R&D Investment and Impact in the Global Construction Industry

R&D Investment and Impact in the Global Construction Industry brings together contributions from leading industry researchers in a diverse group of countries to investigate the role of research and development (R&D) in the construction industry. Investment in R&D is a proven factor in economic growth, and helps develop a more productive and innovative industry. This book explores how policy makers and industry leaders can better target future investment; and how industry and researchers can manage their efforts to improve productivity whilst addressing the environmental and social needs of their communities. Case studies present projects where R&D ideas funded by both the private and public sectors have been translated from research into practice or policy, and examine drivers, successes and barriers to the delivery of R&D outcomes in industry. Based on research from the CIB Task Group TG85 (R&D Investment and Impact) and concluding with a roadmap for maximising the impact of R&D in the future, the book holds valuable lessons for practitioners, policy makers and researchers across the global construction industry.

Dr Keith D. Hampson Over the past 20 years Keith has made a notable contribution to building collaborative innovation networks between industry and research globally. He is committed to crafting a more effective construction industry by promoting better education, applied technology and innovative practices. Keith serves as CEO of the SBEnrc (and previously the Australian Cooperative Research Centre (CRC) for Construction Innovation) and Coordinator of the CIB Task Group 85.

Dr Judy A. Kraatz Judy is an architect with a doctorate in urban development. Career highlights include being a design architect, leading multi-disciplinary teams delivering city-wide solutions, and integrating sustainability into curriculum and practice. Judy is currently a Senior Research Fellow with the SBEnrc, and Coordinator of the CIB Task Group 85.

Adriana X. Sanchez Adriana's experience focuses mostly on sustainable water and transport infrastructure management. Special areas of interest include procurement, risk management and translating policy into tangible outcomes. She has a masters in sustainable resource management, conducted research on four continents and is currently the Commission Secretary for the CIB Task Group 85.

About CIB and the CIB series

CIB, the International Council for Research and Innovation in Building and Construction, was established in 1953 to stimulate and facilitate international cooperation and information exchange between governmental research institutes in the building and construction sector, with an emphasis on those institutes engaged in technical fields of research.

CIB has since developed into a worldwide network of over 5000 experts from about 500 member organisations active in the research community, in industry or in education, who cooperate and exchange information in over 50 CIB Commissions and Task Groups covering all fields in building and construction related research and innovation.

www.cibworld.nl/

This series consists of a careful selection of state-of-the-art reports and conference proceedings from CIB activities.

Open & Industrialized Building *A. Sarja*
ISBN: 9780419238409. Published: 1998

Building Education and Research *J. Yang* et al.
ISBN: 978041923800X. Published: 1998

Dispute Resolution and Conflict Management *P. Fenn* et al.
ISBN: 9780419237003. Published: 1998

Profitable Partnering in Construction *S. Ogunlana*
ISBN: 9780419247602. Published: 1999

Case Studies in Post-Construction Liability *A. Lavers*
ISBN: 9780419245707. Published: 1999

Cost Modelling *M. Skitmore* et al.
(allied series: Foundation of the Built Environment)
ISBN: 9780419192301. Published: 1999

Procurement Systems *S. Rowlinson* et al.
ISBN: 9780419241000. Published: 1999

Residential Open Building *S. Kendall* et al.
ISBN: 9780419238301. Published: 1999

Innovation in Construction *A. Manseau* et al.
ISBN: 9780415254787. Published: 2001

Construction Safety Management Systems *S. Rowlinson*
ISBN: 9780415300630. Published: 2004

Response Control and Seismic Isolation of Buildings *M. Higashino* et al.
ISBN: 9780415366232. Published: 2006

Mediation in the Construction Industry *P. Brooker* et al.
ISBN: 9780415471753. Published: 2010

Green Buildings and the Law *J. Adshead*
ISBN: 9780415559263. Published: 2011

New Perspectives on Construction in Developing Countries *G. Ofori*
ISBN: 9780415585724. Published 2012

Contemporary Issues in Construction in Developing Countries *G. Ofori*
ISBN: 9780415585716. Published: 2012

Culture in International Construction *W. Tijhuis* et al.
ISBN: 9780415472753. Published: 2012

R&D Investment and Impact in the Global Construction Industry
ISBN: 9780415859134. Published: 2014

R&D Investment and Impact in the Global Construction Industry

Edited by Keith D. Hampson,
Judy A. Kraatz and Adriana X. Sanchez

Routledge
Taylor & Francis Group
LONDON AND NEW YORK

First published 2014
by Routledge
2 Park Square, Milton Park, Abingdon, Oxon OX14 4RN

and by Routledge
711 Third Avenue, New York, NY 10017

Routledge is an imprint of the Taylor & Francis Group, an informa business

British Library Cataloguing in Publication Data
A catalogue record for this book is available from the British Library

Library of Congress Cataloging-in-Publication Data
R&D investment and impact in the global construction industry / edited by Keith Hampson, Judy Kraatz and Adriana Sanchez.
pages cm.—(CIB)
Includes bibliographical references and index.
1. Construction industry—Research. 2. Research, Industrial.
I. Hampson, Keith (Keith Douglas) II. Title: R & D investment and impact in the global construction industry.
TH213.5.R155 2014
338.4'7624—dc23 2013045462

ISBN13: 978-0-415-85913-4 (hbk)
ISBN13: 978-1-315-77491-6 (ebk)

Typeset in Sabon
by Keystroke, Station Road, Codsall, Wolverhampton

MIX
Paper from responsible sources
FSC® C013056

Printed and bound in Great Britain by
TJ International Ltd, Padstow, Cornwall

Contents

List of figures

List of tables

Preface

Investment in research and development (R&D) has been recognised by economists globally as a key contributing factor to more productive and innovative industries, and central to economic growth. This applies to the construction industry, with its complex supply chain, and its significant role in shaping national economies, societies and the natural and built environment. However, the manner in which this impact is quantified and measured remains a source of debate, and often fails to effectively inform the development and evaluation of R&D policies and programmes.

Construction is pivotal to ensuring a sustainable future for humanity. In that context, this book has been shaped by questions such as: what difference does R&D investment make across the global construction industry?; how can policy makers and industry leaders better target future investment?; and how can industry and researchers better align and manage their efforts to improve industry-wide productivity and address the broader environmental and social needs of our communities?

The desire of industry, policy makers and analysts, government clients, and researchers to address these issues motivated the formation of the International Council for Research and Innovation in Building and Construction (CIB) Task Group 85 (TG85): R&D Investment and Impact, in 2011. This task group has built a vigorous international network that now benefits from the active involvement of 39 members from 19 countries across six continents. This publication presents contributions from authors representing 14 of these participating countries.

The establishment of TG85 recognises that effective R&D investment strategies have become increasingly important for the construction industry to respond to changing global dynamics. Effectively leveraging R&D investment to deliver industry advancement is a major challenge and this group provides an important international forum in which to improve relationships between industry and research. The receptivity of the construction sector to research-based innovation is increasing, in part due to the efforts of industry-focused R&D partnerships brokered through a range of mechanisms in different countries.

In Australia, by way of example, the formation of the Cooperative Research Centre (CRC) for Construction Innovation in 2001 was an important milestone in a sustained commitment to construction research via collaborative effort. This centre evolved from collaborations initiated in 1994 through the Commonwealth Scientific and Industrial Research Organisation (CSIRO) and Queensland University of Technology (QUT), later formalised through the CSIRO/QUT Construction Research Alliance in 1996. This alliance also included the Royal Melbourne Institute of Technology (RMIT) and the Construction Industry Institute Australia (CIIA). The CRC for Construction Innovation's successor, the Australian Sustainable Built Environment National Research Centre (SBEnrc), now headquartered at Curtin University, was established in 2010, and continues as a key research broker between industry, government and research organisations.

The activities of TG85 have expanded this national undertaking into the international arena, with this task group being a prime example of how successful collaboration can inform the global construction industry.

Contributions from individuals and organisations who strive to make a difference are essential if we are to deliver the benefits of R&D to this industry. We look forward to continuing to work together to better align research policies, funding, and collaborative research teams for a stronger and more productive global construction industry.

Keith D. Hampson
Judy A. Kraatz
Adriana X. Sanchez

Foreword

The construction industry impacts every global society. It provides the setting for modern human activity, ranging from buildings to neighbourhoods to cities, and includes supporting infrastructure such as transport, water supply, telecommunications and energy. It provides constructed facilities where people live and work and is a major contributor to a nation's economy.

Effectively constructing facilities depends on the strength of the industry and the science that underpins it. And it is research and development (R&D) that will create more productive and safer construction operations, processes and techniques. But R&D is not only about discovering new ideas and developing new processes and techniques, it also enhances industry's ability to understand and use innovations that are developed elsewhere.

The International Council for Research and Innovation in Building and Construction (CIB) is the world's foremost platform for international cooperation and information exchange in building and construction research and innovation. It has a global network of over 5,000 experts from 500 organisations active in research, industry or education.

It gives me great pleasure to recommend this unique publication, *R&D Investment and Impact in the Global Construction Industry.* It brings together a set of 14 country contributions from CIB members representing the developed and developing world and evaluates the role and impact of R&D on the performance of projects, companies, governments and society as a whole. The authors then provide a cross-country analysis that distils valuable lessons for industry practitioners, government clients and policy analysts, and scholars of R&D and innovation in the global construction industry.

The seed funding provided by the Australian Sustainable Built Environment National Research Centre (SBEnrc) has been complemented by support from the Australian Research Council (ARC) and a newly focused CIB Task Group TG85: R&D Investment and Impact – a brilliant example of building and maintaining national and international innovation networks.

I commend the work by the three editing authors Keith Hampson, Judy Kraatz and Adriana Sanchez and the 36 contributing country authors in publishing a reference that I expect will be influential in the global R&D and innovation field.

John V. McCarthy AO
Global CIB President (2010–2013)

Acknowledgments

The editors wish to thank all those who have made this publication possible through their contributions and support.

We first wish to thank our international cadre of authors who have contributed to the chapters and generously shared the outcomes of many years of research experience in this field. These contributions have been the foundation, and without them this book would not have been possible. We would also like to expressly thank those who provided an early review of this book's intent.

This publication is the outcome of the International Council for Research and Innovation in Building and Construction (CIB) Task Group 85: Research Investment and Impact, formed in 2011. We therefore extend our sincere thanks to those in the CIB Secretariat who have both facilitated and encouraged the interactions of this Task Group, of which this publication is a major outcome. Acknowledgment is also made of the work of the former Task Groups TG47: Innovation Brokerage in Construction and TG58: Clients and Construction Innovation, on which many personal contacts and friendships have been founded as well as conceptual underpinnings.

The editors received both encouragement and financial support from the Australian Sustainable Built Environment National Research Centre (SBEnrc) and the Australian Research Council (ARC). Without support from these organisations, and national and international associated networks, this publication would not have been realised. Thanks are also extended to Professor Catherin Bull, who has provided guidance and insights to the editors in the course of the Australian research activity which formed the genesis of this publication.

Finally, acknowledgment is also recorded to those who have granted permission to reproduce their material in this book.

Notes on contributors

Dr Miimu Airaksinen, Research Professor, VTT Technical Research Centre of Finland, Espoo, Finland. Miimu completed her PhD in engineering at Helsinki University of Technology in 2003 on buildings moisture and mould problems. Her current research focuses on eco-efficiency in built environment through projects such as IBEN (energy-efficient, intelligent-built environment).

She was recently involved in the development of a strategic research agenda for RYM SHOK (strategic research centre for the built environment in Finland). Previous to her current role, Miimu was involved with several building and city-level projects in energy and eco-efficiency. She is currently a member of the steering committee in E2BA, the European Construction Technology Platform, Energy Efficient Buildings Association. In addition, she is an active member of EERA, the European Energy Research Alliance Smart City steering group.

Dr Frédéric Bougrain, Researcher, Economics and Human Sciences Direction, Centre Scientifique et Technique du Bâtiment (CSTB), Champs-sur-Marne, France. Frédéric works as a Researcher for the Economics and Human Sciences Department at CSTB (a state-owned industrial and commercial research centre under the wing of the Ministry of Housing) in France. His research is concentrated on society issues with a particular focus on innovations in the building and construction industry and energy saving performance contracts.

He previously lectured at the University of Orléans (France) where he defended a thesis on innovation, small and medium-sized enterprises and the consequences for regional technology policy. Frédéric has published papers on public–private partnerships, energy-saving performance contracts, innovation in small and medium-sized enterprises and the social housing sector.

Dr Jan Bröchner, Professor and Chair of Organization of Construction, Division of Service Management, Chalmers University of Technology, Gothenburg, Sweden. Jan holds the Chair of Organization of Construction at Chalmers University of Technology in Göteborg, Sweden. He is an adjunct member, representing the Swedish universities of technology, of the Research Committee of the Development Fund of the Swedish Construction Industry (SBUF).

Currently, he leads the productivity and innovation effect analysis project within Phase II of the 'Bygginnovationen' Programme, supported by the Swedish Governmental Agency for Innovation Systems (VINNOVA) and an industry consortium.

Dr Robert E. Chapman, Chief, Applied Economics Office, Engineering Laboratory, National Institute of Standards and Technology (NIST), Gaithersburg, USA. Robert joined NIST, formerly the National Bureau of Standards, in 1975. In his current position as Chief of the Office of Applied Economics, he leads a group of economists that evaluates new technologies, processes, government programmes, legislation, and codes and standards to determine efficient alternatives and to measure their economic impacts.

Previously, Robert was the Program Manager for the information and outreach branch of the Baldrige National Quality Program, where he provided liaison to a rapidly expanding network of state and local quality award programmes. Between 1988 and 1990, Robert conducted a series of studies on how federal, state and local technology-based programmes were assisting small and medium-sized businesses; these studies were published as a formal report to Congress.

Dr Geert Dewulf, Professor and Research Chair of Planning and Development, University of Twente, Enschede, the Netherlands. Geert is Professor of Planning and Development and Head of the Department of Construction Management and Engineering at the University of Twente, the Netherlands. He is also Vice Dean of the Faculty of Engineering.

In 2012–2013 he was the UPS Foundation Visiting Professor at Stanford University. Before he joined Twente University he worked at the Netherlands Organisation for Applied Scientific Research (TNO) and Delft University of Technology. He holds a PhD in social science from the University of Utrecht. He was a Visiting Fellow at Harvard University in 1990–1991.

Geert has written numerous publications on public–private partnerships, urban planning and infrastructure asset management. His research focuses on public–private governance issues, strategic planning, and infrastructure asset management. He was Scientific President of PSIBouw, the Rethinking Construction Program of the Netherlands (research fund of EUR34 million) and until 2012 Scientific Director of the 3TU (Federation of the Dutch Technical Universities) Center of Expertise on the Built Environment.

Dr Emilia Van Egmond-de Wilde de Ligny, Senior Lecturer and Researcher, Faculty of Architecture, Building Engineering & Planning, Eindhoven University of Technology, Eindhoven, the Netherlands. Emilia carries out analyses and project evaluations, innovation management and feasibility studies for enterprises, branch organisations and governmental and non-governmental organisations.

Her research and teaching activities focus on international technology and knowledge transfers, innovation management, technology and knowledge management, industrialisation and innovation in the construction industry, sustainable building construction and tropical building concepts and technologies.

Dr Francisco Ferreira Cardoso, Professor, Head of Department of Construction Engineering, University of São Paulo, Escola Politécnica, São Paulo, Brazil. Francisco is a Professor at Escola Politécnica, the Engineering School of University of São Paulo, teaching construction technology and management. He is also Vice President of the Undergraduate Committee of Escola Politécnica, Director of the National Association for the Built Environment Technology (ANTAC) and Adviser to the Brazilian Council for Sustainable Construction (CBCS).

Francisco operates in the Brazilian Program of Quality and Productivity in Habitat (PBQP-H) with the Ministry of Cities and is also a member of the National Committee for Technological Development of Housing – CTECH – as well as Researcher for the Foundation for the Technological Development of Engineering (FDTE). Francisco has experience in civil engineering (construction), with an emphasis on competitiveness, quality and production modernisation, production management in construction, and innovation and rationalisation in construction processes.

Dr Keith D. Hampson, CEO, Sustainable Built Environment National Research Centre, Curtin University, Perth, Australia; CIB TG85 Coordinator. Keith has over 30 years of industry, Government and research leadership. He has a Bachelor of Engineering (Hons) from QUT, an MBA, and a PhD from Stanford University focusing on innovation and business performance. He is a Fellow of the Institution of Engineers Australia, a Fellow of the Australian Institute of Company Directors and a Fellow of the Australian Institute of Management.

Keith serves as CEO of the Sustainable Built Environment National Research Centre, successor to the Australian CRC for Construction Innovation, for which he led the bid team in 2000 and was CEO for its nine years of operation.

As Professor of Construction Innovation at Curtin University, he continues to work collaboratively with colleagues across Australia and globally to transform industry performance in sustainability, safety and productivity for a stronger and more competitive industry.

Dr Kim Haugbølle, Senior Researcher, Danish Building Research Institute (SBi), Aalborg University, Copenhagen, Denmark. Kim is an experienced researcher on innovation and sociotechnical change in the construction industry with special emphasis on lifecycle economics, sustainability, procurement and building performance.

Kim has been involved in the coordination and management of several national and international R&D projects, and headed the secretariat of the Danish Building Development Council think tank and a research department. Currently, he is deeply involved in the planning and execution of a new education programme, the BSc (Eng) in Process and Innovation.

Kim is the international co-coordinator of the CIB Working Commission W118 on Clients and Users in Construction as well as a member of the Nordic researchers' network on construction economics and organisation (CREON).

Jingke Hong, MSc, PhD Research Student, Department of Building and Real Estate, Hong Kong Polytechnic University, Hong Kong, People's Republic of China. Jingke obtained his Bachelor in Engineering degree in Project Management from Chongqing University in 2010, and was recommended to Harbin Institute of Technology for postgraduate studies based on his excellent academic performance.

He obtained his MSc degree in 2012. His research field is green building, mainly focusing on evaluating whole lifecycle sustainability performance of building.

Dr Anna Kadefors, Associate Professor, Technology Management and Economics, Chalmers University of Technology, Gothenburg, Sweden. Anna carries out research in the areas of communication, decision making and learning across contractual and organisational boundaries. She has also studied the effects of contracts, leadership and new ICT for collaboration in construction projects.

Anna's key areas of research are procurement and innovation in project-based organisations. Anna is a member of the Royal Academy of Engineering Sciences as well as of several research networks in the area of trust, collaboration and procurement.

Arun Kashikar, Head of R&D, TATA Housing and Development Co., Mumbai, India. Arun has extensive experience in civil and structural design and engineering coordination of projects in various sectors such as oil and gas, chemical and petrochemical, nuclear, industrial and real estate. He also has experience in managing innovations within private industry.

Arun specialises in design management, seismic analysis and design, finite element analysis, analysis and design of RCC and steel structures, value engineering and innovation, quality and safety management systems. As head of R&D of TATA Housing and Development, Arun is responsible for managing innovations for the organisation with the objective of reducing construction time and cost, improving quality and safety and sustainability.

Dr Ole Jonny Klakegg, Professor, Department of Civil and Transport Engineering, Norwegian University of Science and Technology (NTNU), Trondheim, Norway. Throughout his 23 years of work experience, Ole Jonny (MSc, PhD) has alternated between teaching and research at the university, and working as a consultant in project management, building substantial experience including theoretical and practical perspectives.

Ole Jonny shares his time between his current position as Professor of Project Management and his role as R&D Director of Faveo Project Management, the biggest project management consultancy in Scandinavia. He has been involved in a large number of major projects in Norway in both public and private sectors, including building, civil engineering, transport, health, defence and organisational development.

Dr Judy A. Kraatz, Senior Research Fellow, Faculty of Science and Engineering, Griffith University, Brisbane, Australia; CIB TG85 Coordinator. Judy is a Senior Research Fellow working with the Australian Sustainable Built Environment National Research Centre, investigating R&D investment and impact in the built environment.

With over 25 years as a registered architect, Judy has led the delivering of city-wide solutions for public buildings and open places and integrated sustainability into university curriculum, regional initiatives and design and business practice. In 2009 Judy completed a PhD (urban development) at the Queensland University of Technology addressing issues of corporate responsibility in the delivery of major economic infrastructure projects.

As Group Manager Architecture with Brisbane City Council (2001–2005), Judy led a team of up to 40 design professionals delivering urban and social infrastructure across the city. This followed three years as the Programme Director for building courses (building design, survey, construction and project management) being delivered in Australia and Southeast Asia through the Central Queensland University. Prior to that time, Judy had several roles in the Commonwealth Government's building procurement groups, as design architect, change agent and senior manager.

Charles Ma, Research Student, Department of Civil and Environmental Engineering, University of Auckland, Auckland, New Zealand. Charles completed his Civil Engineering degree in 2012 at the University of Auckland. As part of his degree, he completed a research project on innovation in construction focusing on how to improve innovation within construction organisations.

Charles' research interests include strategy, innovation and business management. Charles is currently completing his Bachelor of Commerce degree.

Torill Meistad, MSc, PhD candidate, Department of Civil and Transport Engineering, Norwegian University of Science and Technology (NTNU), Trondheim, Norway. Torill has 20 years of experience from social community research, specialising within industrial development, change and restructuring.

She has a Masters degree in Economics from the Norwegian University of Life Science, and a Masters in Learning in Complex Systems from the University of Oslo and Akershus. Currently, she works at the Norwegian University of Science and Technology, NTNU in Trondheim. Her current field of study is how the construction industry is improving environmental sustainability. In her research, she explores role model building projects with high energy and environmental ambitions, where she focuses on how the participating stakeholders develop, exchange and transfer knowledge throughout the process. Her work is supported by the Research Centre on Zero Emission Buildings (ZEB) at NTNU/SINTEF.

Dr Masi Mohammadi, Professor, Faculty of Technique and Life Sciences, HAN University of Applied Sciences, Arnhem, the Netherlands. Masi lectures and researches at HAN University of Applied Sciences (Netherlands) as a Professor of Architecture in Health, and at the Technical University of Eindhoven as an Assistant Professor with her focus being on smart and sustainable living environments.

She conducts research into the potential and added value of smart innovations in improving the residential comfort, efficiency and sustainability in the built environment. Her main expertise lies in enabling aging-in-place and the acceptance of smart-care innovations through familiarising end users and professionals with smart technology, domesticating it, as well as applying need-based technology in the pre-existing homes.

As the Scientific Coordinator of the Academic Domotics Centre she has not only initiated, coordinated and performed scientific research but also continues to realise experimental and collaborative research in the field of smart healthy environments, together with industrial partners. This platform endeavours to bring cooperation with other knowledge institutes and fuse different stakeholders together so that knowledge transfer is feasible in this area.

Dr Anita Moum, Professor, Faculty of Architecture and Fine Art, Norwegian University of Science and Technology (NTNU), Trondheim, Norway. Anita is a Professor and Senior Adviser for Research at the Faculty of Architecture and Fine Art at NTNU. She is involved in several strategic initiatives that aim to improve building processes and the performance of the architecture, engineering and contractor industry.

Anita completed both her architect degree and PhD at NTNU, the later focused on the use of BIM in collaborative teams. She has been working as an architect and project manager in large-scale projects in Germany for 10 years. Anita was also a Research Manager at the Foundation for Scientific and Industrial Research (SINTEF) Building and Infrastructure 2008–2010, responsible for the development of the building process and architectural research. Her main fields of interests are integrated and collaborative design, delivery processes and project management.

Dr Suvi Nenonen, Research Manager, School of Engineering, Aalto University Espoo, Finland. Suvi holds a Research Manager position at Aalto University where she carries out facility service research activities. Her research group focuses on the business relations between companies, work environment management, construction and real estate practices, as well as contract lifecycle and environmental management.

The group's research activities are characterised by interdisciplinary, close cooperation with business, domestically and internationally networked operations, as well as work-based dissertation research. She has also been coordinating the Built Environment Programme at the Finnish Funding Agency for Technology and Innovation (Tekes) since 2009.

Dr Lúcia Helena de Oliveira, Associate Professor, University of São Paulo, Escola Politécnica, São Paulo, Brazil. Lúcia Helena is an Associated Professor at the Department of Construction Engineering of Escola Politécnica at the University of São Paulo where she teaches and conducts researches on building services.

She is a member of the Commission – W062 Water Supply and Drainage for Buildings of CIB – International Council for Research and Innovation in Building and Construction, and member of the Brazilian Council for Sustainable Construction (CBCS), where she acts as one of the coordinators of the thematic group 'Water'.

Dr Rachel L. Parker, Assistant Dean and Professor, Business School, Queensland University of Technology, Brisbane, Australia. Rachel's research focuses on comparative business systems and the institutional foundations of innovation and industrial competitiveness.

Her work has contributed to understandings of the way in which Australian and international public policy programmes affect firm and industry behaviour and therefore industrial development and transformation. She has published over 40 articles and three books and her publications appear in leading international journals in the field including *Entrepreneurship Theory and Practice*, *Organization Studies*, *Political Studies*, *International Journal of Cultural Policy* and *Work, Employment and Society*. Additionally, Rachel recently worked as a consultant/advisor on knowledge transfer activities for the Department of Innovation, Industry, Science and Research; Queensland Rural Industry Training Council; QMI Solutions and Australian Institute for Commercialisation.

Dr Jeff H. Rankin, Professor and Research Chair in Construction Engineering and Management, University of New Brunswick, Fredericton, Canada. Jeff holds the position of Professor and is the M. Patrick Gillin Chair in Construction Engineering and Management. Previous to joining the Department of Civil Engineering, Jeff was the Executive Director of the Construction Technology Centre Atlantic, an innovation broker for the region's construction industry.

Jeff began in the construction industry as a labourer and his experience has included various project and construction management roles. He has been involved in many types of construction project including commercial high rises and larger projects such as the Confederation Bridge (a 13 kilometre bridge linking two Canadian provinces). While completing his doctorate at the University of British Columbia, Jeff's research centred on the development of integrated and distributed construction management systems based on information standards (a precursor to building information modelling). Currently, Jeff's research programme focuses on improving the performance of the construction industry by strengthening its capacity for innovation, with a specific interest in the appropriate adoption and implementation of information and communication technologies.

Dr Saiedeh Razavi, Assistant Professor and Chair in Heavy Construction, McMaster University, Ottawa, Canada. Saiedeh is the Inaugural Chair in Heavy Construction, and Assistant Professor at the Department of Civil Engineering at McMaster University. She earned her PhD in Civil Engineering at the University of Waterloo and joined McMaster from Concordia University where she was working as a postdoctoral fellow in automation in construction.

Through academic and industry involvement, Saiedeh has gained 14 years of experience in collaborating and leading multi-disciplinary team-based projects in sensing, automation, information technology, intelligent systems, and their applications in construction, transportation, infrastructure management, and logistics. Saiedeh holds academic appointments in the Department of Civil Engineering and at the McMaster-Mohawk Bachelor of Technology–Civil Engineering Technology Program. She is also an associate member of the McMaster's School of Geography and Earth Sciences and an associate editor of the ASCE *Journal of Computing in Civil Engineer*ing. She teaches courses in project management, engineering economics, construction management and optimisations for civil engineering systems.

Dr Aminah Robinson Fayek, University of Alberta, Faculty of Engineering, Department of Civil and Environmental Engineering, Edmonton, Canada; CIB TG85 Coordinator. Aminah is an internationally recognised expert in the development and application of fuzzy logic and fuzzy hybrid modelling techniques for intelligent decision support for the construction industry. She is a Professor in the Hole School of Construction Engineering in the Department of Civil and Environmental Engineering at the University of Alberta, and holds the Ledcor Professorship in Construction Engineering.

Aminah holds the NSERC Senior Industrial Research Chair (IRC) in Strategic Construction Modelling and Delivery. She was also the Associate Industrial Research Chairholder (IRC) in Construction Engineering and Management from 2006–2011 and has a long record of successful university–industry collaborations in construction management research.

Göran Roos, MSc, MBA, Professor and Senior Advisor, Aalto University, Aalto, Finland. Göran chairs the Advanced Manufacturing Council in Adelaide, Australia and is a member of the Prime Minister's Manufacturing Leaders Group; the Economic Development Board; the International Advisory Group (IAG) for DesignGov, the Australian Centre for Excellence in Public Sector Design, and the Manufacturing Sector Advisory Council of the Commonwealth Scientific and Industrial Research Organisation (CSIRO), among other organisations.

Göran is a Stretton Fellow appointed by the City of Playford at the University of Adelaide and Professor in Strategic Design in the Faculty of Design, Swinburne University of Technology, Melbourne, Australia. He is also an adjunct and visiting professor at several Australian and British universities.

Göran is one of the founders of modern intellectual capital science and a recognised world expert in this field as well as a major contributor to the thinking and practice in the areas of strategy and innovation management as well as industrial and innovation policy.

Dr Knut Samset, Professor, Department of Civil and Transport Engineering, Norwegian University of Science and Technology (NTNU), Trondheim, Norway. Knut is Professor of Project Management at the Faculty of Engineering Science and Technology, Norwegian University of Science and Technology.

He is the Director of the Concept Research Programme on Front-End Management of Large Investment Projects, and Founding Director and senior partner of Scanteam, an international consultancy based in Oslo.

His academic background is in both engineering and social science and he holds a PhD in Risk Management. He has extensive experience as advisor to national and international governmental and non-governmental organisations, and has written a number of books on project design, evaluation and project risk. In his current position, he and his team are carrying out pioneering research into ways and means to ensure quality at entry of major projects upstream in order to improve return on investments downstream.

Adriana X. Sanchez, MSc, Research Associate, Sustainable Built Environment National Research Centre, Curtin University, Perth, Australia; CIB TG85 Commission Secretary. Adriana holds an MSc (Hons) in Sustainable Resource Management from the Technische Universität München (TUM), majoring in water, soil and renewable resource management. She has been working at the Australian Sustainable Built Environment National Research Centre (SBEnrc) since January 2012. Prior to this role, Adriana carried out research support activities at the Department of Climate Change and Ecology of the TUM and at Smart Utilities Solution; an environmental consulting firm based in Munich, Germany, with international projects covering tailor-suited solutions in waste-to-energy, water management and off-grid green energy technologies for farmers and production plants. Adriana also worked for the German International Cooperation (GIZ) in a project partially sponsored by Deltares in the development and implementation of a computerised interactive flood model for a local watershed in the Philippines.

Dr Mercia Maria Semensato de Barros, Assistant Professor, University of São Paulo – Escola Politécnica, São Paulo, Brazil. Mercia is an Assistant Professor at the Department of Construction Engineering of Escola Politécnica at the University of São Paulo where she teaches and conducts researches focused on innovation and rationalisation in construction processes and production management. Mercia also operates in the area of rehabilitation of buildings with a focus on the technologies and costs.

She also has worked as a Researcher at the Foundation for Technological Development Engineering (FDTE) and as a Consultant *ad hoc* advising for the Foundation for Research Support of the State of São Paulo (FAPESP) and the Financier of Studies and Projects (FINEP).

Dr Geoffrey Q. Shen, Chair Professor, Head of Department of Building and Real Estate and Associate Dean, Hong Kong Polytechnic University, Hong Kong, People's Republic of China. Geoffrey has a proven track record in academic and research leadership in collaborative working in construction, supported by information and communication technologies. He has led a large number of research projects with total funding of over HKD20 million including five RGC-CERG grants in a row since 2002/03.

Geoffrey has authored over 300 publications including more than 160 papers in academic and professional journals and over 30 scholarly books and research monographs. He has been invited to give keynote presentations in a number of international conferences, has chaired and co-chaired a number of major international conferences, and has served the scientific committee of many international conferences. He teaches extensively in these fields at both postgraduate and degree levels, and has successfully supervised a large number of PhD and MPhil students.

Dr E. Sarah Slaughter, President, Built Environment Coalition, Boston, USA. Sarah is President and Founder of the Built Environment Coalition, a research and education non-profit organisation focused on sustainability and disaster-resiliency advancements to the built environment. She was most recently the Associate Director for Buildings and Infrastructure in the Massachusetts Institute of Technology (MIT) Energy Initiative, Co-Founder and Head of the Sustainability Initiative in the MIT Sloan School of Management.

Previously, Sarah was Founder and CEO of MOCA Systems, Inc., a construction programme management company based on the construction micro-simulation software system she developed in her research as a professor in the MIT Department of Civil and Environmental Engineering. Earlier, she was a Professor in the Department of Civil and Environmental Engineering at Lehigh University. Sarah is currently a member of the

National Academies Department of Defense (DOD) Standing Committee on Materials, Manufacturing, and Infrastructure, and is a National Academy Associate. She currently serves on the Board of Directors for Retroficiency, Inc., and the Charles River Watershed Association.

Dr Alexandra Staub, Associate Professor of Architecture, Pennsylvania State University, Pennsylvania, USA. Alexandra holds a BA from Columbia University in New York, a professional degree in architecture from the University of the Arts Berlin, and a PhD in architecture from the Brandenburg Technical University Cottbus, Germany.

In addition to practising architecture, she has spent 20 years in research and education at university level, both in Germany and the United States. In addition to training students in the research and design process, including questions of technology and sustainability, her research centres around architectural and urban production as a cultural product, examining technical and spatial development in the context of social demands and development. She has published widely in this area, and serves on the editorial board of *Enquiry: The ARCC Journal of Architectural Research* and the bilingual theory-based journal *Wolkenkukuksheim/Cloud-Cuckoo-Land.*

Dr Marit Støre Valen, Associate Professor and Head of Department of Civil and Transport Engineering, Norwegian University of Science and Technology (NTNU), Trondheim, Norway. Marit has over 10 years' experience of carrying out leading research in the areas of real estate management of public and private building portfolios, project management and culture for innovation and interaction processes in the construction industry.

As an Associate Professor in Real Estate Management she has over a decade of experience of teaching and developing education programmes within maintenance and modernisation of buildings and sustainable FM. Additionally, as a head of the department, Marit is responsible for the educational quality of the study programmes and developing research strategies within civil and transport engineering. Under her leadership, the department has a broad cooperation and joint cooperation in education and research activities with the construction industry and oil/gas industry.

Douglas Thomas, Research Economist, Applied Economics Office, Engineering Laboratory, National Institute of Standards and Technology (NIST), Gaithersburg, USA. Douglas is a research economist for the Engineering Laboratory's Applied Economics Office at the National Institute of Standards and Technology. Currently, his activities are focused in two interrelated subjects: manufacturing and construction industry activity and the impact of natural and man-made disasters.

His current research on manufacturing and construction activity includes examining spatial and temporal variations in the quality and quantity of industry activity in relation to the domestic and international economy. It utilises industry data combined with various methods of analysis, including input–output analysis. Douglas' second key area of research examines the impact of natural and man-made disasters, which includes gathering and analysing data on the occurrence and economic impact of disasters.

Dr Russell J. Thomas, Director Fire Research (retired), National Research Council of Canada, Ottawa, Canada. Russ has recently retired from his post as Director, Fire Research, in the Construction Portfolio of the National Research Council (NRC) of Canada.

Prior to becoming the Director of Fire Research some 15 years ago, Russ was responsible for leading the team developing the framework for Canada's new Objective-Based Building and Fire Codes. He came to NRC in the late 1980s from an academic post at the University of Warwick in the UK to set up the Advanced Construction Technology Laboratory where research was undertaken into applying advanced computational techniques and human computer interaction techniques to the construction sector. In addition to running the Fire Research Program, Russ was also responsible for NRC-IRC's Center for Computer-assisted Construction Technologies, based in London, Ontario, where research was undertaken into the development of advanced computational tools for the construction industry.

Terttu Vainio, Lic. Tech., Senior Research Scientist, VTT Technical Research Centre of Finland, Espoo, Finland. Terttu started her career at VTT before graduating and has carried out applied research ever since. She has a Master of Science (Technology) and Licentiate of Science (Technology) pre-doctoral postgraduate degree from Tampere University of Technology.

She is an expert in construction economics, particularly in construction as a part of the national economy, input–output analysis, and money flow analysis. Terttu was recently involved in efforts to investigate energy renovation of building stock and energy renovation as business opportunity.

Dr Suzanne Wilkinson, Professor, Department of Civil and Environmental Engineering, University of Auckland, Auckland, New Zealand. Suzanne is Professor in Construction Management at the University of Auckland. Her research interests focus on disaster recovery and reconstruction, construction contract administration and relationship management for construction projects.

Suzanne completed a Bachelor in Civil Engineering (Honours) and a PhD in Construction Management from Oxford Brookes University, UK. Her recent research book, co-authored with Rosemary Scofield, *Management for the New Zealand Construction Industry* (Prentice-Hall), has been adopted as a standard text at New Zealand universities and used by construction companies in New Zealand. She lectures undergraduates and postgraduate students in project management, construction management, and construction law and administration. Suzanne also acts as an advisor to Government and construction companies on aspects of construction productivity and disaster recovery.

List of abbreviations

A&E	architects and engineers
AAU	Aalborg University
ABNT	Associação Brasileira de Normas Técnicas (Brazilian Association of Technical Standards)
ABS	Australian Bureau of Statistics
ACIF	Australian Construction Industry Forum
ADEME	Agence de l'Environnement et de la Maîtrise de l'Energie (French Environment and Energy Management Agency)
AEC	architecture, engineering and construction
AEGIS	Australian Expert Group on Industry Studies
ANTAC	Associação Nacional de Tecnologia do Ambiente Construído (National Association of Technology of the Built Environment)
ANVAR	National Agency for the Valorisation of Research (France)
ANZSIC	Australian and New Zealand Standard Industrial Classification
APCC	Australian Procurement and Construction Council
ARC	Australian Research Council
ASCE	American Society of Civil Engineers
ASHRAE	American Society for Heating, Refrigerating, and Air-Conditioning Engineers
AUD	Australian dollar
BACnet	building automation and control networks
BBR	Bundesamt für Bauwesen und Raumordnung (Federal Office for Building and Regional Planning) (Germany)
BBSR	Bundesinstitut für Bau-, Stadt- und Raumforschung (Federal Institute for Research on Building, Urban Affairs and Spatial Development) (Germany)
BCITF	Building and Construction Industry Training Fund
BCSPP	Building and Construction Sector Productivity Partnership (NZ)
BEIIC	Built Environment Industry Innovation Council (Australia)
BIC	Byggsektorns Innovationscentrum (Swedish Construction Sector Innovation Centre)

BIM	building information modelling
BMTPC	Building Materials and Technology Promotion Council (India)
BMVBS	German Federal Ministry of Transport, Building and Urban Development
BNDES	Banco Nacional de Desenvolvimento Social (Brazilian Development Bank)
BQR	Rådet för Byggkvalitet (Council for Constructing Excellence) (Sweden)
BRL	Brazilian real
CAD	Canadian dollar
CADD	computer-aided design and documentation
CAPES	Coordenação de Aperfeiçoamento de Pessoal de Nível Superior (Coordination of Improvement of Higher Education Personnel) (Brazil)
CBI	CBI Betong Institutet (Swedish Cement and Concrete Research Institute) (Sweden)
CBIC	Câmara Brasileira de Indústria e Comércio (Brazilian Chamber of Industry and Commerce) (Brazil)
CBRI	Central Building Research Institute (India)
CBS	cybernetic building systems
CCMC	Canadian Construction Materials Centre
CDC	centres for disease control (USA)
CEF	Caixa Econômica Federal (Federal Savings Bank) (Brazil)
CERBOF	Centrum för Energi- och Resurseffektivitet i Byggande och Förvaltning (the Centre for Energy and Resource Efficiency in the Built Environment) (Sweden)
CERF	Civil Engineering Research Foundation (USA)
CESBC	civil engineering surveying, building and construction
CGDD	Commissariat Général au Développement Durable (General Commission for Sustainable Development) (France)
CIB	International Council for Research and Innovation in Building and Construction
WBC13	CIB World Building Congress 2013
CIC	Construction Industry Council (NZ)
CIDC	Construction Industry Development Council (India)
CIFE	Stanford University's Center for Integrated Facilities Engineering (USA)
CII	Construction Industry Institute (USA)
CIPET	Central Institute of Plastics Engineering & Technology (India)
CityU	City University of Hong Kong
CNPq	Conselho Nacional de Pesquisa Científica e Tecnológica (National Council for Scientific and Technological Development) (Brazil)
CNRS	Centre National de la Recherche Scientifique (National Centre of Scientific Research) (France)

CONSIAT	construction systems integration and automation technologies
CONSITRA	Consórcio Setorial para Inovação Tecnológica em Revestimentos de Argamassa (Sector Consortium for Technological Innovation in Mortar Coatings) (Brazil)
CRC	cooperative research centres
CRC CI	Cooperative Research Centre for Construction Innovation (Australia)
CRDs	collaborative research and development grants
CREAHd	Construction Ressources Environnement Aménagement et Habitat durables (Construction Resources Environment and Sustainable Habitat Development) (France)
CRP	Concept Research Programme (Norway)
CRRI	Central Road Research Institute (India)
CRI	Crown Research Institutes (NZ)
CSG	Construction Strategy Group (NZ)
CSIRO	Commonwealth Scientific and Industrial Research Organisation (Australia)
CSIT	computing science and information technology
CSNA	Canadian System of National Accounts
CSO	Central Standards Office (India)
CSTB	Centre Scientifique et Technique du Bâtiment (Scientific and Technical Centre for Building) (France)
CUHK	Chinese University of Hong Kong
CURT	Construction Users Roundtable (USA)
CUSP	Curtin University's Sustainability Policy Unit (Australia)
DDT	Departamento de Desenvolvimento Tecnológico (Technological Development Department) (Brazil)
DIISR	Department of Innovation, Industry, Science and Research, Australia
DKK	Danish krone
DM	German mark
DOE	Department of Energy (USA)
DOT	Department of Transportation (USA)
DTAPP	Digital Technology Adoption Pilot Program (Canada)
DTI	Danish Technological Institute
EA	Engineers Australia
ECIS	Eindhoven Centre of Innovation Studies
ECS	Early Career Scheme
ECTP	European Construction Technology Platform
EEE	electrical and electronic engineering
EIP	European Innovation Partnerships
ENCORD	European Network of Construction Companies for Research and Development
ENOVA	National Norwegian Energy Fund
EPA	Environmental Protection Agency (USA)

EPD	environmental product declaration
EREI	energy and resource efficiency innovation
ESPON	European Spatial Planning Observation Network
ESTCP	Environmental Security Technology Certification Program
EU	European Union
EUR	euro
ExWoSt	Experimenteller Wohnungs- und Städtebau (Experimental Housing and Urban Design) (Germany)
FAPESP	Fundação de Amparo à Pesquisa do Estado de São Paulo (Foundation for Research Support of the State of São Paulo) (Brazil)
FINEP	Financiadora de Estudos e Pesquisas (Financier of Studies and Research) (Brazil)
FM	Frascati Manual
FME	environmentally friendly energy
Formas	Forskningsrådet för Miljö, Areella Näringar och Samhällsbyggande (the Research Council for Environment, Agricultural Sciences and Spatial Planning) (Sweden)
Fraunhofer IRB	Fraunhofer-Informationszentrum Raum und Bau (Fraunhofer Information Centre for Regional Planning and Building) (Germany)
FSES	fire safety evaluation system
GBCA	Green Building Council of Australia
GC	general contractors
GDP	gross domestic product
GERD	gross domestic expenditure on research and development
Glafo	Glass Research Institute (Sweden)
GNP	gross national product
GRF	General Research Fund Scheme
GRI	Global Reporting Initiative
GSA	General Services Administration (USA)
HDI	Human Development Index
HKBEAM	Hong Kong Building Environmental Assessment Method
HKBU	Hong Kong Baptist University
HKD	Hong Kong dollar
HKIEd	Hong Kong Institute of Education
HKU	University of Hong Kong
HKUST	Hong Kong University of Science and Technology
HSCE	Hole School of Construction Engineering (Canada)
HVAC	heating, ventilation and air conditioning
HVAC&R	heating, ventilation, air conditioning and refrigeration
IBGE	Instituto Brasileiro de Geografia e Estatística (Brazilian Statistics and Geography Institute)
ICALL	International Construction Research Alliance
ICC	International Commerce Centre (HK)
ICT	information and communication technology

IFB	Institut für Bauforschung (Institute for Building Research) (Germany)
INPI	Instituto Nacional de Propriedade Industrial (National Institute of Industrial Property) (Brazil)
INR	Indian rupee
INSDAG	Institute of Steel Development & Growth (India)
IQS	IQ Samhällsbyggnad (Centre for Innovation and Quality in the Built Environment) (Sweden)
IPD	integrated project delivery
IPRs	intellectual property rights
IRAP	Industrial Research Assistance Program
IRCs	industrial research chairs
IS	Indian standard
ITF	Innovation and Technology Fund
IWU	Institut Wohnen und Umwelt (Germany)
JPI	Joint Programming Initiative
KIG	Koordinations- og Initiativgruppen for viden i byggeriet (Coordination and Innovation Group for Knowledge in Building) (Denmark)
KPI	key performance indicator
LBL	Lawrence Berkeley Laboratory (USA)
LCA	lifecycle analysis
LU	Lingnan University (HK)
LVL	laminated veneer lumber
M&E	mechanical and electrical
MBIE	Ministry of Business, Innovation and Employment (NZ)
MCTI	Ministério da Ciência, Tecnologia e Inovação (Ministry of Science, Technology and Innovation) (Brazil)
MEC	Ministério da Educação e Cultura (Ministry of the Education and Culture) (Brazil)
Mistra	Stiftelsen för Miljöstrategisk Forskning (Swedish Foundation for Strategic Environmental Research)
MIT	Massachusetts Institute of Technology (USA)
MPI	mechanical, production and industrial
MSI	Ministry of Science and Innovation (NZ)
NACE	Nomenclature des Activités Économiques dans la Communauté Européenne (Nomenclature of Economic Activities in the European Community)
NAICS	North American Industry Classification System
NCB	National Council for Cement and Building Materials (India)
NCSBCS	National Conference of States on Building Codes and Standards
NIOSH	National Institute of Occupational Safety and Health (USA)
NIST	National Institute of Standards and Technology (USA)
NOK	Norwegian krone
NRC	National Research Council of Canada
NREL	National Renewable Energy Laboratory (USA)

NSERC	Natural Sciences and Engineering Research Council of Canada
NSF	National Science Foundation (USA)
NTNU	Norwegian University of Science and Technology
NUTEK	Närings- och teknikutvecklingsverket (National Board for Industrial and Technical Development)
NZD	New Zealand dollar
O&M	operations and maintenance
O&O	owners and operators
OECD	Organisation for Economic Cooperation and Development
PAC	Programa de Aceleração do Crescimento (Growth Acceleration Programme) (Brazil)
PACE	property-assessed clean energy
PAGIT	Programa Andrade Gutierrez de Inovação Tecnológica (Technology Innovation Programme) (Brazil)
PBRF	Performance-Based Research Fund (NZ)
PCA	plan construction and architecture
PI	principal investigator
PIT	Programa para Inovação Tecnológica em Construção (Programme for Technological Innovation in Construction) (Brazil)
PNNL	Pacific Northwest National Laboratory (USA)
PolyU	Hong Kong Polytechnic University
PREBAT	Programme de Recherche et d'Expérimentation sur l'énergie dans le Bâtiment (Research and Experimental Programme on Energy in Building) (France)
PSIBouw	Process and Systems Innovation in the Construction Industry (Netherlands)
PURA	Programa de Uso Racional da Água (Water Conservation Programme) (Brazil)
QA	Quality Assurance System
QDPW	Queensland Department of Public Works (Australia)
QUT	Queensland University of Technology (Australia)
R&D	research and development
RCC	reinforced cement concrete
RDI	research, development and innovation
RDSO	Research Designs and Standards Organisation (India)
RecRes	Canterbury Resourcing Project (NZ)
Resorgs	Resilient Organisations Programme (NZ)
RGC	Research Grants Council (HK)
RISE	Research Institutes of Sweden
RMIT	Royal Melbourne Institute of Technology (Australia)
RTG	Fletcher Building Roof Tile Group (NZ)
RTRC	Railway Testing and Research Centre (India)
RYM	Strategic Centre for Science, Technology and Innovation (SCSTI) of Built Environment (Finland)

S&T	science and technology
SABESP	Companhia de Saneamento Básico do Estado de São Paulo SA (Basic Sanitation Company of the State of São Paulo, S.A.)
SBEnrc	Australian Sustainable Built Environment National Research Centre
SBi	Statens Byggeforskningsinstitut (Danish Building Research Institute)
SBUF	Svenska Byggbranschens Utvecklingsfond (the Development Fund of the Swedish Construction Industry)
SCMD	strategic construction modelling and delivery
SCSTI	Strategic Centres for Science, Technology and Innovation (Finland)
SEA	Swedish Energy Agency
SEK	Swedish krona
SF&S	specialty fabricators and suppliers
SHKP	Sun Hung Kai Properties (HK)
SIC	Standard Industry Classification
SINAT	Sistema National de Avalianção Técnica (Brazilian National System of Technical Evaluation)
SME	small and medium-sized enterprise
SNA	system of national accounts
SNM	strategic niche management
SIO	strategic innovation area
SPI	NSERC's Strategy for Partnerships and Innovation
SSC	steel structure conservation
SSCS	steel structure conservation sector
SSF	Stiftelsen för Strategisk Forskning (Foundation for Strategic Research) (Sweden)
STAN	OECD Structural Analysis Database
STIC	Science, Technology and Innovation Council (Canada)
STIC	Structural Timber Innovation Company Ltd (NZ)
STU	Styrelsen för Teknisk Utveckling (Board for Technical Development) (Sweden)
Tekes	Finnish Funding Agency for Technology and Innovation
UGC	University Grants Committee (HK)
UNB-CEM	University of New Brunswick's Construction Engineering and Management Group (Canada)
UNSD	United Nations Statistical Commission
USD	United States dollar
USITC	United States International Trade Commission
VTI	Transport Research Institute (Sweden)
VTT	Technical Research Centre of Finland
WAG	Western Australian Government
WH&S	work place, health and safety

1 Introduction

Keith D. Hampson, Judy A. Kraatz, Adriana X. Sanchez and Aminah Robinson Fayek

The International Council for Research and Innovation in Building and Construction (CIB),[1] established in 1953, is a global network for exchange and cooperation in research and innovation for the building and construction industry. CIB consists of over 5,000 experts from about 500 member organisations active in academia, research, Government, industry and education. CIB provides support to improve processes and performance in the built environment.

CIB's Task Group 85 (TG85) R&D Investment and Impact was established in Helsinki in October 2011. The task group's objective was to establish an international network to exchange knowledge and develop new understandings related to leveraging R&D investment in the construction industry. An active forum comprising members from 19 countries has been established to strengthen collaboration between private firms, Government agencies and research institutions. Its intent is to inform policies and programmes to enhance R&D investment outcomes. TG85 Coordinators are Keith Hampson, Sustainable Built Environment National Research Centre (SBEnrc), Australia, Judy Kraatz, Griffith University, Australia and Aminah Robinson Fayek, University of Alberta, Canada. The Commission Secretary is Adriana Sanchez, SBEnrc, Australia.

Fourteen country-based chapters have been prepared by task group members. These form the core of this book and report on the construction industry and the state of specific R&D investment in the following countries: Australia, Brazil, Canada, Denmark, Finland, France, Germany, Hong Kong (China), India, the Netherlands, New Zealand, Norway, Sweden and the USA. Chapter 2 provides an introduction to the construction industry and its characteristics and addresses key issues and concepts that will be discussed in the book, including a focus on R&D investment processes and funding. It introduces the impact of R&D[2] on innovation outcomes[3] and mechanisms for disseminating and measuring the impact of R&D.

Each country chapter presents an introduction to the country's construction context, followed by a discussion of R&D trends in the private and public construction sectors. Most chapters present one or more case studies of projects and programmes that illustrate how R&D ideas and products are translated from research into practice or policy. The case studies also present drivers, successes and barriers to the delivery of impactful R&D outcomes. Most chapters present insight into how to measure the success of R&D impact, illustrate the industry impact of their country's R&D and identify the parties that have benefited from such impact. A roadmap or strategy for future investment in R&D is also outlined in several chapters.

The country chapters reflect a developing culture of research and innovation in the global construction industry: Sweden states its ambition to become a prominent research nation; New Zealand identifies R&D as a key driver for innovation, business success and economic growth; Finland discusses the foundations for its ranking as having one of the best innovation systems in the world; Hong Kong (China) outlines investment in academic research since the 1960s to enhance the growth of the special administrative region as an international metropolis. Despite this developing culture for research and innovation and its importance to economic and community health, many countries (including Australia, Canada, Finland, France, the Netherlands, New Zealand and the USA) report insufficient research funding directed to the construction industry.

It is noted that sometimes conflicting objectives arise between industry, Government and researchers due to differing priorities for R&D. In Germany and the USA, academic funding flows to basic and applied research, while industry funding is more heavily focused on development and demonstration phases. Germany's R&D efforts are an example of a Government's interest in overall national reputation, a concern not always central to private industry's interests. A further conflict arises with academic objectives having longer timelines and less commercially focused outcomes while industry seeks a more immediate return on investment (Australia, Canada and the USA).

Key challenges for the construction industry identified from the country chapters include:

- industry diffusion of R&D and innovation;
- tight profit margins, risk aversion and short-term focus curtail private funding;
- the fragmented contracting process (design, bid, build/construct) causes conflicting incentives and inhibits application of advanced knowledge across the supply chain;
- innovation tends to be ad hoc and project based;
- knowledge and skills are lost due to cyclical fluctuations of the industry.

A strong synergy naturally exists between the industry and R&D providers, therefore the challenges that exist in the broader industry impact on the provision of R&D policy and services.

Several country authors have considered future directions and development needs of the construction industry in their country. Some common themes have emerged:

- improve research collaboration and dissemination of findings between industry and researchers;
- develop partnerships to access expertise, resources and experience between the construction industry and sectors servicing the industry;
- improve productivity and reduce costs using advanced information and communication technologies;
- translate research from more technologically advanced sectors to the construction industry;
- expedite practical implementation of research outcomes through pilot trials, field tests and demonstrations.

These themes and challenges are based on the country reports, but assessments are not restricted to these countries alone. They represent the similarities that exist across the globe, and at the same time highlight the diversity that exists based on economic, social and environmental conditions, and across the private and public sectors.

The final chapter draws on the individual country chapters to synthesise and showcase what the future might bring for the construction industry. In particular, it focuses on mechanisms to take advantage of R&D, including opportunities for collaboration and better leveraging new technologies. It addresses approaches to developing infrastructure to support R&D and increasing research impact, the latter by bridging the divide between researchers and industry and disseminating the R&D outcomes.

We trust that this book will lead to maximising the impact of R&D investment in construction.

Notes

1 For more information, please refer to www.cibworld.nl/site/about_cib/index.html.

2 Systematic research and development activity aimed at discovering solutions to problems or creating new goods and knowledge to enable development of new products, processes and services.

3 Innovation is a process that leads to change in a product, service, organisation, industry sector or region as a result of new ideas being developed into something of value.

2 The global construction industry and R&D

Keith D. Hampson, Judy A. Kraatz and Adriana X. Sanchez

Introduction

The built environment provides the setting for modern human activity, ranging from buildings to neighbourhoods to cities, and includes supporting infrastructure such as transport, water supply, telecommunications and energy networks. It is typically the greatest asset of a nation. It is where a nation's population lives and, in industrial societies, where up to 95 per cent of the population works and approximately 80 per cent of national gross domestic product (GDP) is generated. The built environment also encompasses the spaces, namely homes, offices, shopping centres, health facilities and entertainment venues, in which the population spends, on average, 97 per cent of its time.

Increasing urbanisation is now a dominant global trend. The planning, construction and management of sustainable urban settlement is possibly the greatest global challenge of the twenty-first century, especially when coupled with adaptation to the impacts of climate change and resource constraints. Constructing our built environment is truly a cornerstone activity of our society (Newton, *et al.*, 2009).

Defining construction

The United Nations categorises construction activities for buildings and civil engineering works into general and specialised, where:

> *General construction* is the construction of entire dwellings, office buildings, stores and other public and utility buildings, farm buildings etc., or the construction of civil engineering works such as motorways, streets, bridges, tunnels, railways, airfields, harbours and other water projects, irrigation systems, sewerage systems, industrial facilities, pipelines and electric lines, sports facilities etc.
> *Specialised construction* is the construction of parts of buildings and civil engineering works without responsibility for the entire project. These activities are usually specialised in one aspect common to

> different structures, requiring specialised skills or equipment, such as pile driving, foundation work, carcass work, concrete work, brick laying, stone setting, scaffolding, roof covering, etc. Specialised construction activities are mostly carried out under subcontract, but especially in repair construction it is done directly for the owner of the property. These activities include new work, repair, additions and alterations, the erection of prefabricated buildings or structures on the site, repair of buildings and engineering works, and also construction of a temporary nature.
>
> (UNSD, 2008)

There is, however, no globally accepted statistical classification that defines the construction industry. Some countries use narrow definitions of the construction industry, while others choose to include *support* elements in their definitions, making it difficult to directly compare cross-country statistics (Lewis, 2009). For example, to compare France and Australia, the French Statistics Bureau uses NACE Rev 2, the *statistical classification of economic activities in the European Community*, which includes construction of buildings, civil engineering and special construction activities (Eurostat, 2008). The Australian Bureau of Statistics (ABS) uses the Australian and New Zealand Standard Industrial Classification (ANZSIC) which includes building construction, heavy and civil engineering construction and construction services (Trewin & Pink, 2006). NACE classifies *service to buildings and landscape activities* under *administrative and support service activities,* while ANZSIC classifies these services under *construction.*

Regardless of its classification boundaries, the industry is traditionally considered as being *low tech*. Tunzelmann and Acha (2006) define low-tech industries as *mature industries*, in which *technologies and market conditions may change slowly*. Miles (2006) adds that *many services are arguably labouring under a heritage derived from past periods in which few generic technologies found ready application in their activities*. This low-tech definition is based on the industry's low R&D intensity and small numbers of people employed directly in R&D. Notably, however, certain subsectors within the industry, such as building product manufacturers and equipment suppliers, break from this generalisation. Beyond such statistical understanding however, this *low-tech* nature has an important impact on its operation, productivity and ability to engage in R&D. The Australian Expert Group on Industry Studies (Marceau, *et al.*, 1999) identifies activities in the construction arena as a *product system*, as opposed to a cluster, complex or sector, due to: (i) its reach into both services and manufacturing and (ii) the manner in which innovation in this system impacts across products, processes and services, including: elements of goods-producing industries; goods-related service industries; knowledge-based services; in-person services and Government and defence activities. Figure 2.1 shows a map of this product system (adapted from Marceau, *et al.*, 1999).

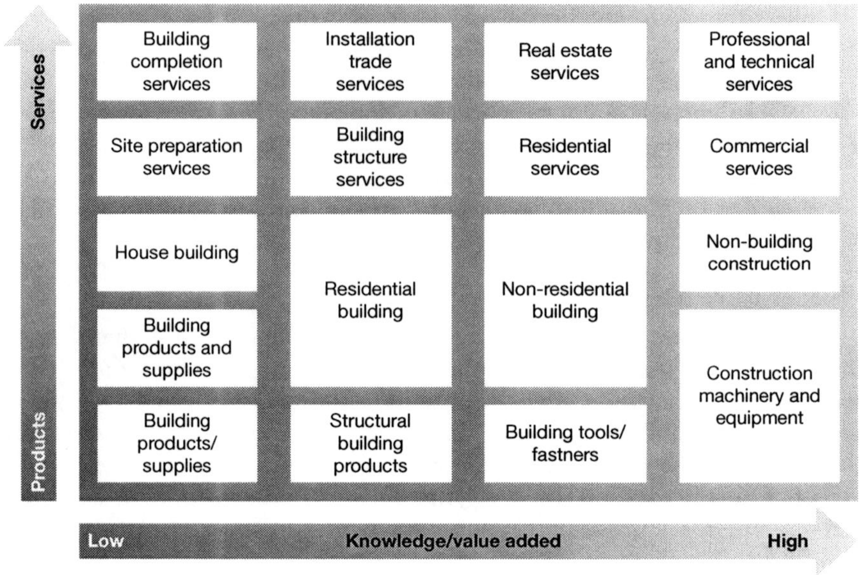

Figure 2.1 Map of the creation-production-distribution chain

Source: Adapted from Marceau, *et al.* (1999)

Country-based differentiation

This book provides a unique consolidation of current conditions and drivers affecting R&D investment across 14 nations, with representation from countries across the spectrum of global economic development.

For example, as will be explained in further chapters, productivity is a major issue affecting the construction industry in Australia, Brazil, Canada, India, the Netherlands, New Zealand, Sweden and the USA. The USA and Canada chapters report decreasing relative productivity levels in their construction industries. The USA has then identified five key areas in which to improve efficiency and productivity, including: deployment of interoperable IT applications across the industry; improved interfacing of people, processes, materials, equipment and information on projects; greater adoption of prefabrication; widespread demonstrations to diffuse knowledge; and targeted performance benchmarking. In Australia, national key performance indicators (KPIs) have been identified: safety; productivity and competitiveness; economic security; workplace capability; and environmental sustainability/eco-efficiency (Furneaux, *et al.*, 2010).

A common theme for research and innovation across participant countries is to reduce resource consumption and CO_2 emissions by creating a more environmentally sustainable built environment using eco-friendly technology and construction processes. The German Government challenged researchers and prefabrication companies to create model homes that generate surplus energy, which can be transferred to, for example, electric motor vehicles. Norway has a similar programme, in which the environmental protection organisation ZERO challenged the construction industry to design and build energy-producing buildings. Finland, Germany and Norway report active research beyond individual buildings towards the development of more sustainable communities.

Information and communication technology (ICT) has been adopted to improve productivity and reduce inefficiencies by Australia, Brazil, Canada, Finland, Hong Kong (China) and the USA. In Australia, public sector construction projects have provided an opportunity to develop building information modelling (BIM) and integrated project delivery (IPD) and associated national guidelines to better integrate the supply chain. In the USA, widespread deployment and use of interoperable IT within the facilities management sector has been identified as having the potential to cut costs by reducing supply chain inefficiencies. In Canada, research into the field of fuzzy hybrid modelling is improving construction project performance to enhance industry competitiveness.

Despite a developing industry culture for research and innovation, many countries (including Australia, Canada, Finland, France, New Zealand, the Netherlands and the USA) report insufficient research funding directed to the construction industry. In Australia, however, there has been a significant recorded increase in private funding while the public sector's contribution has been declining. This underinvestment is counter to the position of the construction industry as a major contributor to national employment and economic growth and as being central to a rising standard of living and improved community sustainability (Hampson, *et al.*, 2013).

The Organisation for Economic Cooperation and Development (OECD) STAN Database for Structural Development provides a valuable source of cross-country analysis of R&D funding. Figure 2.2 presents data relating to R&D investment in the construction industry by country as a proportion of the 16 OECD nations combined: Australia, Belgium, Canada, Denmark, Finland, France, Germany, Italy, Japan, the Netherlands, Norway, Spain, Sweden, Turkey, the UK and the USA.

Table 2.1 presents seven key indicators that are referred to throughout this book to build a better understanding of the construction industry and the impacts of R&D in the diverse nations represented in this publication.

Characteristics of the construction industry

Characteristics and challenges

Construction projects are complex undertakings. Several reasons exist for this complexity, including the need to coordinate numerous participants

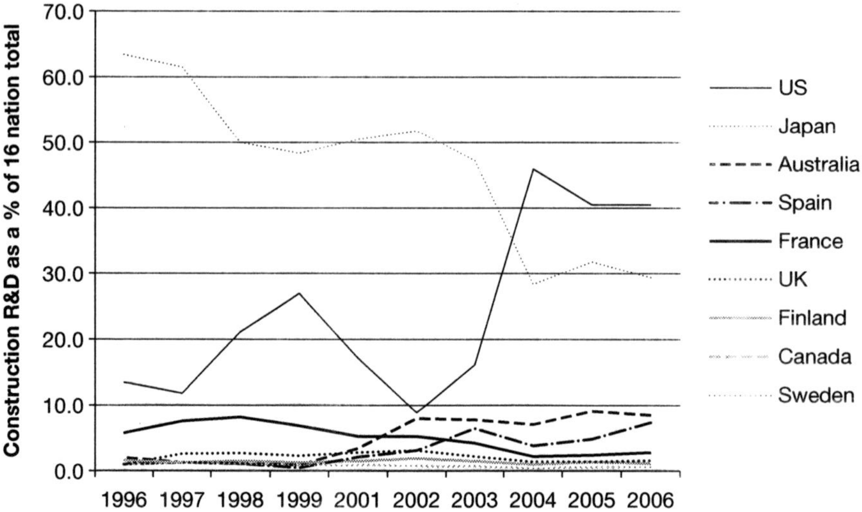

Figure 2.2 R&D in the construction industry as a share of 16 OECD countries

Source: Barlow (2012)

Notes: (i) Derived from OECD STAN. (ii) R&D expenditures in the construction industries are shown as a % of that of 16 OECD nations combined: Australia, Belgium, Canada, Denmark, Finland, France, Germany, Italy, Japan, Netherlands, Norway, Spain, Sweden, Turkey, UK, and the US.

often comprising small and medium-sized enterprises (SME), being a largely site-based industry and the level of specialist knowledge required. Yet, as also highlighted by Dubois and Gadde (2002), the industry has failed to fully integrate technologies and processes from other sectors that can contribute to an improvement in performance and productivity such as: just-in-time delivery; partnering with suppliers; supply chain management; off-site manufacturing; and, most recently, advances in digital information and communication technologies.

Despite its size and importance to the economy, the construction industry has not kept pace with other large industries in productivity improvement. Haskell's (2004) study suggests that this has been less than half that of other industries from 1966–2003 (Jonassen, 2010).

Characteristics of the industry that reflect the need for a change, as defined by Jonassen (2010), include:

- segmentation into silos of interests;
- little alignment of objectives among the silos with each optimised for its own interests;
- poor and discontinuous information flow between parties;

Table 2.1 Country key indicators (bold: highest and lowest)

	Australia	*Brazil*	*Canada*	*Denmark*	*Finland*	*France*	*Germany*
Population (million)	22.3	196.7	34.5	5.6	5.4	65.4	81.8
Population density (people/km^2)	**2.9**	23.2	3.8	131.3	17.7	119.5	234.7
Urban population (% of total)	89	85	81	87	84	86	74
GDP (USD billion)	1,379	2,476	1,736	333	263	2,773	3,600
GDP per capita ('000 USD)	67.0	11.3	52.2	56.2	46.2	39.8	41.5
Gross expenditure on R&D (% of GDP)	2.4	1.2	1.8	3.1	**3.9**	2.3	2.8
Construction industry contribution to GDP (% of GDP)	**6.8**	4.8	6.0	4.7	**6.8**	5.5	4.5
Construction employees (% of workforce)	9.1	7.3	7.4	5.5	7.6	6.8	6.0

	Hong Kong (China)	*India*	*Netherlands*	*New Zealand*	*Norway*	*Sweden*	*USA*
Population (million)	7.1	**1,241.5**	16.7	**4.4**	5.0	9.4	311.6
Population density (people/km^2)	**6,786.6**	417.6	494.9	16.7	16.3	23.0	34.1
Urban population (% of total)	**100**	**31**	83	86	79	85	82
GDP (USD billion)	249	1,873	836	**160**	486	540	**14,991**
GDP per capita ('000 USD)	36.8	**1.5**	46.1	32.0	**99.6**	55.2	50.0
Gross expenditure on R&D (% of GDP)	**0.8**	0.9	1.8	1.3	1.7	3.4	2.9
Construction industry contribution to GDP (% of GDP)	**3.4**	6.0	4.4	4.0	5.4	5.0	**3.4**
Construction employees (% of workforce)	**3.7**	**11.3**	4.3	8.3	7.5	6.9	4.2

Source: Latest data available from ABS (2010); Deloitte (2011); DESTATIS (2013); Government of India (2013); Hong Kong Census and Statistics Department (2013); IBGE (2012; 2013); Insee (2013); Kane (2012); Official Statistics of Finland (OSF) (2013); Statistics Canada (2012); Statistics Denmark (2013); Statistics Netherlands (2013); Statistics Norway (2013); Statistics Sweden (2013); U.S. Department of Commerce (2012); World Bank (2013)

- significant duplication of effort, inefficiency and waste with estimates varying between 30 and 60 per cent;
- current business models and contracts reinforcing barriers to collaboration and integration.

Miozzo and Dewick (2004) regard the industry as an archetypal network system in which a coalition of firms and institutions come together on a temporary basis to undertake a project, thus often creating performance problems due to inadequate inter-organisational cooperation.

Additional challenges include intense competition; limited investment in R&D, partially due to the project-based nature of the industry; limited advantage taken of the potentials of digital modelling and electronic transactions; and increased community expectations regarding environmental and social performance (Newton, *et al.*, 2009).

Such challenges demand both technical and process innovation within the firm and across the supply chain. Transformational change is required in some areas in order to meet the productivity improvements demanded by clients and owners in countries contributing to this publication: whether it is in the way the USA is addressing challenges of climate change and infrastructure security or India is addressing the need to provide housing and transport infrastructure for a growing population.

Another key challenge around the globe is the predominance of micro, small and medium-sized firms in the construction industry. Micro and small firms especially have difficulties committing time and resources to the uptake of new technologies and associated practices limiting the benefits of R&D (Chapter 8). Another consequence of firm size, and in Brazil limited entry barriers, is low skill levels, prevalent in countries including Brazil, India and New Zealand.

Contribution to economy and productivity

The construction industry makes a significant contribution to GDP, capital formation and employment (Hillebrandt, 2000). In addition to the development of new built environment assets, the industry maintains and repairs these assets and produces the building materials that are used to construct them. The industry's significance is not only due to the fact that it provides the buildings and infrastructure on which virtually every other sector depends, but to the fact that it is such a sizeable sector in its own right (OECD, 2008).

Globally, the construction industry is responsible for between 3 to 10 per cent of GDP if only the raw site-based construction activity is considered, but, more importantly, between 10 to 30 per cent of a country's GDP if the broader construction supply network definition is used. Typically, employment is also represented by a similar proportion (DIISR, 1999; Klakegg, *et al.*, 2013). Hillebrandt (2000) explains that the *gross output of the construction industry is the value of all the buildings and works produced by the industry*. The *net* output of construction is:

> [T]he same as value added, that is, it is the value of the contribution of the construction industry itself to the production of buildings and works and excludes the inputs of other industries, such as materials and equipment . . . whatever the measure is used, an industry which produces such a large component of GDP is of great significance for any economy.
>
> (Hillebrandt, 2000)

The ABS reinforces the critical nature of construction with its estimates showing that from an initial AUD1 million of extra output in construction, a possible AUD2.9 million in additional output would be generated in the economy as a whole. This additional output would create nine jobs in the construction industry and 37 jobs in the economy as a whole (ACIF, 2002). A similar leveraging relationship has also been determined in New Zealand, where every dollar spent on construction translates into three dollars of economic activity across the economy (Kane, 2012). An improvement in the performance of this industry can therefore have a significant impact on national GDP and a country's economic health. By way of example, the Built Environment Industry Innovation Council (BEIIC) in Australia calculated that the:

> [A]ccelerated widespread adoption of BIM could boost Australia's economic output (GDP) by 0.2 basis points in 2011. As the difference in adoption of BIM increases over time, impacts on productivity also become larger. This flows on to higher GDP over time. In 2025 GDP is estimated to be 5 basis points higher, when compared with a 'Business as Usual' (BAU) scenario.
>
> (Allen Consulting, 2010)

Economic modelling shows that this economic stimulation through construction process improvement is at least double that of any other industry sector.

Accordingly, this comparative global evaluation of R&D investment and impact in construction is of critical importance in enhancing our understanding of how public and private sector policy and practice can better leverage R&D investments. Given the importance of the link between construction activity and GDP growth for Government policy making, the industry is being used *as a driver of growth, and as a catalyst for other industries to develop* (Ruddock & Ruddock, 2009).

The extensive network of players in this industry's supply chain spans primary, secondary and tertiary sectors. The weight of each part of the chain will vary across countries, according to their level of economic development, with a higher concentration of primary and secondary sector firms in developing countries and more tertiary firms in developed countries (Ruddock & Ruddock, 2009). These differences in part contribute to the distinctions that will become apparent in the following chapters.

Issues related to poor productivity growth are a focus of attention in a number of the following chapters. Challenges highlighted include: deeply embedded local laws, regulations and institutions (Chapter 12); lag in developing labour saving ideas and in finding ways to substitute equipment for labour (Chapter 16); and research to improve construction labour productivity and efficiency through the development of strategic solutions and real-time decision support systems that reduce project execution risks for owners and contractors (Chapter 5). It remains difficult to draw precise comparisons of productivity across countries, given the variations in locale, weather, labour skills, regulations, building complexity, degree of mechanisation, supply chain structures and purchasing power parities, for example.

Small and medium-sized enterprises (SMEs)

SMEs are a key source of growth, employment and innovation in modern economies due to their dynamic and flexible nature and the increasing importance of the ICT sector populated by SMEs (Parker, 2007). Efforts to compile data on the size of the SME sector across countries have been challenged by problems of comparability and consistency. Different countries adopt different criteria – such as employment, sales or investment – for defining small and medium enterprises. Hence, different sources of information on SMEs use different criteria in compiling statistics (Ayyagari, *et al.*, 2005).

Table 2.2 shows employee numbers as a crucial initial criterion for determining in which category a firm falls across the countries represented in this book.

Additionally, many countries have set limits for maximum annual turnover (Denmark, Finland, France, Germany, the Netherlands and Sweden), maximum annual balance sheet (Denmark, Finland, France and Sweden),[1] revenue from exporting services (the USA) and annual revenue (Canada). India classifies SMEs in the service sector only in terms of their investment on equipment, without considering the number of employees: micro enterprises up to INR1 million (USD18,737);[2] small enterprises up to INR20 million (USD374,742); and medium enterprises up to INR50 million (USD936,855) (Government of India, 2006). Other criteria used include: distinction between manufacturing and service firms (Hong Kong (China) and India) (Government of India, 2006; So, 2012); and exporters versus non-exporters (USA) (USITC, 2010).

More broadly, SMEs form a large proportion of each country's construction firms. For example, in New Zealand 90 per cent of all companies employ five or fewer employees (Kane, 2012); in Norway 99 per cent of all firms are classified as SMEs (Norwegian Ministry of Trade and Industry, 2013); in Hong Kong 98 per cent of all firms were SMEs and most

Table 2.2 SMEs by employee numbers

Country/region	*Maximum number of employees*		
	Micro	*Small*	*Medium*
Australia	4	19	199
Brazil	9	49	249
Canada	4	49	499
Denmark	9	49	249
Europe	9	49	249
Finland	–	49	249
France	9	49	249
Germany	9	49	249
Hong Kong (China)	–	49	
India	–	–	–
Netherlands	–	99	
New Zealand	–	20	
Norway	–	99	
Sweden	9	49	249
USA	–	499	

Source: Deutsche Bank Research (2011); DIISR (2011); Environment Canada (2010); EUResearch (2006); IBGE (2006); Innovation and Employment (2013); Insee (2008); Kless (2008); New Zealand Ministry of Business, Innovation and Employment (2013); Norwegian Ministry of Trade and Industry (2013); SMV Portalen (n.d.); So (2012); Statistics Finland (2012); Statistics Sweden (2010); USITC (2010)

had fewer than 10 employees (So, 2012); and in Brazil, 92 per cent of all firms were considered micro firms (IBGE, 2006). This representation is similarly reflected in the construction industry in most countries. In Germany, 92 per cent of all construction workers are employed by SMEs (Kless, 2008), and in Australia, construction businesses with fewer than five employees accounted for 94 per cent of all businesses and over two-thirds of all employees in this industry (de Valence, 2010).

Research and development

R&D is viewed as a crucial investment for the long-term growth of economies. Increased productivity in OECD countries has been a major determinant in economic growth. This productivity increase has been linked strongly to increases in public and private R&D. R&D intensity of countries and their growth in performance is correlated also with the share of research financed by the business sector. In addition, R&D activities contribute to national competitiveness and generate valuable public goods (OECD, 2010).

According to Statistics Canada (2011a) there are two fundamental issues in measuring R&D, the first being defining the content and scope of R&D and, the second, determining the point at which R&D becomes an asset to the economy. Therefore, there have been competing visions of how R&D and its subcomponents should be defined in the national accounts. The most common starting points for these discussions come from SNA93[3] and the OECD Frascati Manual.

Most countries represented in this book use the OECD *Frascati Manual* definition:

> Research and experimental development (R&D) comprises creative work undertaken on a systematic basis in order to increase the stock of knowledge, including knowledge of man, culture and society, and the use of this stock of knowledge to devise new applications. The term R&D covers three activities: *Basic research* is experimental or theoretical work undertaken primarily to acquire new knowledge of the underlying foundation of phenomena and observable facts, without any particular application or use in view. *Applied research* is also original investigation undertaken in order to acquire new knowledge. It is, however, directed primarily towards a specific practical aim or objective. *Experimental development* is systematic work, drawing on existing knowledge gained from research and/or practical experience, which is directed to producing new materials, products or devices, to installing new processes, systems and services, or to improving substantially those already produced or installed. R&D covers both formal R&D in R&D units and informal or occasional R&D in other units.
>
> (OECD, 2002)

Some countries, however, have made some refinements to this definition. For example, Canada uses the SNA and CSNA[4] concepts, which include a fourth activity, *innovation*, which includes the costs associated with marketing new products. It also states that when R&D is purchased either domestically or from an international source, it is considered as a *purchase of fixed capital formation and not as an input (intermediate expense) for production* (Statistics Canada, 2011b). In Norway, R&D includes the *acquisition, combination and use of existing knowledge and skills to prepare plans, projects or proposed new, altered or improved products, services or production processes* (Research Council of Norway, 2002).

Australia defines *R&D core activities* as solely experimental for tax purposes as follows:

- Whose outcome cannot be known or determined in advance on the basis of current knowledge, information or experience, but can only be determined by applying a systematic progression of work that (i) is based on principles of established science; and

(ii) proceeds from hypothesis to experiment, observation and evaluation, and leads to logical conclusions;
- That are conducted for the purpose of generating new knowledge (including new knowledge in the form of new or improved materials, products, devices, processes or services).

(Austlii, 2011)

The OECD uses the Oslo Manual (2005) as the basis to define innovation as the *implementation of a new or significantly improved product (good or service), or process, a new marketing method, or a new organizational method in business practices, workplace organization or external relations* (Canadian Science, Technology and Innovation Council, 2011).

R&D and innovation in construction

The innovation performance of the construction industry has been the focus of significant attention by industry practitioners, Government analysts and policy makers and researchers across the globe. The answer to the industry's continuing problems is said to lie in *building a stronger innovation culture to improve the rate and quality of innovation across the construction system, particularly given increasing client demands for integrated services* (Manley, *et al.*, 2008). Nevertheless, construction continues to underperform significantly compared to other industries in terms of innovation activity. In addition to those broader industry challenges already highlighted, a history of limited investment in R&D and new technologies needs to be addressed to raise the overall capacity for R&D and its adoption in this industry.

R&D investment in construction

Inadequate R&D investment has been cited across the world as one of the weaknesses of the construction industry (Forbes & Ahmed, 2011). In part this may be attributed to the *applied* nature of R&D undertaken in the construction industry, which has traditionally not been captured by country-based statistics. This applied research basis, where the industry takes advantage of knowledge spillovers from other industries, enables benefit to be gained from research undertaken by those more able to sustain a larger R&D investment. For example, some of the digital modelling advances made in the construction industry in the past decade flow on from those made in the aeronautics and manufacturing industries in the later part of the last century. Additionally, most of the improvement in construction productivity in the USA has been the result of R&D work in the manufacturing industry related to construction machinery: *earthmoving equipment has become larger and faster, and power saws have replaced handsaws* (Forbes & Ahmed, 2011).

Public sector funding for R&D is provided through a variety of mechanisms including centrally funded research institutions, contestable funding schemes and tax incentives such as concessions and credits. These are addressed in the following chapters.

In Germany, a long history exists in publicly funded research institutions. The main policy instrument for construction R&D is direct Government subsidies, with major funding coming through the Federal Ministry of Transport, Building and Urban Development (BMVBS) (Chapter 9). Additional funding is also provided through individual States and the European Union. In the US, nearly two-thirds of research expenditures was funded by Federal agencies, while industry contributed 16 per cent, academia 12 per cent, and other organisations 9 per cent (Chapter 16).

Contested funding schemes enable applicants to seek funding for specific, often short-term projects through contested rounds. In New Zealand, such funding is available through TechNZ and New Zealand Science Challenge (MBIE, 2013a, 2013b). In Australia, the Australian Research Council (ARC) offers a range of schemes contested on an annual or bi-annual basis (ARC, 2011). In Canada, the government provides funding through the Industrial Research Assistance Program (IRAP) and the Construction Portfolio (National Research Council of Canada, 2013), and in Hong Kong, the Theme-based Research Scheme focuses on academic research efforts on themes of strategic importance to the long-term development of Hong Kong (UGC, 2013).

Governments in different countries offer various mechanisms to encourage private firms to leverage their R&D investments, ranging from direct grants to R&D tax concessions and incentives. Temporary financial incentives are also used as a mechanism to encourage continued investment during difficult economic times. The OECD (2010) reported that *more than 20 OECD Governments provide fiscal incentives to sustain business R&D, up from 12 in 1995 and 18 in 2004*. The 2010 OECD report *R&D tax incentives: Rationale, design, evaluation* provides a useful outline of different public mechanisms used in those countries. This report and the following chapters provide further detail:

- Australia: collaborative grants for public/private sector collaborations; R&D tax incentives; and encouragement for SME investment (OECD, 2010).
- Brazil: sector-based funds to improve the technological capacity of companies; new innovation laws enabling subsidies to R&D companies and more favourable tax regime with incentives for R&D (Chapter 4).
- Canada: indirect support mechanisms encouraging SME investment via higher tax exemptions though the private sector contributes only a small portion of formal R&D (OECD, 2010).
- China: generous tax reductions for R&D firms located in certain new technology zones or investing in key areas such as biotech, ICT and other *high-tech* fields (OECD, 2010).
- Sweden: the Development Fund of the Swedish Construction Industry (SBUF) provides the largest source of private sector grants with financing

from around 5,000 contractors contributing to the funding of many university-based research projects complementing targeted Government innovation funding (Chapter 15).

An important consequence of investment in construction R&D is to build the absorptive capacity of both individual firms and the industry more broadly (Kraatz & Hampson, 2013) through improving the capabilities of organisations to innovate and develop associated systems. A more holistic approach to R&D investment, supported by Government policy, can thus provide broad benefits to the industry and consumers. For example, the Canadian Federal Government helps companies bring innovative products to market, a process which has impacts across the supply chain, from manufacturers and suppliers through to the end consumers (Chapter 5); and in Sweden, construction firms are actively encouraged to trial new technologies and processes through demonstration projects supported by industry and Government-funded initiatives (Chapter 15).

Collaboration, diffusion and impact

A recurring theme through the following country chapters is the importance of *collaboration*. This focus is evident between industry, Government and researchers in order to strengthen the effectiveness and impact of R&D. This is evident in the broader industry also with a move from traditional adversarial approaches in procurement to a more relationship-based approach through project alliancing and partnering.

This tripartite (or triple helix) approach to collaboration is considered a key to successful R&D and innovation enabling more effective diffusion of research outcomes. The USA identifies public–private partnerships fostered by federal funds as a key in creating roadmaps for more cost-effective construction and a more productive workforce. Canada identifies more practical and tangible benefits when such an approach is taken. Norway highlights the benefit of researchers and industry mutually challenging and inspiring one another to improve and innovate. The Swedish Government has recently brought together public sector and industry leaders to identify innovative projects for R&D funding. Finland describes the need for collaboration between businesses across the value chain and benefits to smaller companies working together and sharing data. Brazil and India cite collaboration between Government and industry to develop more efficient and lower cost technologies to help solve acute housing shortages. Denmark highlights the need for a systematic effort towards improved knowledge dissemination between researchers, the local industry, and international knowledge centres.

Measuring the impact that derives from R&D presents many practical policy challenges. Research Councils UK (2012) describe *academic impact as the demonstrable contribution that excellent research makes to academic advances, across and within disciplines, including significant advances in understanding, methods, theory and application*; and economic and societal impacts as:

> The demonstrable contribution that excellent research makes to society and the economy. Economic and societal impacts embrace all the extremely diverse ways in which research-related knowledge and skills benefit individuals, organisations and nations by:
>
> - Fostering global economic performance, and specifically the economic competitiveness of the United Kingdom;
> - Increasing the effectiveness of public services and policy;
> - Enhancing quality of life, health and creative output.
>
> (Research Councils UK, 2012)

A diverse set of measures are thus required. Garnett, Roos and Pike (2008) report on a case study carried out at Charles Darwin University with the intent of providing a reliable, repeatable and *comprehensive, transparent and agreeable assessment of a university's research value as seen by the stakeholders who sponsor research*. The following chapters address these issues in their national context.

Following on from more detailed accounts of these issues in the country chapters, Chapter 17 then provides a further analysis to improve our understanding of R&D in construction, in order to better match funding strategies to industry needs for a stronger, more productive and more highly valued industry.

Notes

1 GDP contribution by industry was not available for Denmark; therefore gross value added was used instead.
2 Average exchange rate 2012 from Federal Reserve System (2013) Foreign Exchange Rates – G.5A. Available at: www.federalreserve.gov/releases/g5a/current/ (accessed 14 January 2013).
3 System of National Accounts 1993 (SNA93). Available at: http://unstats.un.org/unsd/nationalaccount/sna1993.asp.
4 SNA Rev1 refers to the proposed treatment in the new international SNA guideline. CSNA refers to the treatment in the Research and Development Satellite Account.

References

ABS (2010) *Australian economic indicators, October 2010*, Canberra: Australian Bureau of Statistics.

ACIF (2002) *Innovation in the Australian building and construction industry: Survey report*, Canberra: Australian Construction Industry Forum for the Department of Industry, Tourism and Resources.

Allen Consulting (2010) *Productivity in the building network*, Canberra: Built Environment Industry Innovation Council.

ARC (2011) *Australian government, Australian research council: About ARC.* Available at: www.arc.gov.au/about_arc/default.htm (accessed 12 September 2012).

Austlii (2011) *Commonwealth numbered acts: Tax laws amendment (research and development) Act 2011, Income Tax Assessment Act 1997.* Available at: www.austlii.edu.au/au/legis/cth/num_act/tlaada2011399/sch1.html (accessed 7 June 2013).

Ayyagari, M., Beck, T. & Demirgüç-Kunt, A. (2005) *Small and medium enterprises across the globe*, s.l.: World Bank.

Barlow, T. (2012) *The built environment sector in Australia. R&D investment study: 1992–2010*, Brisbane: Sustainable Built Environment National Research Centre.

Canadian Science, Technology and Innovation Council (2011) *Imagination to innovation. Building Canadian paths to prosperity*, Ottawa: Science, Technology and Innovation Council, Government of Canada.

de Valence, G. (2010) 'Defining an industry: What is the size and scope of the Australian building and construction industry?', *Australasian Journal of Construction Economics and Building*, 10(1/2): 53–65.

Deloitte (2011) *Research and development expenditure: A concept paper*, London: Deloitte Touche Tohmatsu India Private Limited.

DESTATIS (2013) *National accounts, economic indicators.* Available at: www.destatis.de/DE/ZahlenFakten/Indikatoren/Indikatoren.html (accessed 23 April 2013).

Deutsche Bank Research (2011) *SMEs in the Netherlands.* Available at: www.dbresearch.com/PROD/DBR_INTERNET_EN-PROD/PROD0000000000271742/SMEs+in+the+Netherlands%3A+Making+a+difference.pdf (accessed 30 April 2013).

DIISR (1999) *Building for growth: An analysis of the Australian building and construction industries*, Canberra: Department of Innovation, Industry, Science and Resources (DIISR).

DIISR (2011) *Key statistics Australian small business*, Canberra: Australian Government Department of Innovation, Industry, Science and Research.

Dubois, A. & Gadde, L.-E. (2002) 'The construction industry as a loosely coupled system: implications for productivity and innovation', *Construction Management and Economics*, 20(7): 621–631.

Environment Canada (2010) *Small and medium-sized enterprises.* Available at: www.ec.gc.ca/p2/Default.asp?lang=En&n=D35E8873-1 (accessed 13 May 2013).

EUResearch (2006) *SME definition.* Available at: www.euresearch.ch/index.php?id=266&L=2 (accessed 30 April 2013).

Eurostat (2008) *NACE Rev. 2, Statistical classification of economic activities in the European community*, Luxembourg: Office for Official Publications of the European Communities.

Forbes, L.H. & Ahmed, S.M. (2011) *Modern construction, lean project delivery and integrated practices*, Boca Raton, FL: CRC Press, Taylor & Francis.

Furneaux, C., Hampson, K.D., Scuderi, P. & Kajewski, S. (2010) 'Australian construction industry KPIs', in *CIB World Congress proceedings – building a better world*, Barrett, P., Amaratunga, D., Haigh, R., Keraminlyage, K. & Pathirage, C. (eds), paper presented at CIB World Building Congress 2010, Salford, 10–13 May.

Garnett, H.M., Roos, G. & Pike, S. (2008) *Outcomes of higher education: Quality relevance and impact. IMHE Programme on Institutional Management in Higher Education*, Paris: OECD.

Government of India (2006) *Definitions of micro, small & medium enterprises.* Available at: www.dcmsme.gov.in/ssiindia/defination_msme.htm (accessed 30 April 2013).

Government of India (2013) *Data table.* Available at: http://planningcommission.gov.in/data/datatable/index.php?data=datatab (accessed 23 April 2013).

Hampson, K.D., Kraatz, J.A., Sanchez, A.X. & Herron, N.A. (2013) *Investing for impact*, Brisbane: Sustainable Built Environment National Research Centre (SBEnrc).

Hillebrandt, P.M. (2000) *Economic theory and the construction industry*, 3rd edn, Basingstoke: Macmillan Press Ltd.

Hong Kong Census and Statistics Department (2013) *National income, census and statistics department*, Hong Kong: Government of the Hong Kong Special Adminsitrative Region. Available at: www.censtatd.gov.hk/hkstat/sub/sp250.jsp?tableID=036&ID=0&productType=8 (accessed 26 April 2013).

IBGE (2006) *Demografia das empresas* (*Business demography*), Rio de Janeiro: Instituto Brasileiro de Geografia e Estatística.

IBGE (2012) *Censo demográfico 2010, Resultados gerais da amostra* (*Census 2010, overall results of the sample*), Rio de Janeiro: Instituto Brasileiro de Geografia e Estatística.

IBGE (2013) *Instituto Brasileiro de geografia e estatística (IBGE)* (*Brazilian Institute of geography and statistics*). Available at: www.ibge.gov.br/home/estatistica/indicadores/pib/defaultcnt.shtm (accessed 26 April 2013).

Insee (2008) *Small and medium enterprises/SME.* Available at: www.insee.fr/en/methodes/default.asp?page=definitions/petite-moyenne-entreprise.htm (accessed 30 April 2013).

Insee (2013) *Macro-economic database.* Available at: www.bdm.insee.fr/bdm2/index.action?request_locale=en (accessed 23 April 2013).

Jonassen, J.O. (2010) *Report on integrated practice. Changing business models in BIM-driven integrated practice*, Seattle, WA: American Institute of Architects (AIA).

Kane, C. (2012) *Productivity roadmap*, Wellington: Building and Construction Sector Productivity Partnership.

Klakegg, O.J., *et al.* (2013) 'Introduction, context and summary', paper presented at 7th Nordic Conference on Construction Economics and Organisation, Trondheim, 12–14 June 2013.

Kless, S. (2008) *Small and medium-sized enterprises in Germany, Statistisches Bundesamt (DESTATIS).* Available at: www.destatis.de/EN/Publications/STATmagazin/EnterprisesBusinessNotification/2008_8/2020_8SMEs.html (accessed 30 April 2013).

Kraatz, J.A. & Hampson, K.D. (2013) 'Brokering innovation to better leverage R&D investment', *Building Research & Information*, 41(2): 187–197.

Lewis, T.M. (2009) 'Quantifying the GDP-construction relationship', in *Economics for the modern built environment*, Ruddock, L. (ed.), Abingdon: Taylor & Francis.

Manley, K., Marceau, J., Parker, R.L. & Matthews, J.H. (2008) 'The potential contribution of small firms to innovation in the built environment', paper presented at 22nd ANZAM Conference 2008 on Managing in the Pacific Century, Auckland, 2–5 December.

Marceau, J., *et al.* (1999) *Mapping the building and construction product system: Preliminary report*, Canberra: Australian Expert Group on Industry Studies, University of Western Sydney (AEGIS).

MBIE (2013a) *National science challenges*. Available at: www.mbie.govt.nz/what-we-do/national-science-challenges (accessed 9 September 2013).

MBIE (2013b) *TechNZ terms and conditions*. Available at: www.msi.govt.nz/update-me/archive/frst-funding-archive/technz-terms-and-conditions/ (accessed 9 September 2013).

Miles, I. (2006) 'Innovation in services', in *The Oxford Handbook of Innovation*, Fagerberg, J., Mowery, D.C. & Nelson, R.R. (eds), Oxford Handbooks Online.

Miozzo, M. & Dewick, P. (2004) 'Networks and innovation in European construction: Benefits from inter-organisational cooperation in a fragmented industry', *International Journal of Technology Management*, 27(1): 68–92.

National Research Council of Canada (2013) *Industrial research assistance program (IRAP). National research council Canada*. Available at: www.nrc-cnrc.gc.ca/eng/irap/index.html (accessed 9 September 2013).

Newton, P., Hampson, K.D. & Drogemuller, R. (2009) 'Transforming the built environment through construction innovation', in *Technology, Design and Process Innovation in the Built Environment*, Newton, P., Hampson, K.D. & Drogemuller, R. (eds), Abingdon: Spon Research.

New Zealand Ministry of Business, Innovation and Employment (2013) *Small and medium sized enterprises*. Available at: www.med.govt.nz/business/business-growth-internationalisation/small-and-medium-sized-enterprises (accessed 30 April 2013).

Norwegian Ministry of Trade and Industry (2013) *Small and medium business*. Available at: www.regjeringen.no/en/dep/nhd/selected-topics/simplification-for-business/sma-og-mellomstore-bedrifter.html?id=614069 (accessed 30 April 2013).

OECD (2002) *Frascati manual, Proposed standard practice for surveys on research and experimental development*, Paris: Organisation for Economic Cooperation and Development (OECD).

OECD (2008) *Competition in the construction industry*, Paris: Organisation for Economic Cooperation and Development, Directorate for Financial and Enterprise Affairs Competition Committee.

OECD (2010) *R&D tax incentives: Rationale, design, evaluation*, Paris: Organisation for Economic Cooperation and Development (OECD).

Official Statistics of Finland (OSF) (2013) *Annual national accounts*. Available at: www.stat.fi/til/vtp/index_en.html (accessed 23 April 2013).

Parker, R. (2007) 'Innovative methodologies in enterprise research: Tackling the question of the role of the state from a macro and micro perspective', in *Innovative Methodologies in Enterprise Research*, Hine D. & Carson D. (eds), Cheltenham: Edward Elgar Publishing Inc.

Research Council of Norway (2002) *Skattefradrag for kostnader til forskning og utvikling (FoU-fradrag) – gjeldende fra og med inntektsåret 2012 (Tax deduction for expenses on research and development (R&D deductions) – current as of fiscal year 2012)*. Available at: www.forskningsradet.no/servlet/Satellite?blobcol=urldata&blobheader=application%2Fpdf&blobheadername1=

Content-disposition%3A&blobheadervalue1=+attachment%3B+filename%3D%22ForskriftforSkatteFUNN-gjeldendefraogmedinntekts%C3%A5ret2012.pdf%22&blobkey= (accessed 7 June 2013).

Research Councils UK (2012) *Excellence with impact*. Available at: www.rcuk.ac.uk/kei/impacts/Pages/meanbyimpact.aspx (accessed 30 September 2013).

Ruddock, L. & Ruddock, S. (2009) 'The scope of the construction sector: Determining its value', in Ruddock L. (ed.), *Economics for the modern built environment*, Abingdon: Taylor & Francis.

SMV Portalen (n.d.) *Definition af SMV*. Available at: www.smvportalen.dk/Om-smvportalen/definition-af-smv (accessed 30 April 2013).

So, G. (2012) *LCQ3: Measures to assist micro-enterprises and small and medium-sized enterprises*. Available at: www.info.gov.hk/gia/general/201205/30/P201205300299.htm (accessed 30 April 2013).

Statistics Canada (2011a) *Definitions and conceptual issues*. Available at: www.statcan.gc.ca/pub/13-604-m/2007056/s4-eng.htm (accessed 23 May 2013).

Statistics Canada (2011b) *Scope of research and development in Canada*. Available at: www.statcan.gc.ca/pub/13-604-m/2007056/t-c-g/tbl1-eng.htm (accessed 7 June 2013).

Statistics Canada (2012) *CANSIM*. Available at: http://cansim2.statcan.ca/ (accessed 21 November 2012).

Statistics Denmark (2013) *StatBank Denmark*. Available at: www.statbank.dk/statbank5a/default.asp?w=1920 (accessed 26 April 2013).

Statistics Finland (2012) *Small and medium size enterprises*. Available at: www.stat.fi/meta/kas/pienet_ja_keski_en.html (accessed 30 April 2013).

Statistics Netherlands (2013) *Statline, quarterly national accounts; Values*. Available at: http://statline.cbs.nl/StatWeb/?LA=en (accessed 23 April 2013).

Statistics Norway (2013) *Statistics Norway (statistisk sentralbyrå)*. Available at: www.ssb.no/en/statistikkbanken (accessed 22 April 2013).

Statistics Sweden (2010) *De små och medelstora företagens ekonomi 2008 (Small and medium-sized firms' finance 2008)*, Stockholm: Statistiska Centralbyrån (Statistics Sweden).

Statistics Sweden (2013) *National accounts, quarterly and annual estimates, GDP quarterly 1993–2012: 4*. Available at: www.scb.se/Statistik/NR/NR0103/2012K04/Kvartalstabeller_BNP%20kv%204%202012_201309.xls (accessed 23 April 2013).

Trewin, D. & Pink, B. (2006) *Australian and New Zealand standard industrial classification (ANZSIC)*, Canberra: Australian Bureau of Statistics/Statistics New Zealand.

Tunzelmann, N.V. & Acha, V. (2006) 'Innovation in "low-tech" industries', in *The Oxford Handbook of Innovation*, Fagerberg, J., Mowery, D.C. & Nelson, R.R. (eds), Oxford Handbooks Online.

UGC (2013) *Theme-based research scheme*. Available at: www.ugc.edu.hk/eng/rgc/theme/theme.htm (accessed 9 September 2013).

UNSD (2008) *ISIC Rev. 4 code F*. Available at: http://unstats.un.org/unsd/cr/registry/regcs.asp?Cl=27&Lg=1&Co=F (accessed 23 May 2013).

U.S. Department of Commerce (2012) *Bureau of economic analysis, Industry data*. Available at: www.bea.gov/iTable/iTable.cfm?ReqID=5&step=1#reqid=5&step=2&isuri=1&403=1 (accessed 19 April 2013).

USITC (2010) *Small and medium sized enterprises: Overview of participation in U.S. exports*. United States International Trade Commission. Available at: www.usitc.gov/publications/332/pub4125.pdf (accessed 30 April 2013).

World Bank (2013) *The World Bank: Working for a poverty-free world*. Available at: http://data.worldbank.org/indicator (accessed 19 April 2013).

3 Australia – R&D investments in the construction industry

Judy A. Kraatz and Keith D. Hampson

Background

Australia's population in 2012 was approximately 22.7 million people, with annual population growth for the year ended 30 June 2012 at 1.6 per cent. Overseas migration provides an important source of population growth and skilled labour (ABS, 2007). The total land area is 7.7 million sq km with 1.5 million sq km being classified as forested (Montreal Process Implementation Group for Australia, 2008), with a population density of 2.9 people per sq km (World Bank, 2013). The Australian climate spans from alpine to desert, with temperate, subtropical and tropical regions creating diverse climatic conditions and requiring expansive building regulations.

Australians are well educated in general, with those with qualifications steadily increasing. The proportion of people aged 15–64 years who were enrolled in study for a qualification increased from 17 per cent in 2001 to 19 per cent in 2012 (ABS, 2012a). The population is well connected with just over 12.2 million classified as internet subscribers by December 2012 (ABS, 2012b).

The gross domestic product (GDP) in 2012 was estimated at AUD1.4 trillion (USD1.54 trillion) (up from USD1.03 trillion in 2008), with GDP real growth rate (2012 est.) of 3.6 per cent (2.4 per cent in 2008), and GDP per capita (2012 est.) of AUD65,000 (USD68,000) (USD48,000 in 2008) (DFAT, 2012).

The Australian construction industry

In 2008 the cumulative value of site-based residential, non-residential and engineering construction was AUD160 billion (Newton, *et al.*, 2009). This has grown to AUD201 billion in 2012–2013 (Master Builders Australia, 2013) with a forecast value of AUD203 billion in 2015–2016 reflecting current global and domestic challenges.

The construction industry includes three major sectors: residential building; non-residential building such as commercial premises or educational and recreational facilities; and engineering construction including transport, electrical and water infrastructure facilities (de Valence, 2010). In the context of statistical data, the following sectors and fields apply (Barlow, 2012): (i) the building construction sector, *with civil engineering and built environment and design* being the key fields; (ii) the heavy and civil engineering construction sector, with key fields being *civil engineering, interdisciplinary engineering, resources engineering and information and computing sciences*; and (iii) the construction services sector, with activity in fields including *civil engineering, built environment and design, information and computing sciences and mechanical engineering.*

Construction industry activity in 2010 in Australia accounted for 6.8 per cent of GDP (ABS, 2010) with the industry employing 1,033,900 in 2011 through 351,890 firms comprising 17 per cent of Australia's businesses (ABS, 2012b). Additionally small and medium-sized enterprises (SMEs), that is firms with under 200 employees, were responsible for 64 per cent of the value add in the construction industry from 2009 to 2010, with small businesses employing 69 per cent of the industry's workforce (DIISR, 2011). The Australian Bureau of Statistics (ABS) estimates that from an initial AUD1 million of extra output in construction, AUD2.9 million in additional output could be generated in the economy as a whole. This would create nine jobs in the construction industry and 37 jobs in the rest of the economy (ACIF, 2002).

The nature of the Australian construction industry

The Australian Expert Group on Industry Studies (AEGIS, 1999) identified activity in the Australian construction industry as a *product system*, as opposed to a cluster, complex or sector, due to: (i) its reach into both services and manufacturing; and (ii) the manner in which innovation in this system impacts across products, processes and services. This system includes elements of goods-producing industries; goods-related service industries; knowledge-based services; in-person services; and Government and defence activities.

The Australian Royal Commission into the Building and Construction Industry (2002) adopted this approach and de Valence (2010) reinforces the appropriateness of this broader perspective. The former highlights the degree of complexity and the interrelatedness of those involved in the building and construction industry. This report lists the major industry associations involved with over 80 employer and industry associations, organisations and unions named. The de Valence (2010) research reviews *available data on industry size and scope and compares differences between the structure-conduct-performance approach and the alternative industry cluster approach*. This report presents industry-related data to 1997, which demonstrates the need for this inclusive approach. Further to this, Hampson and Manley (2001) detail the key players in the Australian built environment including research institutions; standards and regulatory bodies; training providers; industry associations; and the like. They provide statistics on industry output, contribution to GDP, employment, income, value of work and employment by subsector.

The Australian Department of Innovation, Industry, Science and Research (DIISR, 2010) identifies an overall decline in spend on science and innovation as a percentage of GDP in Australia since 1993–1994 of 22 per cent. Australia's expenditure on R&D as a percentage of GDP is 2.4 per cent, compared to that of Germany and the United States, which are 2.8 and 2.9 per cent respectively, and Denmark, Finland and Sweden, which are each more than 3 per cent (Chapter 2). To address this, the Australian Government has identified a number of key initiatives including a target of 25 per cent increased business engagement in innovation over the next 10 years; doubling the tax incentive for small business, a critical component of the building and construction industry in Australia; supporting targeted responses to climate change; improving innovation skills and capabilities in the workplace; and maintaining a focus on business innovation through Government-sponsored industry innovation councils such as the past Built Environment Industry Innovation Council (BEIIC) (DIISR, 2010). Informing this is the Australian Government commitment to an increased *use of metrics, analysis and evaluation to inform policy development and decision making* (DIISR, 2010). More specifically, Hampson and Manley (2001) report on the relatively poor innovation record of the Australian industry with an R&D expenditure of 1.4 per cent compared to the share of site-based construction activity in total output of 6.5–7 per cent of GDP.

Productivity growth in this industry continues to lag behind that of the rest of the economy (Property Council of Australia, 2009). To address this, the Australian Procurement and Construction Council (APCC) and the Australian Construction Industry Forum (ACIF) identified a set of national key performance indicators (KPIs) to track industry productivity performance. These indicators were further developed by the Australian Cooperative Research Centre for Construction Innovation (CRC CI) (Furneaux, *et al.*, 2010). These KPIs relate to safety; productivity and competitiveness; economic security; workplace capability; and environmental sustainability/eco-efficiency. Examples of poor performance in these areas that illustrate the extent of the industry problem include:

- Deaths in construction increased from 3.14 deaths per 100,000 workers in 2004 to 4.27 in 2008 (CFMEU, 2010). This compares to an overall fatality rate of 2.7 deaths per 100,000 workers across all industries. Data for 2012 reveal a similar rate of 4.26 fatalities per 100,000 workers being *nearly twice the national fatalities rate of 2.23* (Safe Work Australia, 2012).
- Less than average productivity growth: were productivity growth to match that of the market sector, economic modelling shows that the accumulated gain in real gross domestic product between 2003 and 2010 would approximate AUD12 billion (Royal Commission, 2002).

- Poor documentation contributing an additional 10 to 15 per cent or more to project costs in Australia (Engineers Australia, 2005) with *substandard project documentation equating to an estimated financial loss of AUD12 billion nationwide annually* (EA, 2005).

These examples highlight the need for those participating in R&D to establish a more focused investment platform to address the complex challenges facing this industry.

R&D investment in Australia (1992–2010)

In 2011 the Australian Sustainable Built Environment National Research Centre (SBEnrc) commissioned an audit and assessment of R&D investment in the Australian construction industry. Findings of this study (Barlow, 2012) provide a valuable overview of investment and significant trends in both private and public sector investment in this country.

Shifts in investment

A substantial increase in private sector investment occurred between 1992 and 2010, while public sector investment over this same period decreased as a proportion of total spending (Figure 3.1). In the early 1990s, Australian

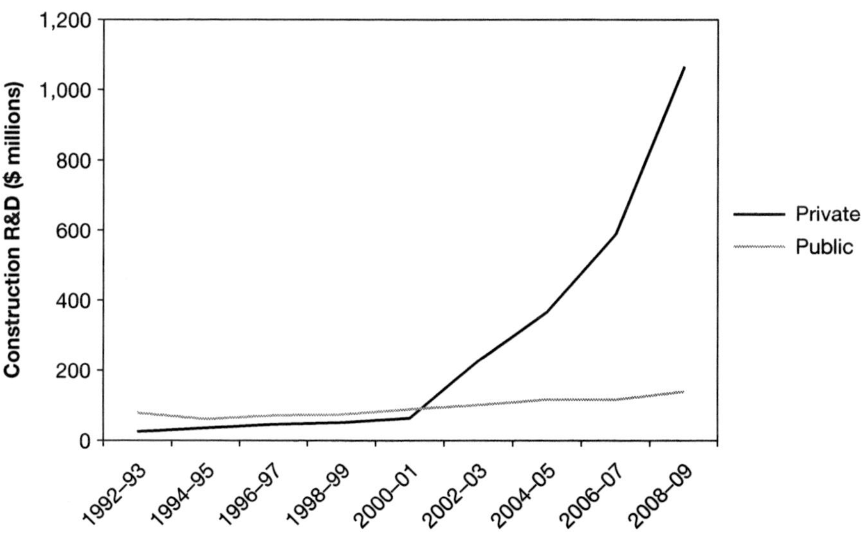

Figure 3.1 Private versus public R&D on construction, Australia

Source: Barlow (2012)

Note: (i) Derived from ABS 8112 and Barlow 2012. (ii) Shows R&D expenditures by sector focused on the socioeconomic objective *construction*. (iii) *Public R&D counts* R&D from the university sector and from State and Federal government agencies

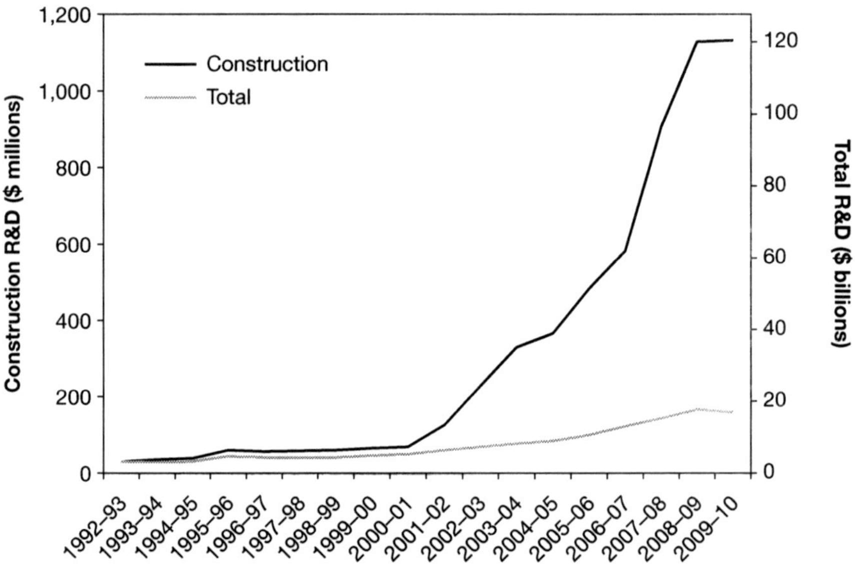

Figure 3.2 Growth in construction R&D relative to total business R&D, Australia

Source: Barlow (2012)

Note: (i) Derived from ABS 8109. (ii) Compares business R&D expenditures focused on the socioeconomic objective *construction* (left axis) with total business R&D expenditures (right axis). (iii) The right axis has been adjusted so that the growth rates of both curves from 1992 are comparable

public institutions were spending three times more on construction related R&D than private sector Australian organisations did. Yet, by 2008 Australian businesses were spending eight times as much on construction-related R&D as public research institutions.

Additionally, a greater percentage of construction research is being undertaken within the built environment sector when compared to total business R&D (Figure 3.2).

Conversely, it is also the case that the Australian Government R&D agencies have a reduced emphasis on construction R&D as a proportion of total spending. Between 1992 and 2008, Government agency spending on construction R&D fell from 2.2 per cent to 0.5 per cent of total Government sector R&D expenditure (Figure 3.3).

Further to this, R&D activity within the Australian construction industry has grown in comparison with selected other Organisation for Economic Cooperation and Development (OECD) nations, based on the OECD Structural Analysis (STAN) Database (Barlow, 2012). Over the past decade, Australian businesses have dramatically increased their share of global construction R&D.

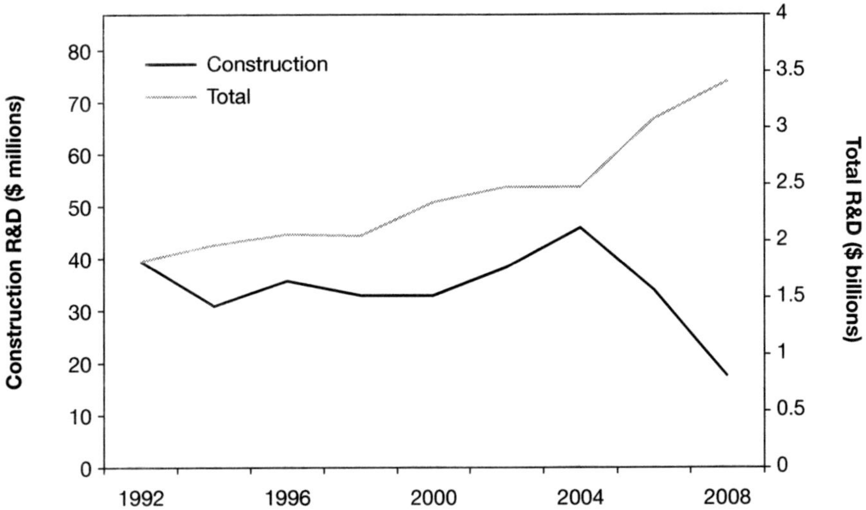

Figure 3.3 Government agency R&D focused on construction, Australia

Source: Barlow (2012)

Note: Derived from ABS 8109 and Barlow 2012. (ii) Compares Government intramural R&D expenditures focused on the socioeconomic objective *construction* (left axis) with total Government intramural R&D expenditures across all objectives (right axis). (iii) The right axis has been adjusted so that the growth rates of both curves from 1992 are comparable

Case studies

A lead SBEnrc project, Leveraging R&D Investment for the Australian Built Environment, has been investigating R&D investment in the Australian construction industry from 1990 to 2010. The overarching goal of this project has been to understand how to maximise the benefits of R&D to Australia's infrastructure and building industry through better matching funding and research strategies to industry needs. The research aims to build new understandings and knowledge relevant to R&D funding patterns, research team formation and management, future industry needs and R&D strategy. To this end, a research method was designed to illustrate the nature of such investments; drivers, successes and barriers to investment; organisational capabilities that contributed to outcomes; and outcomes and impacts of these initiatives. These issues were examined through a series of four nation-wide case studies. Sources of data included: 35 formal interviews (late 2011 to early 2012) for three public sector case studies; and a survey with 61 private sector participants (in late 2012) and six follow-up interviews (early 2013) to build an understanding of private sector investment.

Case study: pursuing the green building agenda in Western Australian (2001–2012)

The Western Australian Government (WAG) has taken a lead role in driving and delivering on green building initiatives, especially from 2001 (Figure 3.4). This process has been characterised by both formal and informal R&D activities including the development of policy guidelines and regulations, and strong links with external agencies. Significant political and bureaucratic support in its early stages was received from: (i) the Premier's office (2001–2006) through the establishment of the Sustainable Policy Unit in 2002; (ii) staff from across 42 divisions who contributed to the State Sustainability Strategy (2003); (iii) WAG departments including Works and Housing, Planning and Infrastructure, and Education; and (iv) allied agencies including LandCorp. This recent focus has been underpinned by a long-term awareness of such issues, through, for example, the publication of *Energy management in the design of new buildings* in 1980 (Western Australia Architectural Division Energy Management Committee, 1980).

Key drivers were also provided by nation-wide initiatives including: National Strategy on Energy Efficiency, an initiative of the Council of Australian Governments; the Energy Roundtable; Online System for Comprehensive Activity Reporting (OSCAR); the Australian Building Codes Board and developments in the environmental and sustainability provisions of the Building Code of Australia; and the Solar Cities programme.

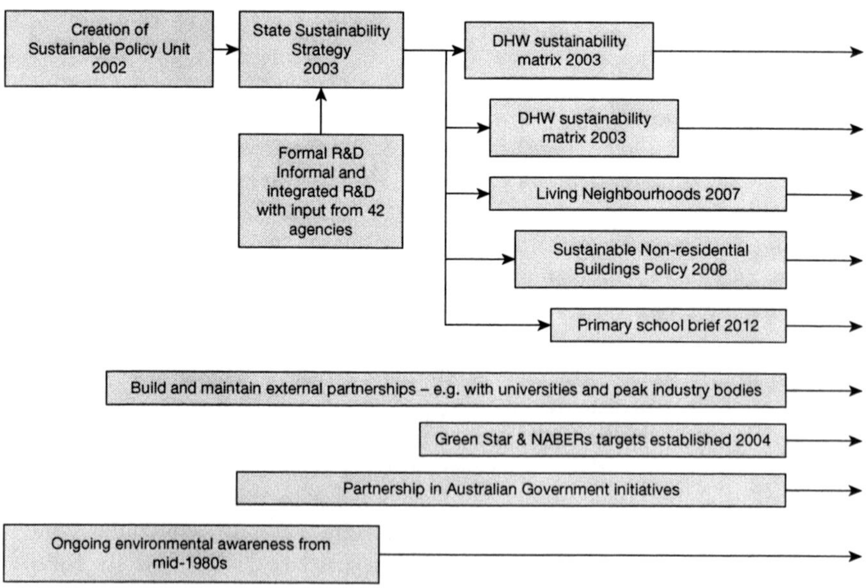

Figure 3.4 Western Australian Government's pathway to greener buildings

In 2003, the establishment of National Australian Built Environment Rating System (NABERS, formerly known as Australian Building Greenhouse Rating) and Green Star rating scheme through the Green Building Council of Australia (GBCA) also provided crucial tools to enable Government and industry to quantify outcomes. Through mandating the use of these tools, WAG also provided a critical lever for achieving enhanced environment and social outcomes in the built environment. Additional leverage has been achieved through the establishment of key relationships with external parties including: other State and local planning authorities; research institutions; industry and supply chain; and industry associations.

Based on interviews conducted in 2011, WAG demonstrated embedded organisational capabilities that facilitated this approach. These included:

- ongoing development of new products and processes including new policies, guidelines and undertaking pilot projects;
- organisational learning such as through 50 PhDs that informed the development of the State sustainability strategy;
- engaging with external R&D agencies such as the CRC CI and Curtin University's Sustainability Policy Unit (CUSP);
- interaction with industry organisations such as the GBCA;
- focusing on financial benefits and cost reduction, for example energy savings.

Other commonly reported benefits of innovation included the creation of knowledge networks and the empowerment of innovation leaders. Outcomes have included:

- the *State sustainability strategy* and associated documents, regulations and policies that informed organisational learning;
- R&D engagement, both formal and informal, with a cross-spectrum of academic institutions and innovation brokers;
- cost benefits such as enhanced business case development, requests for proposals and cost savings associated with increased resource efficiency;
- greater knowledge creation, exploitation and flows including in a project context;
- uptake of new metrics for target setting, assessing and reporting on innovation and performance.

Case study: digital modelling in Queensland public buildings (2005–2012)

This case study explored the evolution of digital project delivery processes in the Queensland Department of Public Works (QDPW) Project Services from: initial implementation of computer-aided design and documentation (CADD) in the mid-1980s; to experimentation with and implementation of building information modelling (BIM) from the mid- to late 2000s; to recent moves (2010–2012) towards integrated project delivery (IPD).[1]

Project Services adopted a strong vision to advance development across the supply chain (Figure 3.5). This was characterised by a focus on developing more efficient delivery mechanisms through the use of new technology enablers, coupled with process changes including pilot projects, strong researcher engagement and targeted industry leadership and partnerships.

QDPW Project Services implemented an integrated and informal R&D process with the incremental adoption of new technologies and work practices. Internal proof of concept was achieved on a project-by-project basis, which was complemented with formal R&D engagement through core involvement with CRC CI from 2001 to 2009, and now with SBEnrc. Complementing this, they have been involved with a number of Australian Research Council (ARC) Industry Linkage projects led by the Queensland University of Technology (QUT) and the Royal Melbourne Institute of Technology (RMIT). Key outcomes include:

- Mareeba Court House and Police Station (2006): first BIM pilot;
- Queensland State Archives (2006): 4D model developed;
- North Lakes Police Station (2008): BIM approach further developed;
- structural steel design provided to the fabricator from Project Services;
- Dandiiri Contact Centre (2008): 4D model developed including energy modelling. The building was awarded the highest environmental performance of any Australian building under construction at that time

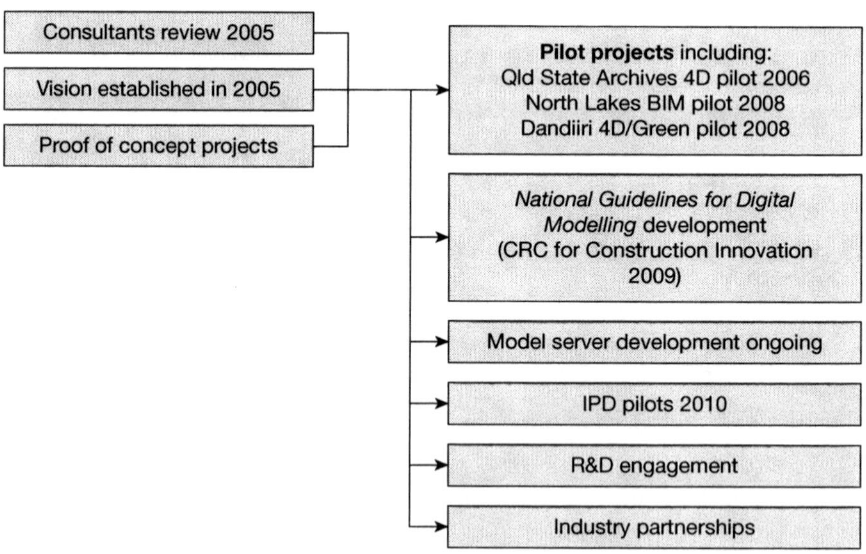

Figure 3.5 Digital modelling pathway to innovation

- development of national guidelines for digital modelling (2009) with the CRC CI.

Industry-wide impacts include QDPW Project Services being acknowledged as a national and international leader in this field characterised by strong research partnerships, industry consultation and engagement. There has been extensive dissemination of leading work practices to other industry researchers, external contractors, suppliers and vendors. This has led to the demonstration of significant productivity benefits for the industry as highlighted by the BEIIC in its 2010 report (Allen Consulting Group, 2010).

Case study: private sector R&D investment (2005–2012)

The intent of this case was to further develop the narrative regarding the nature of construction industry R&D investment throughout the past decade; building on the audit and investment of past investment, and the three prior public sector case studies. With a focus on private sector engagement, issues identified to be addressed included:

- *Nature of the investment*: process differences between large, medium-sized and small firms; focus/area of research; nature of change (transformative or incremental; technical, process and/or coupled);
- *Motivation for investment*: reasons for investment; impact of changes to tax incentives and concessions; and demand-side changes;
- *Mechanisms*: role of strategic relationships; and dissemination avenues and mechanisms;
- *Impact of investments*: for example, on quality, return on investment, safety, skills and supply chain; short- to medium- or long-term impacts.

The findings of the first phase of this case study illustrate the differences that exist between three of the subsectors, namely, consultants, contractors and suppliers (each self-identified), and firm size. The survey included 61 participants: 20 consultants, 21 contractors and 20 suppliers; 41 per cent were from large firms; 36 per cent from medium-sized firms and 23 per cent from small firms. Of these, 55 per cent self-identified as being building-related firms including architects and planners; 23 per cent as civil firms; 7 per cent delivering services in both areas; and 13 per cent in services engineering. Of the 20 consultants surveyed, 17 are SMEs, as were 10 of the 21 contractors; and 12 of the 20 suppliers self-reported as large firms. A second phase, including interviews with a cross-section of respondents, is yet to be finalised.

A significant increase in firms introducing new services, products and/or processes as a result of R&D activities was recorded. From a pre-2005 level of 32.8 per cent this rose to 71.6 per cent (2009–2011) (Figure 3.6).

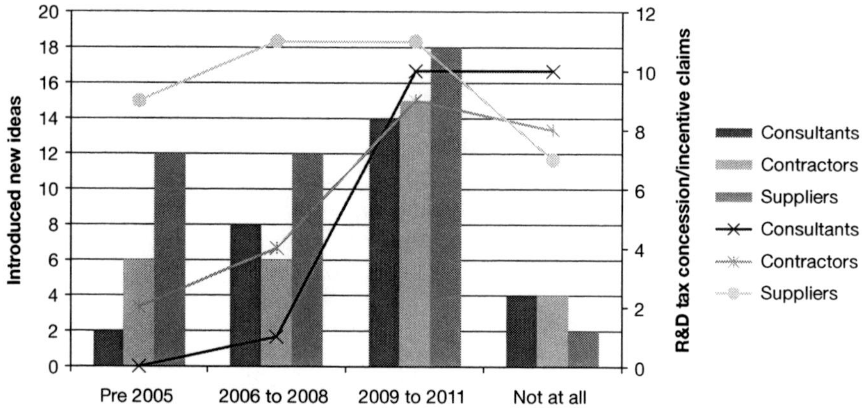

Figure 3.6 Introduction of new ideas correlated with R&D tax concessions/incentive claims

Other findings from this case study include:

- *A difference in key drivers* was found between subsectors: for both consultants and suppliers, market-related issues such as being at the cutting edge, improving responsiveness to customer, and increasing market share; productivity, namely, increasing efficiency, improving quality and improving IT capabilities and capacity; and profit, namely, increasing revenue and reducing costs; ranked highly as key drivers. For contractors, workplace, health and safety (WH&S) followed by productivity, market-related and environmental issues received the top rankings. Drivers were consistent when looked at by size of firm.
- *Internal resources were by far the main sources of new ideas*: consultants turn to websites, journals, conferences and seminars; contractors to their clients, partners and suppliers; and suppliers to a cross-section of all these sources.
- *Dissemination of outcomes* for all subsectors was among the immediate network, including partners, consultants, subcontractors and service providers, via internal training and industry associations and awards for suppliers.
- *Changes made as a result of R&D* activity were different in each subsector where changes implemented by consultants were predominantly related to service delivery; by contractors to their processes, namely, construction, operations and management; and by suppliers to each of these aspects.
- *Impact* of the R&D activities: all subsectors reported improved market-related and project outcomes, followed by profit-related impacts for consultants and suppliers, and process changes for contractors.

Alignment between the drives and changes/impacts is also apparent from the findings.

A roadmap for the Australian construction industry – Construction 2030

A *roadmap for R&D priorities for Australia's built environment* (Bok, *et al.*, 2012) was developed as a part of the aforementioned SBEnrc project to build an understanding on the likely future landscape of this industry. This roadmap (C2030) builds on *Construction 2020* (Hampson & Brandon, 2004), which was developed through extensive, nation-wide industry consultation, leading to nine industry visions for future practice.

The intent of C2030 was to establish a set of priorities that respond to likely industry future scenarios to guide investment. The research team developed a set of macro social and environmental drivers with the potential to impact the construction industry in the future. A map of emergent trends and decision-making scenarios was then developed, which related to likely future uncertainties around climate change, skills, economy, attitudes, policies, governance, energy and technology. Industry leaders then tested these through a series of national workshops in which possible technology capabilities were matched to these scenario conditions. An expert review was then undertaken as to the potential timing and impact of these emerging technologies on the industry.

From this, two categories of required research area were identified: (i) areas that require adaptation to local conditions or partnering with other industries; and (ii) areas that are critical to the industry and require an active research investment (Table 3.1).

Challenges and recommendations

The culmination of this recent SBEnrc research has been the identification of: (i) challenges; (ii) future models for industry R&D activity; and (iii) policy recommendations to inform ongoing discussion in this arena. These are presented in the context of a *vision* for the future that is underpinned by a culture of self-improvement, mutual recognition, respect and support and include:

- a *national industry steering body* that defines long-term strategic industry R&D priorities, and funds associated research in public organisations;
- Government procurement equipped to support construction innovation and supply matching funds for strategic R&D;
- *research institutes with world-leading interdisciplinary capabilities* to provide expertise relevant to the goals of the Australian construction industry.

Table 3.1 Summary of R&D priorities for Australia's property and construction industry

Research area	*Description*	*Industry need*
CONDUCT ACTIVE RESEARCH		
1. Model-based facility lifecycle business models	Model-based IT has potential to facilitate changes in how business is structured/value captured. Enabling alternative business models will be crucial to commercialising critical technologies and solutions	Key link between capital asset and more effective asset delivery and management Collaborative processes supported by robust facility lifecycle management tools
2. Intelligent infrastructure and buildings	Electronics, sensor and communication, analysis and network applications that improve: key aspects of infrastructure and buildings; occupant welfare; and lifecycle sustainability	Enhance control, automation, integration and communication of facility using long-life sensor systems. Enable longer view with reduced lifecycle costs
3. Solutions for a more sustainable built environment	Different types of solution can enhance sustainability throughout project lifecycle. Create incentives for the development and use; many solutions dependent on novel systems, standards, tools, and financial and business models	Adapt to changing business conditions including market and regulatory environment Greening existing and future built environment and adapt to climate change
CONDUCT RESEARCH FOR LOCAL CONDITIONS		
4. ICT for radical redesign	ICT critical to facilitate improved design including need to disseminate information; support new materials, trends, processes and asset management. Predictive tools and optimisation techniques for integrating product and process design at asset level to urban or network level required	Respond to climate change at facility, precinct and regional design levels. Find new energy balances in the design of built environment systems brought about by changes to energy generation
5. Biotech for tree-based materials	Research underway into materials, products and processes based on trees including: UV; moisture and decay resistance; insulation; conduction performance; nano-cellulose-metal composites	Respond to societal expectations, climate change and skills shortages. Possibilities for new materials with customised properties and more effective processes
6. Educational curricula	Need for lifelong learning, shifts in business models, advanced ICT and sustainability presents challenges and opportunities. Includes education in technical, operational and management areas	Integrated teaching in use of new approaches and technologies. Stronger integration of research and teaching, and customised career-long education

Source: Condensed from Bok, *et al.* (2012)

Challenges for the Australian construction industry

The following challenges have been identified as the culmination of two years of research including discussions with State and territory chief scientists across Australia in 2012:

- *Government spend*: between 1992 and 2008, Government agency spending on construction R&D fell from 2.2 per cent to 0.5 per cent of total Government sector R&D expenditure. This has occurred despite the Australian construction industry growing in terms of gross value added significantly faster than the Australian GDP in the last two decades.
- *Timeframes*: there is a mismatch in the nature of research objectives sought from public sector funding, often medium to long term, and from private sector funding, often short term. *A mechanism to encourage/enable public organisations to build greater long-term strategic capabilities is required.*
- *Fragmented nature of the industry*: historically, there has been no coherent strategic planning within the industry due to: fragmentation; the project-centred nature of the industry; and limited capacity for diffused, organisational or industry learning. *Industry will only act strategically to define common problems if it has a clear incentive to contribute and if there is a well-defined structure to define such problems.*
- *Industry structure*: the SME nature of the industry restricts the capacity of most firms to invest directly in long-term R&D. Furthermore, SMEs lack channels to access new ideas when not directly involved in research. *There is a need to provide resources to support both R&D and training for SMEs.*
- *Government risk aversion*: public sector clients are often risk averse, seeking the lowest conforming tender with the least possible risk. This can also restrict innovation in asset construction and maintenance where alignment between research and procurement is lacking. *Governments need to: encourage innovation through the procurement process; establish generic standards for public good and industry outcomes; and introduce an R&D component to all projects.*
- *Public sector expertise*: in most cases, there is greater technical excellence in the private sector than in the public sector, partially due to the level of R&D investment in the respective sectors. Furthermore, there is little incentive for researchers to engage in industry collaborations due to the greater prestige offered by national competitive grant-funded research and lower perceived value from industry-relevant research. *An opportunity exists for practitioner/researcher exchange to build a shared understanding of a culture of innovation unhindered by traditional models. This would lead to: a rise in interdisciplinary approaches; and potentially unorthodox solutions to industry challenges.*

Future models for industry R&D engagement

Several models for engagement have been considered.

Industry-sponsored research councils

Globally, several industries have formed their own research funding bodies. Typically, such bodies have started as industry initiatives without Government funding or leadership, but often go on to leverage Government funds. A key feature of this model is the research focus on strategic, precompetitive research of the sort that would benefit all members of a consortium collectively, rather than individual firms. These consortia may also broker research partnerships between individual companies and research providers in the public sector to address specific challenges faced by a particular company. This activity is enhanced by the existence of a broad capability and an established relationship built up through the sponsorship of previous strategic, precompetitive research.

Government-mediated industry R&D

Australia has a strong tradition of assisting industries with high SME involvement in fund raising for R&D through levies on industry activity, with funds distributed according to priorities determined by an industry board. The *Building and Construction Industry Training Fund* (BCITF) is one such model that invests in skills development. This fund could theoretically be modified through legislative amendments to foster a viable and industry-responsive research fund. Government can also play a role in encouraging industry-based organisations to associate and develop a shared vision. One such example is that of BEIIC, which had been tasked with advising the Australian Government on innovation challenges. A smaller and more focused organisation such as this, led by the industry, could play a significant role in driving a future Australian public–private research agenda.

Government R&D tax programmes

The Australian Government offers R&D tax concessions and incentives to promote innovation. These include:

- The *research and development tax concession* introduced in 1986 to encourage Australian industry to undertake such activities. It aimed to make eligible companies more internationally competitive by encouraging innovative products, processes and services and by promoting technological advancement and strategic R&D planning.

- This concession was replaced on 1 July 2011 by the Australian Government's AUD1.8 billion *R&D tax incentive* to provide tax offsets to encourage more companies to engage in R&D.

The Australian Government has also announced that it will introduce R&D tax incentive quarterly credits for SMEs from 1 January 2014 (ATO, 2011).

Government grants

Historically, the Australian Government has sought to build partnerships between industries and public sector researchers via longstanding public granting schemes. The ARC Linkage scheme and the Cooperative Research Centres (CRC) programme are two such schemes. Recently, these schemes have failed to foster substantial public–private partnerships, despite the extremely strong growth in private sector internal investment on R&D. For example, there is currently no CRC that serves the strategic needs of the construction industry as a whole. The ARC Linkage programme has also experienced a steady decline in the funds granted with 2012 funding being 25 per cent lower than in 2006, in June 2012 Australian dollars. Additionally, success rates after the peer review process remain relatively low averaging 43 per cent over the last 7 years. This becomes a significant disincentive to industry participation if grant applications with willing industry partners are rejected, potentially reinforcing a culture of poor engagement.

Government agency research

Traditionally, Governments maintained internal R&D capabilities related to the built environment. However, as Governments have progressively reduced their internal design and construction activities, they have also reduced the internal investment in R&D. Within the Commonwealth Scientific and Industrial Research Organisation (CSIRO), Australia's largest state-owned industrially oriented research organisation, the realignment of internal priorities has led to a further steep reduction of such R&D. Other countries offer examples of Government agencies that have prioritised construction research as being integral to economic growth, and used the strength of their national institutions to reinforce the capabilities of their local construction industry.

Policy recommendations

For industry, including Government as a developer and asset manager

It is in the best interest of the construction industry to engage with public research, and lead and invest in its own research and innovation. The following actions could potentially provide a link between industry and public sector priorities, and to serve the long-term interests of the industry itself:

- Establish a national industry steering body to define long-term (10–15 years) R&D priorities for the construction industry, to be revised annually.
- Disseminate these priorities throughout Government and public sector research organisations to help align the research priorities and capacity building activities with the long-term strategic interests of the industry.
- Provide a new funding stream, derived in part from industry sources, to be distributed directly by the proposed industry steering body in order to provide incentives to public research organisations to grow capacity that is aligned with the long-term industry needs.

For Government as a client, regulator and investor

All levels of Government must actively ensure that the public infrastructure investment is effectively delivered:

- Public procurement should establish systematic and internationally consistent standards that will drive innovation in the industry and public investment in infrastructure. These activities should be paralleled with investment in relevant R&D capability.
- State Governments, through existing levy mechanisms funding trade training, should allocate a proportion of these funds towards long-term strategic R&D determined by the State chapters of the proposed national industry steering body.
- Federal funding for centres of excellence, CRCs, ARC Linkage funding, and CSIRO internal funding should reflect the long-term strategic priorities identified by the proposed national industry steering body; and offer incentives for public research bodies to align their research capabilities with the needs of regional industry.

For public research organisations

Universities and Government research agencies are highly reactive to external financial incentives. Recognising this constraint, public organisations can nonetheless provide leadership through:

- Senior decision makers in Government agencies and universities integrating industry priorities and regional industry capability into their internal investment allocation in order to ensure that capabilities within public organisations match long-term industry needs.

- Public research organisations building ongoing strategic partnerships with members of the construction industry while retaining a focus on leading-edge practice and driving transformational change.
- Public research organisations building outstanding centres emphasising interdisciplinary models, social as well as technical, to ensure results are globally connected.

Note

1 IPD requires team collaboration across the project supply chain, including design consultants, contractors and subcontractors (CRC for Construction Innovation (2009), *National Guidelines for Digital Modelling*).

References

Allen Consulting Group (2010) *Productivity in the buildings network: Assessing the impact of Building Information Models*, Sydney: Built Environment Industry Innovation Committee.

Australian Bureau of Statistics (ABS) (2007) *Training for a trade*, Canberra: Commonwealth of Australia.

Australian Bureau of Statistics 8112.0 (2009) *Research and Experimental Development, All Sector Summary, Australia, 2008–09, Figure 3.1*. Available at: www.abs.gov.au/ausstats/abs@.nsf/mf/8112.0 (accessed 01 February 2011).

Australian Bureau of Statistics (ABS) (2010) *1350.0 – Australian economic indicators*. Canberra: Commonwealth of Australia. Available at: www.abs.gov.au/AUSSTATS/abs@.nsf/DetailsPage/1350.0Dec%202010?OpenDocument (accessed 23 August 2013).

Australian Bureau of Statistics (ABS) (2012a) *6227.0 – Education and work, Australia*, Canberra: Commonwealth of Australia. Available at: www.abs.gov.au/ausstats/abs@.nsf/mf/6227.0 (accessed 23 August 2013).

Australian Bureau of Statistics (ABS) (2012b) *8153.0 – Internet activity, Australia*. Available at: www.abs.gov.au/ausstats/abs@.nsf/Products/8153.0~December+2012~Chapter~Type+of+access+connection?OpenDocument (accessed 23 August 2013).

Australian Construction Industry Forum (ACIF) (2002) *Innovation in the Australian building and construction industry: Survey report*, Canberra: Australian Construction Industry Forum for the Department of Industry, Tourism and Resources.

Australian Expert Group on Industry Studies (AEGIS) (1999) *Mapping the building and construction product system in Australia*, Sydney: University of Western Sydney.

Australian Taxation Office (ATO) (2011) *Research and development tax incentive quarterly credits*. Available at: www.ato.gov.au/General/New-legislation/In-detail/Direct-taxes/Income-tax-for-businesses/Research-and-development-tax-incentive-quarterly-credits/ (accessed 23 August 2013).

Barlow, T. (2012) *The built environment sector in Australia. R&D investment study: 1992–2010*, Brisbane: Sustainable Built Environment National Research Centre.

Bok, B., Hayward, P., Roos G. & Voros, J. (2012) *Construction 2030 – Executive summary*, Brisbane: SBEnrc. Available at: www.sbenrc.com.au/research/developing-innovation-and-safety-cultures/leveraging-rad-for-the-australian-built-environment (accessed 20 March 2013).

Construction Forestry, Mining and Energy Union (CFMEU) (2010) *National Security Dave Noonan*. Available at: www.WAToday.com.au (accessed 01 February 2011).

CRC for Construction Innovation (2009) *National Guidelines for digital modelling*, Brisbane: CRC for Construction Innovation.

Department of Foreign Affairs and Trade (DFAT) (2012) *Australia: Fact sheet*, Canberra: Commonwealth of Australia.

Department of Innovation, Industry, Science and Research (DIISR) (2010) *Powering ideas: An innovation agenda for the 21st century*, Canberra: Commonwealth of Australia.

Department of Innovation, Industry, Science and Research (DIISR) (2011) *Key statistics: Australian small businesses*, Canberra, Commonwealth of Australia.

de Valence, G. (2010) 'Defining an industry: What is the size and scope of the Australian building and construction industry?', *Australasian Journal of Construction Economics and Building*, 10(1/2): 53–65.

Engineers Australia (EA) (2005) *Getting it right the first time*, Brisbane: Engineers Australia.

Furneaux, C., Hampson, K.D., Scuderi, P. & Kajewski, S. (2010) *Australian construction industry KPIs*, Canberra: Australian Construction Industry Forum.

Hampson, K.D. & Brandon, P. (2004) *Construction 2020: A vision for Australia's property and construction industry*, Brisbane: CRC for Construction Innovation.

Hampson, K.D. & Manley K.M. (2001) 'Construction innovation and public policy in Australia', in *Innovation in construction: An international review of public policies*, Manseau, A. & Seaden, G. (eds), London: Spon Press.

Master Builders Australia (2013) *Building and construction recovery on the horizon*, Canberra: Master Builders Australia.

Montreal Process Implementation Group for Australia (2008) *Australia's state of the forests report: Five-yearly report 2008*, Canberra: Commonwealth of Australia.

Newton, P., Hampson, K.D. & Drogemuller, R. (2009) 'Transforming the built environment through construction innovation', in *Technology, design and process innovation in the built environment*, Newton, P., Hampson, K.D. & Drogemuller, R. (eds), Abingdon: Spon Research.

Property Council of Australia (2009) *Construction sector productivity: KPI framework*, Sydney: PCA.

Royal Commission into the Building and Construction Industry (2002) *Overview of the nature and operation of the building and construction industry*, Canberra: Commonwealth of Australia.

Safe Work Australia (2012) *Construction fact sheet*, Canberra: Commonwealth of Australia.

Western Australia Architectural Division Energy Management Committee (1980) *Energy management in the design of new buildings: Guidelines and standards*, Document 1, Perth: Architectural Division, Public Works Department of Western Australia on behalf of the Energy Management Committee.

World Bank (2013) World DataBank. Available at: http://databank.worldbank.org/data/views/reports/tableview.aspx (accessed 20 March 2013).

4 Brazil – an overview

Mercia Bottura de Barros, Francisco Ferreira Cardoso and Lúcia Helena de Oliveira

Introduction

Brazil is the world's fifth largest country, with a total land area of 8.5 million sq km, divided into 26 States and a federal district. According to the census of 2010, conducted by Instituto Brasileiro de Geografia e Estatística (IBGE, Brazilian Statistics and Geography Institute), there are 190.8 million inhabitants, corresponding to the fifth largest population in the word (IBGE, 2011).

In economic terms, the gross domestic product (GDP) of Brazil varies between the sixth and seventh position worldwide, competing with the United Kingdom. Nevertheless, the Human Development Index (HDI) of Brazil in 2011 was 0.72 which places it at the 84th position among 187 countries assessed (PNUD, 2011). Inflation has been controlled with a ceiling of 6.5 per cent achieved in 2011, and a continuing decline to 5.8 per cent in 2012 (Banco Central do Brasil, 2012). Thus, the country now has a solid self-sufficient economy with great growth potential.

Although the distribution of wealth remains uneven, there has been an improvement in the economic conditions of the general population in recent years (IBGE, 2012). There has been a significant drop in the unemployment rate, especially in the construction industry, which fell from 10 per cent in 2003 to 2 per cent in 2012 (CBIC, 2012). In 2010 construction supply chain workers represented 14 per cent of the economically active population (FIESP, 2012).

GDP is increasing, there are international reserves of more than USD230 billion, the financial system is stable, the number of registered workers is rising and social inequality has fallen six percentage points since the beginning of the decade (IBGE, 2012). In 2010 the construction industry grew 11.6 per cent, compared to 2009, a rate higher than the national industry average of 10.1 per cent (DIEESE, 2011).

A major contributor to the development of the construction industry is the *Programa de Aceleração do Crescimento* (PAC, Growth Acceleration Programme), launched by the Federal Government in 2007. The objectives of this have been: investment in infrastructure; promotion of credit and financing; improvement of the investment environment; relief and improvement of the tax system and other long-term fiscal measures. This programme, in addition to providing direct funding for the production of infrastructure and housing, stimulates the economy mainly through the resources provided by the Banco Nacional de Desenvolvimento Social (BNDES, Brazilian Development Bank) (Ministério do Planejamento, 2012).

The BNDES is a federal public company associated with the Ministry of Development, Industry and Foreign Trade. Its goal is to provide long-term enterprise financing, contributing to the country's sustainable development. The BNDES is one of the largest development banks in the world and has been financing large-scale industrial and infrastructure projects, playing a significant role in the support of investments in agriculture, commerce and the service industry, as well as all sized private businesses (BNDES, 2012).

In its *Corporate Plan 2009/2014*, the BNDES elected local and regional development, innovation and environmental development as the most important aspects of economic growth in the current context. These aspects must be promoted and emphasised in all projects supported by the bank. One effect of the PAC and the BNDES investments has been the growth of the construction industry by creating indirect jobs linked to other productive sectors and trade in building materials (BNDES, 2012).

Between 2002 and 2007, real estate loans increased from BRL4.8 billion (Brazilian real) (USD2.5 billion) to BRL25 billion (USD12.8 billion). In this scenario, construction company turnover, for example, will have more than doubled from BRL53.5 billion in 2007 to BRL129.6 billion in 2030 (Ernest & Young Terco, 2011a, 2011b).

The investment directed to *mega sporting* events, such as the 2014 FIFA World Cup and 2016 Olympic Games, will benefit and support the sustainable development of the country. It is estimated that, in total, BRL142.4 billion (USD 72.9 billion) will be invested in the country from 2010 to 2014, generating 3.6 million jobs a year and BRL63.5 billion (USD32.5 billion) in income (Ernest & Young Terco, 2011a). This will certainly impact the construction market.

Brazil has always had great need of construction, especially residential buildings, and this quantity of available housing required has been increasing by around 16 per cent annually (CAIXA, 2012). The problem has been that this demand was not financially viable, as consumers did not have the capital to invest; more than 80 per cent of the demand is concentrated in families whose monthly salaries are under USD3,000 (CAIXA, 2012).

In the new Brazilian context, it is possible to address this need if the housing units are produced at a large scale, with low production costs and within short timeframes, which are the main characteristics of industrialised construction. Thus, the main investments oriented to innovation have been directed to this construction segment.

Summarising, this new phase of industrial growth is being consolidated within the construction industry, linked to the modernisation of the production process involving among others: applying new project tools such as building information modelling (BIM); increased use of mortar and concrete pump and spray equipment; and higher usage of precast concrete components and cranes.

In order to meet the intense demand, the construction industry has sought to use technology that would allow higher productivity in shorter timeframes, seeking thereby lower costs for both buildings and infrastructure. In this context, the use of prefabricated components has intensified and these are being produced in factories or on construction sites.

The main structural materials used in infrastructure projects are concrete and steel. In this segment, the use of reinforced and pre-stressed concrete structures is high, with either precast or cast on-site components.

Large multi-unit housing developments are present in many Brazilian cities. The buildings are housed in condominiums ranging from five to 40 floors, with four to six dwelling units per floor. The construction technology employed in the production of these buildings has been defined mainly according to their height and the availability of local resources. For buildings of up to 18 floors, the use of structural masonry or massive concrete cast on site has been common. Steel structures associated with masonry cladding or lightweight panels (drywall) have also been used. For buildings with greater height, this intensifies the use of reinforced concrete structures, elements associated with the sealing of masonry and internal partitions of lightweight panels (drywall). In the production of buildings, the diversity of materials and components is large, varying according to the subsystem (Table 4.1).

Table 4.1 Building subsystems and key materials and components

Subsystems of building	*Examples of key materials and components*
Foundation	Reinforced concrete or steel
Structure	Reinforced concrete, steel or structural masonry
Wall systems	Ceramic or concrete masonry, reinforced concrete walls or drywall
Windows	Wood, PVC, steel, aluminium or glass
Water supply and drainage systems	PVC pipe and fittings, copper, PPR (polypropylene copolymer random type 3), PEX (cross-linked polyethylene) and galvanised steel
Electrical systems	Copper wires and cables and polymer or steel conduits
Coating	Rendering, plastering, ceramic and rock tiles, wallpaper, blanket vinyl
Roofing	Ceramic, concrete or steel tiles, bitumen felt, fibre cement sheet
Water proofing	Bituminous, asphalt or polymer skins
Painting	PVA and acrylic paints

The technologies associated with the requirements of environmental sustainability, such as those enabling the reduction of energy and water consumption, must be highlighted. Technologies aimed at reducing greenhouse gas emissions and other environmental impact factors have also been the subject of research and development (R&D) activities through gathering information from research centres and private companies. It is worth emphasising the benefits gained by involving the cement production chain in the development of eco-efficient cements (Damineli & John, 2012).

Innovations have not been restricted to the field of technology. There have also been organisational innovations introduced in the management systems of companies. The use of software to support the design and planning has intensified. For example, the use of BIM has become commonplace in large Brazilian design offices, and construction and real estate companies. In general however, new technologies are not easily incorporated into the production system of construction companies, requiring developments to be undertaken by internal R&D within the organisation in partnership with the suppliers or, in some cases, with the support of the university faculties.

The construction industry is very fragmented, mainly because there are few entry barriers. The laws and legislation to create construction companies are very weak and there is no restriction to access construction technologies. Therefore, large or small contractors work on very similar projects. Even in major infrastructure works, small specialised companies are contracted and coexist with large contractors to perform specific activities.

The main players in the market are:

- *Public*: Caixa Econômica Federal (CEF, Federal Savings Bank): a state bank acting as fiscal agent and administering Government funds allocated for infrastructure and housing; City Hall and municipalities: define the rules of urban planning and construction that the specific projects must meet.
- *Private*: investors, entrepreneurs, real estate companies, designers, materials industries, builders and private banks.

There are national and state regulations related to issues of health and safety, and accessibility, which are mandatory in addition to the technical standards of the Associação Brasileira de Normas Técnicas (ABNT, Brazilian Association of Technical Standards). Furthermore, there are urban and construction rules established by local Governments which therefore may vary from city to city. There is no *code of practices* system that uniquely records good practices in design, construction and maintenance.

Innovative construction solutions are evaluated by the Sistema Nacional de Avaliação Técnica (SINAT, National System of Technical Evaluation), a mechanism for assessing innovative products, introduced in 2008 and linked to the Ministério das Cidades (Home Office). This system provides a preliminary evaluation of innovative products and systems, characterised by the absence of specific prescriptive technical standards. The evaluation is performed in accordance with guidelines proposed by the Performance Standard NBR 15575 (ABNT, 2008). This establishes the requirements and minimum criteria to be met by buildings and parts. Until December 2012 the SINAT had evaluated and approved 13 new construction systems aimed at producing affordable housing and had established nine sets of guidelines for technical assessment of products or systems (Ministério das Cidades, 2012).

Historical R&D investment (1995–2010)

According to the Ministério da Ciência, Tecnologia e Inovação (MCTI, Ministry of Science, Technology and Innovation), considering all economic sectors, Brazil has invested approximately 1 per cent of its GDP in R&D in recent years (2000 to 2010), where 55 per cent of the resources have come from the public sector and 45 per cent from private sources (MCTI, 2012a).

Public sector spending is targeted mainly on academic R&D. According to Cruz (2007), in 2000 Brazilian universities received 68 per cent of Government funding for R&D, while only 0.12 per cent was allocated to the private sector. Moreover, the demand from the private sector for R&D development in universities at this time was very limited; only 2.16 per cent of the funding for R&D in universities came from the private sector.

MCTI (2012a) statistics show this percentage is substantially lower than the United States' 7.5 per cent, or the average of 5 per cent of the countries of the Organisation for Economic Cooperation and Development (OECD). Furthermore, MCTI (2012a) data have revealed that Brazil is also different to other major emerging countries, particularly China and South Korea, where there is strong private sector investment. The average value of 0.42 per cent of GDP in the last 11 years invested in R&D by Brazilian companies represents only 24 per cent of that invested by companies in South Korea. However, Brazil still enjoys higher investment than other countries such as Russia where a mere 0.25 per cent is invested (MCTI, 2012a).

R&D trends in the public sector

The efforts of the Brazilian Government in the pursuit of innovation have been constant. In 2004 the Presidency of the Republic sanctioned law no. 10,973, known as the *Innovation Act*, which provides incentives for innovation, and scientific and technological research in the production environment. This law seeks to achieve national technological autonomy and industrial development. Possibly due to this law, research institutions affiliated to the MCTI tripled the number of patent applications at the Instituto Nacional de Propriedade Industrial (INPI, National Institute of Industrial Property) in recent years from 52 (between 2000 and 2005) to 161 (between 2006 and 2011).

Focusing on the development of the construction industry, the Brazilian Government has maintained some actions using mainly fostering organisations such as Financiadora de Estudos e Pesquisas (FINEP, Financier of Studies and Research) and the Fundação de Amparo à Pesquisa do Estado de São Paulo (FAPESP, Foundation for Research Support of the State of São Paulo), allocating resources for R&D for both to research institutions and the private sector. Additionally, research academic institutions receive resources in the form of scholarships from other sources of Government funding. The main activities of these agencies are detailed in Cardoso, *et al.*, 2000.

FINEP

FINEP's funds originate mostly from the MCTI, created from revenues derived primarily from contributions gained from both the results of the exploitation of natural resources belonging to the state and a proportion of the *excise tax*. Of the 16 ministerial R&D funds, two granted important resources to the construction industry, with the goal of contributing to the resolution of the Brazilian housing problem and the modernisation of the industry, while integrating environmental concerns. In the last decade, the total funds allocated to the construction industry were BRL21.8 million (USD11.2 million) (MCTI, 2012b).

There are no statistical data about the impact of these actions on the private sector or the specific impact over the technological innovation. However, the main qualitative results were published in *Habitare – results of impact 1995/2007* (FINEP, 2008). According to this report, the programme focused on affordable housing and has indeed contributed to the technological improvement of the industry. This programme has been not only a financial source for R&D and the development of innovations, but also a great opportunity for the concentration of effort from Government bodies, research institutions, non-Governmental organisations and private enterprises.

FAPESP

FAPESP is an independent public foundation with the mission to foster research and the scientific and technological development of the State of São Paulo. Other Brazilian States have similar bodies. FAPESP is highlighted because the volume of resources contributing to the research is significantly greater than that of similar agencies. Its annual budget in 2011 was BRL1.03 billion (USD528.5 million); the engineering field received 11 per cent of this amount.

Despite significant FAPESP investment in engineering R&D, the resources for the construction industry have declined over the years (Figure 4.1); it has been estimated that 0.5 per cent was invested in 2011. Low investment

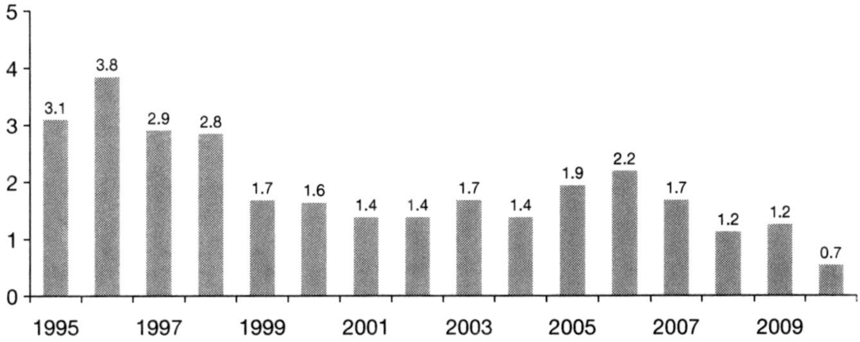

Figure 4.1 Invested resources by FAPESP in R&D in civil engineering, 1995–2010, Brazil

Source: Internal unpublished FAPESP data

stems from reduced demand for research resources, which can be explained by the intense production activity that absorbs graduates from university courses leading to diminishing human resources for research.

It is very difficult to objectively evaluate FAPESP's actions in terms of contributions to innovation in the industry. The grants allocated to young scientists (50 per year) are investments for the future; the results of those allotted to master of science (30 per year), doctoral (55), and postdoctoral (five) students are more obvious, even if FAPESP does not have indicators to measure their impact.

CNPq

The Conselho Nacional de Pesquisa Científica e Tecnológica (CNPq, National Council for Scientific and Technological Development) is a foundation linked to the MCTI to foster research. Its mission is to promote scientific and technological development and to support research necessary for the social, economic and cultural progress of the country.

CNPq performs two basic activities: (i) promoting and fostering the development of scientific and technological research; and (ii) training human resources for research in all areas of knowledge. The funds provided by CNPq to promote scientific and technological development have grown in recent years. In 2012 they surpassed BRL1.73 billion (USD885.8 million) and around 2 per cent of this amount (BRL34.6 million) was invested in the construction engineering area (CNPq, 2012).

CAPES

The Fundação Coordenação de Aperfeiçoamento de Pessoal de Nível Superior (CAPES, Foundation for the Coordination of Improvement of

Higher Education Personnel) is a public body linked to the Ministério da Educação e Cultura (MEC, Ministry of the Education and Culture). Its main objective is to help MEC regarding master degree policies and to coordinate and stimulate the formation of highly qualified professionals by means of scholarships, grants and other mechanisms (CAPES, 2010).

A major programme launched in 2011, the Ciência sem Fronteiras (Science without Borders), seeks to promote the consolidation, expansion and internationalisation of science and technology, innovation and competitiveness through international exchange of Brazilian undergraduate and postgraduate students. This programme will provide approximately 100,000 scholarships over four years (Ministério da Educação, 2012). Its objective is to promote technological development and stimulate innovation processes in Brazil through international mobility of teachers, undergraduate and graduate students, postdoctoral fellows and Brazilian researchers; stimulating the integration of research carried out in Brazilian institutions with best international practices. Its priority areas incorporate all engineering disciplines, industry focused on products and processes for technological development and innovation, new construction technologies, and training of technologists.

It was not possible to obtain CAPES statistics on scholarships specifically granted for construction engineering. However, they can be estimated by adopting the same percentile of incidence of that area in relation to the total amount allotted by CNPq, estimated at 2 per cent. Considering the total investment of CAPES in 2011 was around BRL2.7 billion (USD1.4 billion), this estimation would result in an annual investment of BRL54.1 million (USD27.7 million) for the construction engineering area (CAPES, 2010).

Trends in the private sector

Since 2000 there have been some initiatives initiated by the Federal Government targeting the private sector. For example, the creation of sector funds and the *Innovation Law* has helped to improve the technological capacity of companies in Brazil. According to Cruz (2010), the Innovation Law created the possibility for economic subsidies to R&D companies. In addition, the law has provided for the establishment of a favourable tax regime with incentives for R&D resulting in Act 1196 in 2005, known as *Lei de Bem* (the *Good Law*) which has been used by some companies in the construction industry.

The grant programme, initiated in 2006 by FINEP, has offered about BRL386.8 million (USD198 million) annually to subsidise R&D companies. These resources represent only 0.07 per cent of Brazilian GDP, well below the 0.23 per cent of GDP practised by Canada or the USA 0.20 per cent (Cruz, 2010).

The major challenge for companies in the construction industry is to continually improve the quality of their products and the productivity of their production process. This has been a challenge since the mid-1990s, when the scenario was of no public investment, low-income consumers and therefore increased competition between firms.

Since then, some construction companies and the real estate market decided to invest in R&D to meet challenges more effectively. Through a process of organisational restructuring of their departments, they created specific standalone R&D departments with a focus on innovation. The academic researchers have influenced this process, supporting some companies that, at the time, achieved significant results in terms of process rationalisation, based on technology and construction management.

Presently in Brazil, construction companies with outdated technology are taking the opportunity of bringing their work procedures to the same standards as those of national companies that have reached higher technological and organisational development. Furthermore, opportunities such as analysing and implementing the standards achieved in more developed countries have been brought in by the academic institutions through specialisation courses offered primarily by public universities.

Despite public sector efforts in search of innovation, the construction industry still has one of the lowest levels in terms of programmes or resources invested in innovation processes. Innovation research conducted by the IBGE in 2008 highlighted the construction industry among the segments with lowest technological content, where innovation rates did not reach 25 per cent (IBGE, 2008).

According to data from the Câmara Brasileira de Indústria e Comércio (CBIC, Brazilian Chamber of Industry and Commerce), the predominance of innovation in the construction companies has been mainly a consequence of innovation developed and implemented in other segments such as: (i) materials and components industry; (ii) design process; and (iii) management process. There are also a number of innovation initiatives carried out by the market leading construction companies that are not widespread among smaller companies. Moreover, there is little intervention from research institutions, including academia, in the dynamics of generation, implementation and absorption of innovations developed in projects throughout the various segments of the construction industry (CBIC, 2010).

Case studies

The three case studies presented in this chapter are being carried out by different segments of the private construction sector. The first involves a supply chain dedicated to cement mortar, the second regards a large contractor with expertise in infrastructure and the third refers to a real estate company.

Case study: Sector Consortium for Technological Innovation in Mortar Coatings – CONSITRA

The *Consórcio Setorial para Inovação Tecnológica em Revestimentos de Argamassa* (CONSITRA, *Sector Consortium for Technological Innovation in Mortar Coatings*) is a multi-institutional, multi-disciplinary research programme. It is structured with the goal of *developing new technologies in mortar coatings that are more reliable, highly productive, durable and competitive in the various situations of the Brazilian market* (FAPESP, 2008). Its professional input consists of representatives of entities from all stages of production of mortar coatings: conception, design, production, application and technological control.

The first research phase took place from 2005 to 2008, with BRL671 million (USD344 million) funding, contributed jointly by the private sector (67 per cent) and FINEP (33 per cent). The second phase of the CONSITRA is underway (2012–2015), with allocated resources of about BRL1 billion (USD500 million) granted exclusively from the private sector.

The focus of the first stage was the formulation of mortar, while the development themes of the second phase are: equipment for mixing mortar, guidelines for design, technology to increase productivity and lower production costs, maintaining or improving the quality of the coating.

The first phase resulted in significant advances for programme participants; new formulations of mortars resulted in lower production cost and consequently, a reduced price for the market. The second phase of the programme aims to achieve economic benefits and to improve the coating quality.

Case study: Andrade Gutierrez Technological Innovation Programme – PAGIT

Andrade Gutierrez is a large Brazilian construction company that operates in most markets of the construction industry. Andrade Gutierrez created the *Programa Andrade Gutierrez de Inovação Tecnológica* (PAGIT, *Technology Innovation Programme*) from the opportunity of gaining revenue from tax incentives Federal law 11,196, the *Good Law*. The main incentive offered to Brazilian companies to promote technological innovation is low-interest loans or grants that can be in the form of reimbursements of up to 80 per cent of the total amount invested.

The PAGIT started in 2008 and, up until now, has used only one type of incentive, whose potential return on investment is 20 per cent (Table 4.2). Initially, the projects were proposed aiming at innovation across all areas of the company. However, after the first years of the programme it was identified that the focus should be the area of engineering that represented 69.3 per cent of tax incentives obtained in 2011.

Table 4.2 Number of projects, investments and incentives per year of the PAGIT

Year	*Number of projects*	*Investments ('000 BRL)*	*Investments ('000 USD)*	*Incentive value obtained ('000 BRL)*	*Incentive value obtained ('000 USD)*
2008	7	2,112	1,081	431	221
2009	55	7,527	3,853	1,536	786
2010	94	16,384	8,386	3,211	1,644
2011	123	27,286	13,966	5,566	2,849

Source: Programme coordination team

Over the years that the PAGIT has been active, the innovation curve increased despite growing at a lesser rate in the fourth year. The results over the four years' experience have shown the need to develop a strategic plan for science and technology (S&T) to identify the objectives for innovation. According to the programme management team, ongoing investments should be focused on large projects, particularly within engineering. The leaders of PAGIT believe that *innovation is a process that has a beginning with no end and what really matters is to create a culture of innovation.*[1]

However, until now, no partnerships with universities or research centres have been formed. There is only one committee, comprised of representatives of the main areas of the company, which meets once a month and aims to foster a culture of innovation.

Case study: Tecnisa Technological Development and Research Programme

Tecnisa is a medium-sized Brazilian real estate company founded in 1977. It has become a recognised brand for quality and efficiency in building construction processes, customer relations and excellent products. The company has built over five sq km of housing and has more than 12,000 customers. With the support of academic researchers, they search for innovations bringing added value through customer satisfaction and business profitability.

It is one of few construction companies that have *a technological development programme* focused on innovation conducted by a specific standalone department: Departamento de Desenvolvimento Tecnológico (DDT, Department of Technological Development).

The DDT adopts a management model incorporating the areas of technical assistance and quality and also participates in the strategic processes of the company. The creation of product differentials based on acoustic comfort and the responsibility for integration of the environmental sustainability concept in its projects are other examples of their work.

The R&D process of the DDT involves capturing, formulating and selecting ideas for innovation; and carrying out research, analysis and development through to implementation. In this process, the role of experts by area of knowledge is privileged, driven by the area manager with the support of external consultants, usually from academia.

Involvement with the previous stages of the projects should guide the work of the DDT team in the coming years, influencing the phase of purchasing land and defining products in order to create the most suitable production conditions. Moreover, the concept of sustainability with a focus on energy efficiency and rational use of water in the lifecycle of the building has gained ground in internal studies.[2]

R&D and impact measurement

Measuring the impact of R&D on technological innovation is not easy. However, some indicators can be mentioned. Public investments, made through development agencies such as FINEP, FAPESP, CAPES, and CNPq, for example, have resulted in an increase in the number of patents. In addition, the grants are allotted to young scientists as investments for the future. Research outcomes from funds awarded to master of science, doctoral, and postdoctoral programmes are more obvious, even if indicators to measure their impact are not available.

Moreover, there are investments in R&D that have resulted in direct benefits to the Brazilian society. For example, the *Programa de Uso Racional da Água* (PURA, *Water Conservation Programme*) was created in 1996 aiming to reduce the potable water demand with a focus on technology, legislation and education. Initially, the programme was implemented in a campus of the University of São Paulo, yielding a consumption reduction of 48 per cent. Subsequently, the programme has been implemented across the city of São Paulo presenting results of around 30 per cent reduction in average annual consumption per household (SABESP, 2011).

The first phase of CONSITRA brought significant results especially to manufacturers of mortars who have improved their products and reached new market niches. There are no figures supporting such gains. However, had they not occurred, there would not have been a second phase. In this new phase, it is expected that the return will be even greater, particularly for the construction segment, due to lower production costs and also to the end users with improved product quality. In Tecnisa, the main returns earned in recent years have resulted from initiatives such as: optimisation of the relationship between the strength of concrete and the rate of steel in concrete structures, system enhancement coatings and waterproofing, viability of precast and prefabricated concrete, and rationalisation projects under development. The financial result is an average annual savings of about BRL1.5 million (USD0.77 million) from the discounted costs involved in maintaining staff and research. When investing in a technological innovation, the financial return of Andrade Gutierrez is automatically 20 per cent coming from Government reimbursement due to the Innovation Law. Further results arising from the innovation itself are more difficult to assess.

R&D priority areas for future investment

In order to study, analyse and define guidelines for the development, dissemination and evaluation of technological innovations in construction for permanent and continuous improvement, actions have been undertaken by the *Programa para Inovação Tecnológica em Construção* (PIT, *Programme for Technological Innovation in Construction*). This is an initiative from CBIC and other principal organisations of the construction industry, including the Associação Nacional de Tecnologia do Ambiente Construído (ANTAC, National Association of Technology of the Built Environment), which brings together researchers linked to universities and research institutions.

Initiated in October 2007, the programme seeks to reach the broadest possible spectrum of businesses and professionals, as well as the related technical associations, from the major players of the construction industry.

The programme began with the identification of the main barriers to the dissemination of innovation within the industry, with emphasis on the gap between academia and market; lack of integration; difficulty to access existing knowledge; conservatism and short-term vision; limitations of the legal basis of the stimulus to R&D; inappropriate mechanism for assessing research; unavailability of data supporting R&D; and a combination of inadequate data and limited application of regulations and standards.

Subsequently, CBIC and its partners have defined priorities that have unfolded in nine innovation projects: taxation compatible with industrialisation and innovation; development of technical standards; feasibility of innovation in public construction projects; national construction code; diffusion of innovation; capability innovation; R&D for innovation in construction; knowledge for innovation; and modular coordination.

Considering the proposed projects, it can be observed that there are problems in Brazil with regards to training professionals for innovation, and difficulties in disseminating innovation. Moreover, the transfer of accumulated knowledge, developed in universities and research centres for the market, has been insufficient through the lack of interactions between the parties involved. Furthermore, the mechanisms allowing researchers to work with companies looking for innovation are not sufficiently known, such as those arising from the Innovation Law or from funding agencies such as FINEP and BNDES.

Faced with these difficulties, the ANTAC was asked to identify the actions necessary to bring together academia, private companies and public agents so as to develop and disseminate innovation in the future.

They defined five priority areas: systems and construction processes and production management; quality materials and construction; water, energy and comfort; design, use and operation; and cities, with a focus on urban infrastructure, housing and real estate management. This work was performed through five workshops, conducted between October 2011 and October 2012, which were attended by 325 invited participants, including 88 professors and senior researchers from 29 Brazilian institutions and one foreign researcher.

The following priority research projects were defined in the workshops:

- building management systems, components, systems and processes;
- database of technical information and focusing on environmental performance;
- methodology of *simplified lifecycle assessment*, eco-friendly materials and components, templates and tools for performance evaluation;
- design methodologies and intervention in order to integrate urban morphologies with infrastructure systems and housing, information technology accessible and appropriate for participatory activities and decision making;
- models of production schedules, housing policy proposals and public housing;
- planning tools and formatting of real estate products.

Considering these priority areas, several actions must occur within a 10-year timeframe, for example: implementation of new R&D models demanding closer ties between academia, private and public sectors; better integration between all shareholders within the industry, battling adversity and overcoming fragmentation of the production chain; greater dissemination of R&D results and improvements in system performance reporting; more and permanent financing for R&D; and increased financing for training human resources dedicated to innovation.

The strategy for the implementation of such models in developing R&D should involve actions allowing the main barriers (listed above) to be overcome, highlighting the gap between academia, industry and public agents. Only the joint efforts of these agents may result in the transformation of academic R&D into effective innovation for the industry and Governmental initiatives.

Additionally, mechanisms must be implemented for increasing R&D fundraising to be applied specifically to the area of technology for the built environment. For example, the creation of a specific Sector Fund of Science and Technology linked to the MCTI.

It is important to continue developing actions aimed at improving the Innovation Law (law no. 10,973) with a focus on the construction industry and its wider deployment at companies. This has the potential to lead to closer relations with academia working towards developing a body of research generating technological innovation and processes.

Final considerations

Brazil is in a phase of immense economic and social development. This fact reflects on the construction industry by an increasing demand for large-scale production, increasing production costs, increasing labour cost, and driving the search for greater efficiency and automation within production processes. All of which remains thoroughly favourable towards modernisation and continuing to demand a focus on innovation.

The participation of the construction industry in the GDP has maintained an average of 6 per cent, constituting one of the highest in the single sector economy. In addition, if the entire construction industry, including the materials and components industry, is considered, that figure reaches almost 15 per cent of GDP (IBGE, 2010).

The efforts of the Brazilian public sector in the pursuit of innovation have been constant, albeit mainly in the form of scholarships.

Important actions by the Brazilian Government have been the creation of sector funds and the Innovation Law, which have helped to improve the technological capacity of companies in Brazil.

In order to improve the results of R&D activities, one of the principal strategies being incentivised is the strengthening of the relationship between academia and the public and private sector agents.

Notes

1 The case study information was provided to the authors by the PAGIT team of Andrade Gutierrez Company.
2 The case study information was provided to the authors by the DDT manager of Tecnisa.

References

ABNT (2008) *NBR 15575 Desempenho de edifícios partes 1 a 6 (NBR 15575 building performance parts 1 to 6)*, Rio de Janeiro: Associação Brasileira de Normas Técnicas (ABNT).

Banco Central do Brasil (2012) *Histórico de metas para a inflação no Brasil (History of inflation targets in Brasil)*. Available at: www.bcb.gov.br/Pec/metas/TabelaMetaseResultados.pdf (accessed 20 January 2013).

BNDES (2012) *The BNDES, Banco nacional de desenvolvimento social (The Brazilian Development Bank)*. Available at: www.bndes.gov.br/SiteBNDES/bndes/bndes_en/Institucional/The_BNDES/ (accessed 21 January 2013).

Brasil (2010) *Ministério da educação (Ministry of education)*. Available at: www.capes.gov.br/images/stories/download/Livros-PNPG-Volume-I-Mont.pdf (accessed 20 January 2013).

CAIXA (2012) *Demanda habitacional no Brasil* (*Housing demand in Brazil*). Available at: http://downloads.caixa.gov.br/_arquivos/habita/documentos_gerais/demanda_habitacional.pdf (accessed 21 January 2013).

CAPES (2010) *Plano nacional de pós-graduação (PNPG) 2011–2020, Volume 1* (*Postgraduate national plan 2011–2020, Volume 1*), Fundação Coordenação de Aperfeiçoamento de Pessoal de Nível Superior (Foundation for the Coordination of Improvement of Higher Superior). Available at: www.capes.gov.br/images/stories/download/Livros-PNPG-Volume-I-Mont.pdf (accessed 20 January 2013).

Cardoso, F.F., Rezende, M.A., Barros, M.B.D. & Oliveira, R. (2000) 'Public policy instruments to encourage construction innovation: Overview of the Brazilian case', in *Joint Meeting of the CIB Working Commissions W55 and W65 and Task Groups TG23, TG31 and TG35*, Reading: University of Reading, Departament of Construction Management and Engineering.

CBIC (2010) *Projeto inovação tecnológica* (*Technology innovation project*), Câmara Brasileira da Indústria da Construção (Brazilian Chamber of Construction Industry). Available at: www.cbic.org.br/comissoes-e-foruns/comissao-de-materiais-tecnologia-qualidade-e-produtividade/projetos/pit/pagina/pr (accessed 20 September 2012).

CBIC (2012) *Camara Brasileira da indústria da construção* (*Brazilian chamber of construction industry*). Available at: www.cbicdados.com.br/menu/emprego/emprego-formal-caged (accessed 21 January 2013).

CNPq (2012) *Indicadores de pesquisa* (*Research indicators*), *Conselho Nacional de Pesquisa (National Council for Scientific and Technological Development*). Available at: www.cnpq.br/web/guest/indicadores1 (accessed 20 January 2013).

Cruz, C. (2007) 'Ciência e tecnologia no Brasil' ('Science and technology in Brazil'), *REVISTA USP*, March/May: 58–90.

Cruz, C. (2010) *Ciência, tecnologia e inovação no Brasil: Desafios para o período 2011 a 2015* (*Science, technology and innovation in Brazil: Challenges for the period 2011–2015*). Available at: http://interessenacional.uol.com.br/2010/07/ciencia-tecnologia-e-inovacao-no-brasil-desafios-para-o-periodo-2011-a-2015/ (accessed 23 September 2012).

Damineli, B.L. & John, V.M. (2012) 'Developing low CO_2 concretes: Is clinker replacement sufficient? The need of cement use efficiency improvement', *Key Engineering Materials*, 517: 342–351.

DIEESE (2011) *Estudo setorial da construção* (*Study of the construction sector*), Departamento Intersindical de Estatísticas e Estudos Econômicos (Inter-union Department of Statistics and Economic Studies). Available at: www.dieese.org.br/esp/estPesq56ConstrucaoCivil.pdf (accessed 21 January 2013).

Ernest & Young Terco (2011a) *Sustainable Brazil: Housing market potential.* Available at: www.ey.com/Publication/vwLUAssets/Housing_market_potentials_Publica%C3%A7%C3%A3o/$FILE/Housing%20market%20potentials.pdf (accessed 2 January 2013).

Ernest & Young Terco (2011b) *Sustainable Brazil: Social and economic impacts of the 2014 World Cup*. Available at: www.ey.com/Publication/vwLUAssets/Sustainable_Brazil_-_World_Cup/$FILE/copa_2014.pdf (accessed 21 January 2013).

FAPESP (2008) *Paredes sem fissuras (Walls without cracks)*, São Paulo: FAPESP.

FIESP (2012) *ConstruBusiness, 10° congresso Brasileiro da construção* (*Constru Business, 10th Brazilian construction congress*), São Paulo: Federação das

Indústrias do Estado de São Paulo (Federation of São Paulo State Industries). Available at: http://hotsite.fiesp.com.br/construbusiness/2012/doc/CB2012Port.pdf (accessed 21 January 2013).

FINEP (2008) *Habitare: Resultados de impacto 1995–1997* (*Habitare: Impact results 1995–1997*), Rio de Janeiro: Financiadora de Estudos e Pesquisas.

IBGE (2008) *IBGE divulga os resultados da PINTEC 2008* (*IBGE releases the results of PINTEC 2008*), Rio de Janeiro: Instituto Brasileiro de Geografia e Estatística (Brazilian Statistics and Geography Institute). Available at: www.pintec.ibge.gov.br/index.php?option=com_content&view=article&id=56:ibge-divulga-os-resultados-da-pintec-2008&catid=7:noticias&Itemid=10 (accessed 21 January 2013).

IBGE (2010) *Pesquisa anual da industria da construção 2010* (*Annual survey of the construction industry in 2010*), Rio de Janeiro: Instituto Brasileiro de Geografia e Estatística.

IBGE (2011) *Sinopse do censo demográfico 2010* (*Synopsis of the 2010 census*). Available at: http://pt.scribd.com/doc/73362536/Resultados-Censo-2010 (accessed 21 January 2013).

IBGE (2012) *Indicadores. Pesquisa mensal de emprego* (*Indicators. Monthly employment survey*), Rio de Janeiro: Instituto Brasileiro de Geografia e Estatística. Available at: ftp://ftp.ibge.gov.br/Trabalho_e_Rendimento/Pesquisa_Mensal_de_Emprego/fasciculo_indicadores_ibge/2012/pme_201212pubCompleta.pdf (accessed 21 January 2013).

MCTI (2012a) *Indicadores nacionais de ciência, tecnologia e inovação* (*National science, technology and innovation indicators*), Ministério da Ciência, Tecnologia e Inovação (MCTI, Ministry of Science, Technology and Innovation). Available at: www.mcti.gov.br/index.php/content/view/2076/Recursos_Aplicados.html (accessed 1 September 2012).

MCTI (2012b) *Financiadora de estudos e pesquisas, Fundo nacional de desenvolvimento científico e tecnológico, relatório de gestão do exercício de 2011* (*Financier of studies and research, national fund for scientific and technological management report year 2011*). Available at: http://download.finep.gov.br/processosContasAnuais/relatorio_gestao_ fndct_2011.pdf (accessed 23 September 2012).

Ministério da Educação (2012) *The program, Ciência sem fronteiras* (*The programme, Science without borders*). Available at: www.cienciasemfronteiras.gov.br/web/csf-eng/ (accessed 23 September 2012).

Ministério das Cidades (2012) *Sistema nacional de informações das cidades* (*National information system of cities*). Available at: www.cidades.gov.br/pbqp-h/projetos_sinat.php. (accessed 21 January 2013).

Ministério do Planejamento (2012) *Sobre o PAC* (*About PAC*). Available at: www.pac.gov.br/sobre-o-pac (accessed 20 November 2012).

PNUD (2011) *Relatório do desenvolvimento humano de 2011* (*Human development report 2011*). Available at: http://hdr.undp.org/en/media/HDR_2011_PT_Complete.pdf (accessed 21 January 2013).

SABESP (2011) *Relatório de sustentabilidade 2011* (*Sustainability report 2011*). Available at: http://blog145.tempsite.ws/relatorio-de-sustentabilidade-2011 (accessed 20 July 2012).

5 Canada – innovation through collaboration

Aminah Robinson Fayek, Jeff H. Rankin, Saiedeh Razavi and Russell J. Thomas

The national context

Construction is one of the largest industries in Canada, although the industry faces certain challenges that are complicated by factors that include globalisation, technological advances and the large geographical size of the country. Despite the industry's need for innovation to address these challenges, construction's investment in research and development (R&D) activities is low when compared to both the industry's other expenditures and the R&D spending by other industrial sectors. This chapter provides an overview of R&D investment and impact in the Canadian construction industry. Three case studies illustrate how R&D is carried out in collaborative environments involving Government, universities and colleges and industry. Due to the diversity of this industry, each model of R&D differs in terms of the composition of participating parties and the structure of its interactions, although all utilise similar mechanisms that will be highlighted. The chapter concludes by proposing a roadmap and strategies for future R&D.

The construction industry historically has contributed 5–8.5 per cent of the country's gross domestic product (GDP). Figure 5.1 depicts the annual GDP of the construction industry in Canada between 2002 and 2011 (Statistics Canada, 2012a). In 2011 the Canadian industry represented CAD76.5 billion (Canadian Dollars)[1] in economic activities: 6 per cent of Canada's GDP and a significant growth from the sector's CAD57.8 billion in 2002. The annual GDP at basic price, by different industry sectors including construction, is presented for 2007–2011 in Table 5.1 (Statistics Canada, 2012j; 2012c).

According to Industry Canada (2012a), in 2011 over 278,000 Canadian firms were involved in the construction industry. Of this number, 60.4 per cent were considered micro firms with one to four employees while 28.7 per cent were considered small establishments with five to 99 employees, 0.9 per cent medium-sized firms with 100–499 employees, and 1 per cent large firms with more than 500 employees. Industry Canada (2012b) groups all businesses with fewer than 500 employees using the

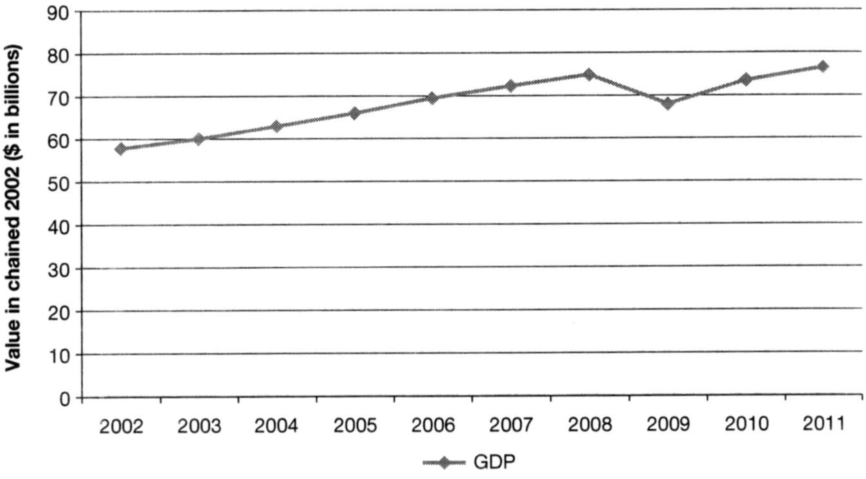

Figure 5.1 GDP construction, 2002–2011, Canada

Source: Statistics Canada (2012d)

term small and medium-sized enterprises (SMEs). Statistics Canada reported that the average total annual revenue for SMEs in the construction industry was CAD409,800 in 2008. In the same year the average total expenses for the same sector and establishment sizes were CAD369,200 (Industry Canada, 2012a).

Labour productivity plays an important role in construction economics. According to Statistics Canada, construction labour productivity declined at an average yearly rate of 0.7 per cent between 2002 and 2011 despite increasing by 0.4 per cent between 2009 and 2011. These changes in productivity are presented and compared to the Canadian economy as a whole in Figure 5.2 (Statistics Canada, 2012d).

By the same token, annual capital investment in construction shows significant growth over the past decade (Statistics Canada, 2011b). The annual capital investment increased from CAD3.5 billion in 2002 to CAD6.1 billion in 2011. On average, during this period investment in machinery and equipment increased at a compound annual rate of 6.2 per cent and construction investment increased at an annual rate of 7.3 per cent (Statistics Canada, 2011b).

Audit and analysis of R&D investment

R&D spending by industry sectors in Canada was expected to be CAD15.5 billion in 2012. The intended R&D spending in 2012 was 0.9 per cent

Table 5.1 Gross domestic product at basic prices by industry, 2007–2011, Canada

Gross domestic product at basic prices, by industry					
Industry	*2007*	*2008*	*2009*	*2010*	*2011*
	Millions of chained dollars (2002)				
All industries	1,218,981	1,229,786	1,193,211	1,233,930	1,266,590
Goods-producing industries					
Agriculture, forestry, fishing and hunting	27,570	30,008	28,082	28,486	29,105
Mining and oil and gas extraction	57,776	56,538	52,125	54,967	57,422
Manufacturing	181,348	171,785	150,431	158,326	162,143
Construction industries	72,330	74,875	68,011	73,467	76,515
Utilities	31,598	33,044	32,191	32,624	34,028
Services-producing industries					
Transportation and warehousing	57,708	57,884	55,338	57,569	59,804
Information and cultural industries	44,568	44,940	44,848	45,240	45,922
Wholesale trade	70,107	69,628	65,268	68,822	70,713
Retail trade	71,733	73,293	72,774	75,634	77,239
Finance and insurance, real estate and renting and leasing and management of companies and enterprises	240,577	245,547	251,128	257,488	264,193
Professional, scientific and technical services	59,246	60,209	59,623	59,948	61,566
Administrative and support, waste management and remediation services	30,799	31,025	29,860	30,329	30,747
Public administration	69,136	71,447	73,742	75,390	76,333
Educational services	58,413	60,140	61,219	62,539	63,142
Health care and social assistance	76,715	78,715	80,888	82,761	84,479
Arts, entertainment and recreation	11,087	11,215	11,272	11,359	11,226
Accommodation and food services	26,531	26,846	26,094	26,611	27,340
Other services (except public administration)	31,442	32,039	31,920	32,329	33,090

Source: Statistics Canada (2012g, 2012i)

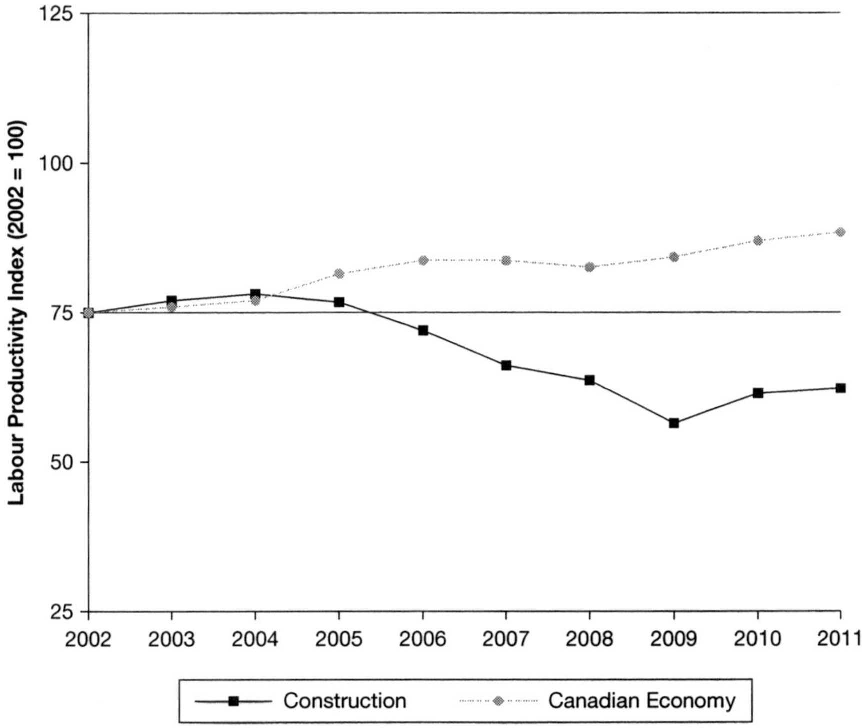

Figure 5.2 Construction labour productivity index, 2002–2011, Canada

Source: Statistics Canada (2012h)

more than in 2011, but less than the CAD16.8 billion recently reached in 2007 (Statistics Canada, 2012e; 2012f). In 2012 the manufacturing sector shared 47 per cent of the total industrial R&D expenditure, the highest rate of R&D spending by an industry sector. While projected R&D spending in the manufacturing sector was CAD7.6 billion in 2012, projected R&D spending for the construction industry was only CAD101 million, less than 1 per cent of the industry's total expenditure. Figure 5.3 presents total R&D expenditure by main industrial groups (Statistics Canada, 2012e; 2012f). In 2010, for which the latest provincial data are available, the provinces of Ontario (CAD6.8 billion) and Quebec (CAD4.7 billion) received the majority of the industrial R&D spending in all sectors in Canada.

Gross domestic expenditure on research and development (GERD) represents the total R&D conducted in a country. The ratio of GERD to gross domestic product (GDP) is a standard measure for a country's R&D effort. From 2001–2010, Canada's GERD to GDP rose to a high of just above 2 per cent in 2006 before it dropped to 1.96 per cent in 2007 and to

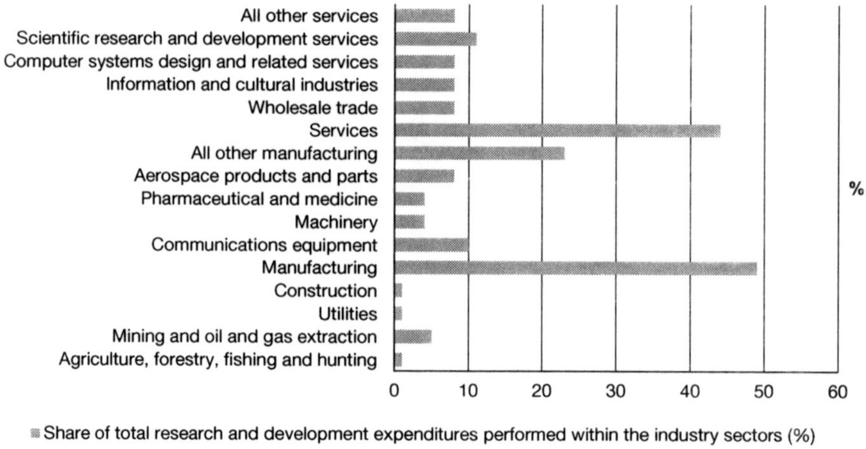

Figure 5.3 Canadian industrial research and development spending by main industrial groups (2012 intentions)

Source: Statistics Canada (2012e, 2012f)

1.81 per cent in 2010 (Statistics Canada, 2012c). For each year in this period, the business sector was the greatest contributor of R&D funding, followed by the Federal Government. Business was also the sector with the greatest R&D performance (CAD15 billion), followed by higher education, which performed over CAD11 billion (Science Technology and Innovation Council, 2011).

The trend of the total R&D expenditure by the construction industry in Canada for the years 2008–2012 shows a decline in R&D spending in construction from CAD122 million in 2008 to CAD101 million intended spending in 2012 (Statistics Canada, 2012e; 2012h). Detailed R&D expenditure in construction for this period is presented in Table 5.2 (Statistics Canada, 2012e; 2012h). In this table, the total R&D expenditure is divided into current and capital. The expenditure information is also provided separately for some categories of current expenditure. The trend shows that almost 70 per cent of the R&D spending is on wages and salaries of R&D personnel.

Construction R&D in Canada is funded by three major sources: the public sector, the private sector and academia (Xu, 2010). While the private sector is the major funding source for R&D in most industry sectors, historical data reveal that only a small portion of construction's private sector undertakes formal R&D activities (Xu, 2010).

The Government of Canada has contributed significantly to R&D expenditure in this industry. The National Research Council (NRC)

Table 5.2 Construction R&D in Canada, 2008–2012

Research and development performed by the construction industry					
Expenditures	*2008*	*2009*	*2010*	*2011*	*2012*
	CAD millions				
Total R&D expenditures	122	119	100	109	101
Current expenditures	118	114	96	105	X
Wages and salaries	86	88	76	83	77
Other current expenditures	32	26	20	22	X
Capital expenditures	4	5^{E}	4	F	X
	Full-time equivalent				
Total R&D personnel	2,806	2,134	1,711	..	..
Professionals	968	869	976	..	..
Technicians	1,210	943	620	..	..
Other support staff	628	322	114	..	..

X: suppressed to meet the confidentiality requirements of the *Statistics Act*
E: use with caution
F: too unreliable to be published
..: not available for a specific period of time

Source: Statistics Canada (2012b, 2012e)

of Canada's Construction Portfolio is one of the main Governmental organisations that supports research and the dissemination of knowledge. The Government also directly provides funding through other channels, most notably through the *Industrial Research Assistance Program* (IRAP). The Government of Canada's programmes to support research in the higher education sector include, but are not limited to, the Canada Foundation for Innovation and the Natural Sciences and Engineering Research Council of Canada (NSERC). In 2000 the Government of Canada also established the Canada Research Chairs programme, which is designed to attract, support and retain excellent researchers and to enhance the capacity of universities to produce and apply new knowledge (Canada Research Chairs, 2012).

NSERC is one of Canada's largest sources of funding for public–private R&D partnerships. For the fiscal year of 2011–2012, for which the latest data are available, NSERC planned to invest more than CAD347 million, equivallent to 31.6 per cent of its annual budget, to foster innovation (Moss, 2011). This investment enables industry members to work with academic researchers to find solutions to industry problems through various initiatives. In 2009 NSERC's *Strategy for Partnerships and Innovation*

(SPI) added a number of new initiatives to stimulate new partnerships between universities and industry. This included Engage Grants, 6-month CAD25,000 research grants to help foster a new relationship and solve a company-driven problem, which have resulted in 1,700 new companies participating in R&D projects with universities across Canada. NSERC's new Research Partnerships programmes include: (i) the *Strategic Partnerships Program*; (ii) the *Industry-Driven Collaborative Research and Development Program*; (iii) the *Commercialisation Program*; (iv) the *Training in Industry Program*; and (iv) the *College and Community Innovation Program* (Natural Sciences and Engineering Research Council of Canada, 2012). From 2011 to 2012, the Strategic Partnerships Program had the highest budget (44 per cent of the total) followed by the Industry-Driven Collaborative Research and Development Program (34 per cent).

This trend is projected to change with the renewed focus on industry-driven collaborative R&D and the increased investment in this area. NSERC received an additional CAD15 million in the 2012 Federal Government budget in support of SPI (Government of Canada, 2012). Investment in the Training in Industry Program is projected to grow to 50 per cent of NSERC's innovation budget for the fiscal year of 2015–2016 (Moss, 2011). This programme includes interaction grants, engage grants, partnership workshop grants, collaborative research and development grants (CRDs) and industrial research chairs (IRCs) (Natural Sciences and Engineering Research Council of Canada, 2012).

The following section describes three R&D groups that deliver innovation to the Canadian construction industry. The case studies highlight funding sources, research themes, methodologies, working relationships, dissemination activities and impacts.

Case studies in Government, university, and industry partnerships

Federal Government leadership

The Government of Canada's research programmes

There are two main channels through which the Canadian Federal Government supports industrial research in the construction industry: through IRAP, and through working and collaborating with the NRC and its construction portfolio. IRAP is administered by the NRC and provides grants and support to SMEs so that they may undertake research in developing a product, process or service (National Research Council of Canada, 2012b). On other occasions, companies work directly with researchers in the NRC to undertake work that meets the needs of the industry as a whole.

Evaluating and creating markets for innovative products

In many cases where research topics have a broad impact on industry, industry members work with the NRC to undertake precompetitive-type research. Results are then shared widely to facilitate the creation or broadening of a market. The primary output of this research programme is findings that can be used to support proposed changes to the national model codes or to various national or international standards. By establishing performance-based metrics that reflect the performance of the minimally performing prescriptive solution permitted within the existing code, this kind of work can open up new market areas to which industry can introduce new and innovative products. In the Canadian objective-based codes system, the existing prescriptive solutions are seen as only one of a number of different possible solutions that would meet the relevant specific objective. The establishment of performance criteria for such a specific objective permits a much larger number of innovative solutions that can be demonstrated to meet or exceed the performance criteria.

In other cases, specific companies approach the NRC via the Canadian Construction Materials Centre (CCMC) to seek an evaluation for their innovative product, whether construction materials, products, systems or services, where there is no existing recognised standard. In such cases, it is often necessary to undertake some research to establish the appropriate test methods and procedures required to demonstrate that the product meets or exceeds the performance expectations established by the relevant code objectives. Once a technical guide has been established and the proponent has been able to demonstrate that the innovative product meets or exceeds the performance standards established in the guide, then they are provided with an evaluation report and the product is placed in the registry of evaluated products. Authorities having jurisdiction recognise the product evaluations as indicating that the product meets or exceeds the objectives of the code.

By undertaking research that allows for changes in national or international product evaluation standards, the Canadian Federal Government, through IRAP and the CCMC, helps construction innovators enter or broaden the industry's competitive market, and facilitates industry and end user access to better products.

NSERC Industrial Research Chair in Strategic Construction Modelling and Delivery

NSERC Industrial Research Chairs Programme

NSERC invests in Canadian science and technology research and promotes it so as to encourage further funding from industry sources. NSERC's IRC programme awards 5-year appointments to Canadian university researchers performing state-of-the-art research that has the potential to be exploited for the economic, social or environmental benefit of the participating industrial organisations and of Canada (Natural Sciences and Engineering Research Council of Canada, 2009). CRDs, awarded for 3- to 5-year periods, facilitate industry access to the specialised knowledge and resources available at Canadian universities and train students in skills needed by industry (Natural Sciences and Engineering Research Council of Canada, 2009). Funding for IRCs and CRDs is industry driven; industry partners commit to support a research programme and provide funding that is matched, at different levels, by NSERC. The industry partners must have the internal capacity to exchange ideas and truly collaborate, as opposed to simply being interested in the outcomes of the research and must have the capacity to exploit or disseminate the final research results.

Industrially relevant research at the Hole School of Construction Engineering

Operating out of the University of Alberta's Department of Civil and Environmental Engineering, the Hole School of Construction Engineering (HSCE) is composed of six faculty members and is the largest construction group of its kind in Canada. Three faculty members each hold IRCs while two others each hold CRDs. Since 1997, the HSCE has received a total of CAD3.88 million in NSERC funding for IRCs and CAD4.33 million for CRDs, with roughly equivalent amounts provided by industry partners in matching funding (Moss, 2011). The current NSERC commitment is CAD3.1 million over 5 years for the three IRCs and CAD1 million over 5 years for the two CRDs (Moss, 2011). The IRCs and CRDs are distinct yet complementary in nature, allowing the researchers in the HSCE to collaboratively develop integrated strategies that address complex issues facing the construction industry today. The IRC and CRD focus areas are illustrated in Figure 5.4.

The NSERC IRC in Strategic Construction Modeling and Delivery

The IRC in Strategic Construction Modeling and Delivery (SCMD) has the overall goal of delivering innovative decision support systems and improved project practices to help Albertan and, ultimately, Canadian construction industries improve their project performance and increase their competitive position globally. This goal is being achieved by advancing the field of fuzzy logic and combining it with other artificial intelligence and simulation techniques to create workable, applicable approaches for solving construction-related problems in three focus areas: labour productivity analysis and modelling, structuring projects and teams for improved

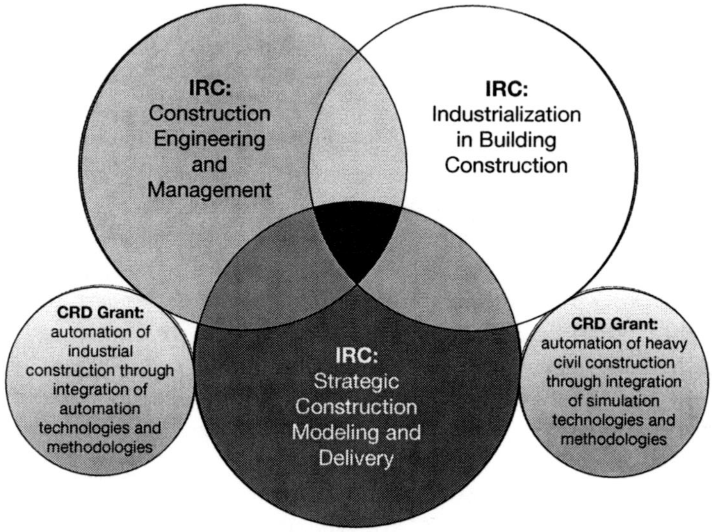

Figure 5.4 Collaborative structure of the Hole School of Construction Engineering, Canada

performance and reducing owner and contractor risk through qualification. Another important objective of the IRC in SCMD is to train highly qualified academic and industry personnel through their involvement in the proposed research.

Partners and mechanisms for collaboration

The IRC in SCMD is funded by NSERC, the University of Alberta and a consortium of Alberta-based construction industry stakeholders. The involved owners and owner associations range in size from 1,100 to 10,000 employees; partner contractors and contractor associations together represent 1,300 companies employing over 70,000 people; and partner labour associations and unions represent a combined total of 125,000 workers. Representatives from each industry partner comprise the management and technical advisory committees. The former oversees and guides the direction of the chair's research by defining areas of focus, providing strategic support and managing industry financial contributions. The last helps facilitate the industrial application of the IRC's research findings and new technologies. These multiple and diverse perspectives provide opportunities for truly innovative research and comprehensive solutions with the potential to make a significant impact on the construction industry as a whole.

Products and mechanisms for dissemination

The province of Alberta is expected to have the second highest non-residential construction investment in Canada between 2012 and 2020 (Construction Sector Council, 2012), with Alberta oil sands representing a price-adjusted CAD364 billion worth of investment between 2012 and 2035 (Burt, *et al.*, 2012). All partners in the IRC are directly involved in oil sands development, are major energy producers for Alberta or are otherwise impacted by the development of the oil sands, so the IRC's programme affects a significant portion of the Canadian economy. Large cost and schedule growth of projects, challenges in labour productivity, skilled labour shortages and international competition have threatened future investment in the Canadian construction industry, particularly the oil sands. The IRC in SCMD works to improve construction labour productivity and efficiency through the development of strategic solutions and real-time decision support systems that reduce project execution risks for owners and contractors.

Ideas are translated into outcomes and disseminated as results in a number of ways. Academic results are disseminated through journal publications, conference and poster presentations, and, in some cases, commercialisation of the research results. Translating results into tangible outcomes for industry is achieved through the following mechanisms:

- The HSCE invites industry partners and other researchers to the annual Innovation in Construction Forum.
- Industry partners attend training workshops for products developed under the IRCs and CRDs.
- Newsletters and websites disseminate research results widely.
- Researchers participate in industry events and committees to help shape industrial, as opposed to academic, research and to further disseminate the R&D results of the IRCs and CRDs.
- The IRC in SCMD undertakes two types of project: individual projects specific to each partner's needs and collaborative projects involving multiple partners in the IRC. Products delivered to the organisations have an immediate impact on their practices. Products include software that implements the advanced models developed in the research programme, internal technical reports of the research results, internal presentations, and process improvement solutions. For collaborative research, similar products are delivered to the construction industry at large, providing tools to develop industry-wide best practices.
- Students of the programme become highly qualified personnel who transfer advanced knowledge and techniques to their eventual employers.
- Leading members of the construction industry are exposed to the latest research methods and technology, enabling them to improve their organisation's practices.

Successes and barriers to R&D impact

The IRC has successfully addressed several barriers that could potentially limit R&D impact. First, staff turnover and corporate restructuring can result in the loss of a champion at the partner organisation to facilitate the research. Furthermore, advanced academic objectives, required by both NSERC and the university, have longer timelines, whereas businesses require more immediate demonstration of a return on investment of resources. Accordingly, as the IRC's programme advances the academic areas of research, it maintains strong ties with senior management to ensure continuity in the research and demonstrates immediate, tangible, short-term value to continuing and new personnel.

Second, with many other available employment opportunities, it can be difficult to recruit and retain research staff and students. The close ties with industry that the IRC programme has to offer are critical in attracting high calibre individuals to the programme.

Third, while financial contributions from partner organisations are essential to establish an IRC, in-kind support, namely, physical resources, data and personnel time, can be equally as important. Time commitments from personnel can be especially difficult to secure, so the IRC's on-site student researchers must actively contribute as team members of the hosting organisation. This involvement helps students develop effective, *one-of-a-kind* research solutions; the individualised results help secure further in-kind commitment.

The IRC in SCMD provides a venue for industry partners, who are normally in a competitive situation or are different parties in the construction supply chain, to interact and share perspectives with each other in a research setting. Ultimately, projects undertaken by the IRC in SCMD have succeeded because they have been structured to mutually benefit the graduate students, researchers and industry partners.

A regional research and innovation cluster for SMEs

The University of New Brunswick's Construction Engineering and Management Group

The context for this case study is a research and innovation cluster within the Maritime Provinces of Canada. In this region, the industry is comprised primarily of SMEs, and although national organisations do have a presence in the region, none has central headquarters there. There are fewer than half the number of medium-sized organisations per capita in this region in comparison with other more populous regions (the provinces of Ontario, Quebec, and Alberta), and only two large organisations in total (Statistics Canada, 2011a). As a result, the research topics undertaken must have a multi-party perspective in order to involve a critical mass of industry participants.

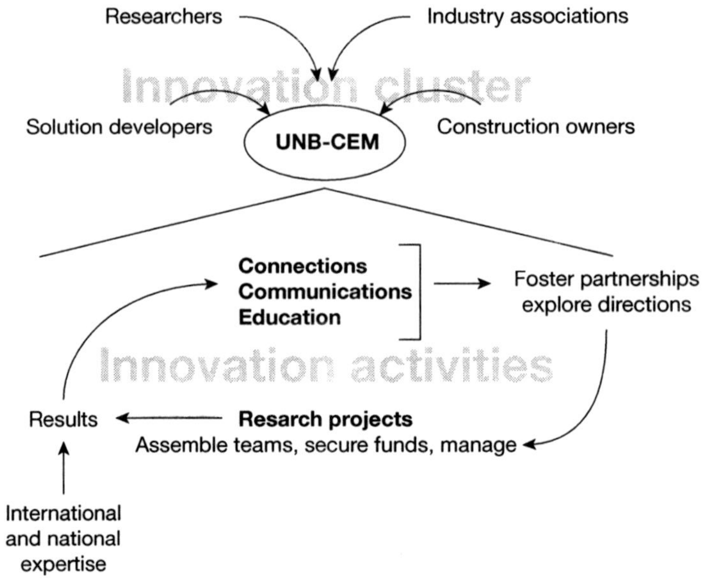

Figure 5.5 Participants and activities of the UNB-CEM Group's research and innovation cluster

In Figure 5.5, the participants are displayed as an innovation cluster, where the University of New Brunswick's Construction Engineering and Management (UNB-CEM) Group is the entity that brokers innovation. The UNB-CEM Group consists of two primary faculty members, a group of undergraduate and graduate research students and an industry advisory board, and is supported in part through the M. Patrick Gillin Chair in Construction Engineering and Management, an endowed chair. Although individual organisations are engaged through the advisory board, the primary mechanism for interacting with the region's industry is through industry associations, namely contractor, consulting and architectural associations; safety associations; sector councils; and public owners of infrastructure.

The UNB-CEM Group's research activities are focused on finding better ways to support an industry that is progressively more collaborative in its methods and as a result requires improved communication and management of information and knowledge. The overlapping research themes of the UNB-CEM Group are: the measurement of industry performance; the adoption of management and technological innovations; and the development of advanced information and communications technologies. The majority of UNB-CEM Group research involves active industry partners and has a track record of commercialisation. Where applicable, the UNB-CEM Group leverages global expertise through cross-disciplinary collaborations with other researchers and technology developers.

Mechanisms for research and innovation

The lower half of Figure 5.5 depicts the typical cycle of research and innovation activities that guides the UNB-CEM Group's innovation brokering activities. The activities of technology connections, technology communications and technology education can be considered collectively as knowledge dissemination activities.

Technology connections link researchers with practitioners. This mechanism typically follows a sequence that begins with presenting practical research results, followed by providing examples of successful implementations and lessons learned, gaining feedback from industry participants, setting new directions for research, and establishing training needs of the industry. Researchers and practitioners are brought together at roundtables, meetings, regional conferences and workshops.

Technology communications form a decision support network for the practitioners. Publications like briefing notes or best practices guidelines raise awareness of technology-related activities within the industry by simplifying the search for who is doing what, where and how they can be reached; they also also raise awareness of available technologies and communicate their potential for successful application towards improving industry productivity.

Technology education activities focus on project management and technology application as they apply to the construction industry. The intent is not to duplicate what is currently offered by other training organisations, but rather to identify gaps and to provide alternative delivery mechanisms that are best suited to industry requirements identified as a by-product of technology connections. Activities include promoting education in technology, developing course content and determining the best learning or training model.

These efforts ultimately aim to develop strategies to overcome the regional shortage of skilled workers, which is not only a problem at a trade skill level but also at the management and supervisory levels.

The Digital Technology Adoption Pilot Programme

A recent research project demonstrates the mechanisms through which research and innovation occur in the cluster. The cluster recognised that one of the dominant impediments to increasing the rate of innovation and improving productivity in the Canadian construction industry is the lack of access to knowledge in forms appropriate to support decision making or realise the capacity necessary to successfully adopt and implement digital technologies. The UNB-CEM Group secured support for a technology-based research project through NRC's IRAP. One of IRAP's current programmes is the *Digital Technology Adoption Pilot Program* (DTAPP), which aims to increase SME adoption of digital technologies and skills through the development of educational resource networks (National Research Council Canada, 2012a). UNB-CEM's research project consists of three phases: a series of diagnostic assessments; a series of digital adoption and implementations; and dissemination and awareness activities.

Diagnostic assessments examine organisations' operational processes and capacities, and identify areas having the potential to improve firm performance and productivity through successful technology adoption. Data collection is conducted through face-to-face interviews with key individuals identified by each construction organisation. The interviews follow a scripted data collection survey and subsequent verification of organisational practices. Results of the data collection for each construction organisation are used to conduct an assessment of organisational practices, and are analysed to identify opportunities for improvement, highlighting areas suitable for the adoption of digital technologies. Results of the analysis are presented as preliminary recommendations for each construction organisation. For example, an opportunity may be identified for an organisation to improve its construction materials management planning practices.

Preliminary recommendations are validated through workshops with members of each individual construction organisation. Participants validate the results and select the most relevant opportunities for the organisation to pursue. An adoption and implementation plan is then developed for each construction organisation. The adoption and implementation plan addresses organisational capacity and is structured in a way that is conducive to further DTAPP support and case studies.

The data collected using the standard assessment framework for each organisation are aggregated to provide benchmarking information on the performance of the industry with respect to key organisational practices and the capacity to adopt digital technologies. The benchmark provides a basis of comparison against which individual organisations can assess their performance before participating in a technology adoption and implementation project; following project completion, the benchmark allows organisations to evaluate project impact.

Successes and barriers to R&D impact

In general, the challenges encountered in this case study are not unlike those discussed in the previous case study concerning the IRC in SCMD. However, more specific to this scenario and to the research project example described are the limited mechanisms currently available to disseminate the results broadly and in a form that would efficiently stimulate more innovation.

However, the project does result in knowledge in a practical form: findings that can be used by subsequent potential adopters, namely, industry practitioners, for decision-making purposes.

When aggregated, the results of the case studies provide tangible evidence of the common issues that need to be addressed at an industry level. In addition, the series of case studies are also available for subsequent awareness and training activities such as creating technology connections between researchers and practitioners, supporting technology practitioners through communications, and introducing technologies to the industry through technology education, in the form of lessons learned and best practices for the local, regional, and national construction industries. Thus, in reference to Figure 5.5, the loop is closed with R&D dissemination, as the results feed into additional opportunities and potential expansion into other geographical and sector areas.

R&D impacts

Organisations and contractors that partner with Government or university-affiliated research groups experience immediately applicable, tangible benefits from their involvement in a research programme. In each of the case studies outlined above, collaborative arrangements ensure R&D is targeted at improving products or practices of specific companies; findings can then be used to benefit the construction industry as a whole. For example, by developing metrics, standards, and policies that potential buyers can rely on, the Canadian Federal Government helps companies bring innovative products to market.

This process not only helps manufacturers and suppliers, but it also helps industry users deliver better products to end consumers. The IRC in SCMD works with individual companies and organisations to better their competitive positions by developing and implementing tailor-made solutions. In partnering with owner associations, contractor associations and labour organisations, the IRC in SCMD's programme ensures its R&D has an even more widespread impact on the construction industry; research is both carried out and disseminated through the extensive membership of these groups. The UNB-CEM Group continuously builds on relationships with industry connections, cultivating awareness of both its research results and its members' expertise for the betterment of a unique regional industry. One mechanism the UNB-CEM Group uses is benchmarking, a natural extension to the improvement activities described in this case study. Each of the case studies featured in this chapter delivers innovation that has, or has the potential to have, far reaching impacts.

Working to tackle industry-wide issues from interconnected R&D hubs is both effective and economical. When the cost of funding R&D is shared between industry partners of varying sizes and Governments of differing jurisdictions, R&D resources, experts and findings become more accessible to a wider array of industry members. By cultivating long-term, mutually beneficial, collaborative working relationships between researchers and industry representatives, the Canadian Government, the IRC in SCMD and the UNB-CEM Group all enhance the competitiveness and viability of the Canadian construction industry as a whole.

Strategy for future investment

The Canadian construction industry has greatly benefitted from advancements in R&D, as outlined in the preceding case studies. The path forward to address current and future issues facing the Canadian construction industry is summarised in Figure 5.6. The industry and its researchers require improved mechanisms to reduce the risks inherent in pursuing new ways of working. Support could take the form of piloting innovative solutions on public projects while efficiently capturing lessons learned and sharing them industry wide. Research findings must be captured in a way that demonstrates their impact on practice, and in a format that supports

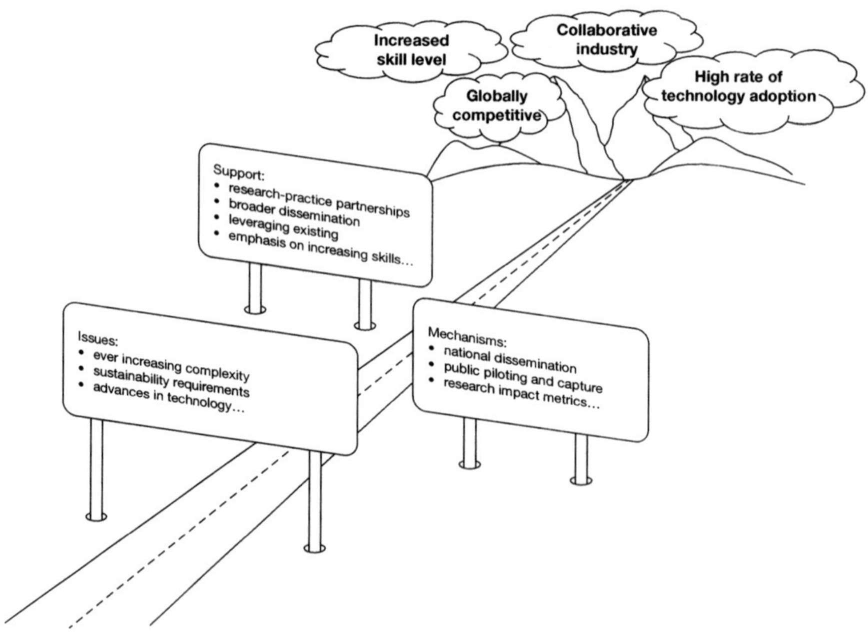

Figure 5.6 Roadmap for future research and development in the Canadian construction industry

and reduces the risk of technologies for those adopting them. Supporting these mechanisms requires further enhancements of the programmes that support research partnerships with an emphasis on increasing the skills of those conducting research and those practicing within the industry. Researchers must look for ways to not only generate new knowledge, but also to leverage existing knowledge by developing effective communications infrastructure.

Research–industry partnerships have been successful in ensuring R&D outcomes are put into practice. However, findings and innovations still require a more effective mechanism for national dissemination if they are to reach a broader audience, particularly SMEs. A comprehensive information infrastructure would enable researchers to better share knowledge across Canada's vast geography. Currently, R&D in the Canadian construction industry is fragmented; Government agencies, university researchers, funding agencies and other stakeholders may be simultaneously dedicating funds and efforts toward similar projects with no awareness of R&D contexts at other levels of Government, other institutions, and/or in other regions. Increasing contextual awareness would result in more efficient data collection as well as faster industry-wide adoption of new methods, tools, and practices.

However, such adoption remains voluntary until R&D is translated into policy. Public owners can be important drivers of this mechanism; by incorporating R&D findings into company policies, they can lead competitors by exemplifying new industry standards. Safety management practices, sustainability guidelines and building information modelling (BIM) techniques have all entered the industry in this way. Unfortunately, due to the speed at which construction takes place and the absence of innovation as a management priority, there is often a lack of desire to implement significant, long-term change. As the earlier case studies illustrate, continuing to create and nurture industry–research partnerships is crucial to ensure that R&D findings are actually implemented so that their effects can be measured and demonstrated to other industry members.

Developing appropriate metrics to measure the impact of R&D on the Canadian construction industry would better enable researchers to demonstrate the value of their work to public owners. Academic research output has some established metrics, while funding agencies make individualised attempts to evaluate projects from their own perspective. However, the industry itself lacks a suitable R&D impact evaluation system. Measuring the leading indicators of the industry's ability to innovate or capacity to adopt new ideas is one possible direction for a metrical system to take. Once presented with figures that demonstrate the measurable value of construction R&D, industry members may be encouraged to further invest in R&D endeavours and implement findings; competitors may also be motivated to pursue innovation and invest themselves. Such funding is needed to overcome the organisational overheads associated with data collection and further research on appropriate R&D impact metrics (Rankin, *et al.*, 2008).

As the construction industry becomes more globalised in nature, projects rapidly increase in societal, political, legal, economic and financial complexity. This complexity is compounded by simultaneous rapid advances in information and communications technology, increasing volumes of available information, and the analytical challenges that follow (American Society of Civil Engineers, 2007). Since dealing with these challenges often requires a multi-disciplinary approach, the ability to communicate, collaborate, and build relationships outside areas traditionally associated with the construction industry has become a skilled trade itself. As the urgency increases to address such diverse concerns as sustainability, energy efficiency and health and safety, the infrastructure that facilitates R&D, its implementation and its dissemination will play a role of ever increasing importance in the resilience of the local, regional and national economies of Canada. It is therefore imperative to invest in construction R&D in Canada, and not only financially. Personnel in diverse roles representing academic institutions, Governments, private companies, trade associations, and labour unions must collaborate if Canadian construction efforts are to remain globally competitive. Together, they must ensure that R&D becomes and persists as an industry priority.

Acknowledgements

The authors would like to thank Andrea Churchill Wong for her editorial work on this chapter. The NSERC Industrial Research Chair in Strategic Construction Modeling and Delivery gratefully acknowledges the support of the Natural Sciences and Engineering Research Council of Canada and the Chair's industrial partner organisations. The University of New Brunswick's Construction Engineering and Management Group members thank their industry advisory board and the National Research Council Canada's *Industrial Research Assistance Program*.

Note

1 According to Federal Reserve System (2013) Foreign Exchange Rates – G.5A (available at: www.federalreserve.gov/releases/g5a/current/ (accessed 14 January 2013)), the average exchange rate in 2012 was USD1 = CAD0.9995. Therefore, all CAD values are equivalent to roughly the same amount in USD.

References

American Society of Civil Engineers (2007) *ASCE policy statement 465: Academic prerequisites for licensure and professional practice*, Reston, VA: American Society of Civil Engineers.

Burt, M., Crawford, T.A. & Arcand, A. (2012) *Fuel for thought: The economic benefits of oil sands investment for Canada's regions*, Ottawa: Conference Board of Canada.

Canada Research Chairs (2012) *Canada research chairs*. Available at: www.chairs-chaires.gc.ca/ (accessed 20 December 2012).

Construction Sector Council (2012) *Total non-residential investment (excluding maintenence) for Canada and provinces, 2012–2020*, Construction Forecasts. Available at: www.constructionforecasts.ca/ (accessed 20 December 2012).

Government of Canada (2012) *Budget 2012*. Available at: http://budget.gc.ca/2012 (accessed 1 November 2012).

Industry Canada (2012a) *Canadian industry statistics: SME benchmarking construction (NAICS 23)*, Industry Canada. Available at: www.ic.gc.ca/cis-sic/cis-sic.nsf/IDE/cis-sic23bece.html (accessed 20 December 2012).

Industry Canada (2012b) *SME research and statistics: Key small business statistics – July 2012*, Industry Canada. Available at: www.ic.gc.ca/eic/site/061.nsf/eng/02715.html (accessed 14 February 2013).

Moss, P. (2011) *NSERC's strategy for partnerships and innovation . . . after two years*, Edmonton: Natural Sciences and Engineering Research Council of Canada.

National Research Council Canada (2012a) *Industrial research assistance program: Digital technology adoption pilot program (DTAPP)*, National Research Council Canada. Available at: www.nrc-cnrc.gc.ca/eng/irap/dtapp/index.html (accessed 20 December 2012).

National Research Council of Canada (2012b) *About NRC industrial research assistance program*, National Research Council of Canada. Available at: www.nrc-cnrc.gc.ca/eng/irap/about/index.html (accessed 20 December 2012).

Natural Sciences and Engineering Research Council of Canada (2009) *Natural sciences and engineering research council of Canada*. Available at: www.nserc-crsng.gc.ca (accessed 20 December 2012).

Natural Sciences and Engineering Research Council of Canada (2012) *Connect. Collaborate. Prosper*. Available at: www.nsercpartnerships.ca/ (accessed 20 December 2012).

Rankin, J., Fayek, A. Robinson, Meade, G., Haas, C. & Manseau, A. (2008) 'Initial metrics and pilot program results for measuring the performance of the Canadian construction industry', *Canadian Journal of Civil Engineering*, 35(9): 894–907.

Science, Technology and Innovation Council (STIC) (2011) *State of the nation 2010: Canada's science, technology and innovation system*. Available at: www.stic-csti.ca/eic/site/stic-csti.nsf/eng/h_00038.html (accessed 20 December 2012).

Statistics Canada (2011a) *Canadian industry statistics: Establishments construction (NAICS 23)*, Ottawa: Industry Canada. Available at: www.ic.gc.ca/cis-sic/cis-sic.nsf/IDE/cis-sic23etbe.html (accessed 20 December 2012).

Statistics Canada (2011b) *Fixed capital flows and stocks* [2002–2011], Catalogue no. 15-549-XIE, Ottawa: Statistics Canada. Available at: www.statcan.gc.ca/pub/15-549-x/15-549-x2007001-eng.pdf (accessed 20 December 2012).

Statistics Canada (2012a) *Gross Domestic Product by Industry: Provinces and Teritories [2002 to 2011], component of Catalogue no. 11-001-X*, Ottawa: Statistics of Canada. Available at: www.statcan.gc.ca/daily-quotidien/120427/dq120427a-eng.pdf (accessed 20 December 2012).

Statistics Canada (2012b) *Gross Domestic Product by Industry [2002 to 2011]*, Catalogue no. 15-001-X, Ottawa: Statistics of Canada. Available at: www.statcan.gc.ca/pub/15-001-x/15-001-x2012009-eng.pdf (accessed 20 December 2012).

Statistics Canada (2012c) *Gross domestic product (GDP) at basic prices, by North American industry classification system (NAICS), monthly (dollars)*, CANSIM Table 379-0027, Ottawa: Statistics Canada Database. Available at: http://cansim2.statcan.ca/ (accessed 20 December 2012).

Statistics Canada (2012d) *Indexes of labour productivity and related variables, by North American industry classification system (NAICS), seasonally adjusted, quarterly (index, 2007=100)*, CANSIM Table 383-0012, Ottawa: Statistics Canada Database. Available at: http://cansim2.statcan.ca/ (accessed 20 December 2012).

Statistics Canada (2012e) *Business enterprise research and development (BERD) characteristics, by industry group based on the North American industry classification system (NAICS) in Canada, annual (dollars unless otherwise noted)*, CANSIM Table 358-0024, Ottawa: Statistics Canada Database. Available at: http://cansim2.statcan.ca/ (accessed 20 December 2012).

Statistics Canada (2012f) *Business enterprise research and development (BERD) characteristics, by industry group based on the North American industry classification system (NAICS), provinces and territories, annual (dollars unless otherwise noted)*, CANSIM Table 358-0161, Ottawa: Statistics Canada. Database. Available at: http://cansim2.statcan.ca/ (accessed 20 December 2012).

Statistics Canada (2012g) *Gross domestic expenditures on research and development in Canada (GERD), and the provinces*, Catalogue no. 88-221-X, Ottawa: Statistics Canada. Available at: www.statcan.gc.ca/pub/88-221-x/88-221-x2012001-eng.pdf (accessed 20 December 2012).

Statistics Canada (2012h) *Industrial research and development: Intentions 2012*, Catalogue no. 88-202-XIE, Ottawa: Statistics Canada. Available at: www.statcan.gc.ca/pub/88-202-x/88-202-x2012000-eng.pdf (accessed 20 December 2012).

Statistics Canada (2012i) *Gross domestic product by industry: Provinces and territories* [2002–2011], component of Catalogue no. 11-001-X, Ottawa: Statistics Canada. Available at: www.statcan.gc.ca/daily-quotidien/120427/dq120427a-eng.pdf (accessed 20 December 2012).

Statistics Canada (2012j) *Gross domestic product by industry* [2002–2011], Catalogue no. 15-001-X, Ottawa: Statistics Canada. Available at: www.statcan.gc.ca/pub/15-001-x/15-001-x2012009-eng.pdf (accessed 20 December 2012).

Xu, K.J. (2010) *Roadmap to research and innovation in the Canadian construction industry: Towards a change management strategy*, Ottawa: Canadian Construction Association.

6 Denmark – building/housing R&D investments

Kim Haugbølle

Background: the national context

From the early 1990s onwards a new business policy perspective gradually emerged in Denmark called *resource areas* (Danmarks Statistik, 2001; Erhvervsfremme Styrelsen, 1993). Contrary to most other business policy studies, resource areas include the four industries: primary industry, such as raw material extraction; manufacturing industry, such as production of building components; supporting industry such as production and leasing of building machinery; and service industry such as contractors and consultants. As a consequence, building is no longer simply viewed in terms of contractors and consultants, but embraces all industries contributing to the production of the built environment.

During the 1990s and early 2000s, a range of policy reports addressed the challenges of the Danish building/housing resource area. One of the more prominent reports was the policy report *The future of building – from tradition to innovation* by the Building Policy Task Force established by the Ministry of Housing and Urban Affairs and the Ministry of Business Affairs (By- og Boligministeriet & Erhvervsministeriet, 2000). This task force identified a range of problems for this resource area and proposed a number of initiatives within four areas related to: (i) the role of building clients; (ii) competitiveness; (iii) cooperation; and (iv) innovation.

In relation to innovation, the Building Policy Task Force noted that the development of research and development (R&D) investments and patents was decidedly negative within the resource area. The number of patents had decreased, and the level was generally lower with respect to other Organisation for Economic Cooperation and Development (OECD) countries. Corporate R&D investment had fallen in the past 10 years, and in terms of trade and industry investments in R&D the Danish building industry invested around 20 per cent below the level of other OECD countries. Thus, the task force proposed, *inter alia*, that the Government should take the initiative to scrutinise the existing knowledge system for building and develop a national action plan for building/housing research. Based on this recommendation, the Government established a new task force with a dedicated focus on building research to develop proposals for an action plan for Danish public building/housing research. This was named the Task Force for Building Research.

The Task Force for Building Research was not the first to address building-related R&D. Building-related R&D has been the attention of a range of surveys and policy studies in Denmark during the past decades, similar to other countries such as the UK (see, for example, Fairclough, 2002). These studies include, among others, questionnaire surveys of building-related research in Denmark (Boligministeriet, 1993; Christoffersen & Bertelsen, 1990), the use of technological services in the building sector (Bang, 1997), production, use and dissemination of research-based knowledge in the building sector (Dræbye, 1997), and overview of research on cities, housing and building (Det Offentlige Forskningsudvalg for Byer og Byggeri, 2000). In summary, the previous studies, surveys and policy analyses can be characterised by:

- Common questions: (i) *How much should different research themes take up?* And (ii) *How can knowledge dissemination improve from research institutes to companies?* Other pressing questions are, however, left in the dark, such as: *Do we have sufficient building research? How can we improve the capabilities of companies to adopt new research results?*
- Inward looking perspective: mappings and analyses adopt an inward looking perspective with little comparison to other sectors and countries, link to broader R&D policy issues, or utilisation of theoretical contributions on what fosters R&D investments and how the R&D community operates in general.
- Vague definitions: in general, the very definitions of what to include in the studies are vague with respect to the subject area, namely, housing, building, construction, planning to mention some; research areas such as technical science vs. social science; type of effort, namely research, development, dissemination among others; and indicators of input such as funding, and output such as articles and tools.
- Weak evidence for conclusions: the available resources in the individual surveys and analyses have apparently been relatively modest. It is characteristic of several of the mappings and analyses that the database for firm observations and conclusions is small and inadequate.

Taking the work of the Task Force for Building Research (established June 2001) as the starting point, this chapter will: (i) describe the organisational setup of Danish building and housing research; (ii) quantify the Danish public and private building R&D investments; and (iii) provide an update on recent developments in the Danish building R&D environment.

Adopting a resource area perspective, this chapter provides a comprehensive quantitative survey of R&D investments coupled with a qualitative analysis of professional competences and profile of individual research institutes. The institutional profiles were based on an initial screening of all relevant research institutes followed by a more detailed mapping of the most prominent research institutes. The survey was conducted on the basis of existing sources that were publicly available especially public research statistics, annual reports and websites of research institutions and funding agencies. In addition, special surveys in the research statistics of the Analyseinstitut for Forskning (Research Centre for Studies in Research) were conducted. It should be noted though that drawing firm conclusions on the level of building-related R&D expenditures warrants careful consideration with regard to the scope of the study, the definition of the construction industry, the difficulties on how to classify and register research activities adequately under different research purposes, to mention some. Data stem from a previous study extending to 2001. Although the ambition has been to update the estimate, this has proved cumbersome due to an organisational restructuring of the research system in 2007, changes in the reporting of research statistics and a change in the policies of universities on reporting.

Structure of Danish building and housing research

The starting point for the Task Force for Building Research was the view that the organisational setup of Danish public building/housing research was too fragmented. Although this research was spread over many different actors in 2002, most of the activity was concentrated on five core research and educational institutions with building/housing research as their primary activity:

- Aalborg University (AAU, primarily the Institute for Building Technology);
- Technical University of Denmark (mainly BYG-DTU);
- School of Architecture in Copenhagen (KARCH);
- Aarhus School of Architecture (AAA);
- Danish Building Research Institute (SBi).

Although the majority of the other smaller or more peripheral institutions remained in place, significant changes in the organisational setup of universities and Governmental research institutes took place due to the national university reform in 2007. The hitherto 12 universities and 15 Governmental research institutes merged into eight universities and a few Governmental research institutes. During this process, the national building research institute merged with Aalborg University, but has remained an independent faculty of the university. At the Technical University of Denmark, the technical and management part of the building technology department was separated. The construction management group became part of a newly established department of engineering management.

Along with the five core public research institutions, the technological service provider Danish Technological Institute (DTI, notably the Construction Division) does building-related R&D. In line with OECD definitions, DTI is considered in statistical terms a private institution and not a public institution despite the public support granted by the Government: about 10 per cent of turnover. In addition to the core institutions, a variety of other R&D institutions regularly carry out building-/housing-related research and development, but not with building/housing research as their primary activity. There are also several supporting institutions in the form of knowledge brokers, which do not conduct research themselves, but provide support for, initiate and coordinate research and development, disseminate knowledge or otherwise act as catalysts in or for change.

Finally, the main sources of public funding for the building/housing area's R&D are a range of ministries, research councils, the European Union, private not-for-profit foundations and various bodies of the Nordic Council of Ministers.

Building/housing-related R&D investments

Public R&D funding

One way of estimating national Government R&D expenditure for building/housing is by research purpose (Table 6.1). First, the statistics on public research budget for 2007–2013 show that funding for the research purpose *housing and physical planning* is very small compared to the total budget. Second, the appropriations for the research purpose *industry, handicraft and building and civil engineering* is much higher and almost doubled in 2008 due to a new funding scheme towards globalisation. Since this research purpose is primarily centred on industry in general, it is not possible to give an accurate estimate of the building/housing proportion. Third, the research purpose *general scientific development* represents more than half of total public R&D expenditures, which also contains R&D expenditures relevant to building/housing.

With the methodology applied it is not possible to more precisely distinguish R&D expenditure directed specifically at building/housing. As statements of public R&D expenditure by research purpose cannot give an exact picture of the public R&D efforts in building/housing, the estimate will instead be based on the turnover of the five core public research institutions plus other public research institutions.

The level of R&D by the five core public institutions in 1999 was approximately DKK175 million (USD30 million) plus DKK55 million (USD9.5 million) for the DTI (Table 6.2). It should be noted that, at present, several institutions only make aggregated financial indicators publicly available.

In general research statistics (Analyseinstitut for Forskning, 2001a and 2001b), the R&D efforts by other public research institutions for research

Table 6.1 Public R&D expenditures distributed on research purpose (million DKK), Denmark

	2007	*2008*	*2009*	*2010*	*2011*	*2012*	*2013*
Farming, forestry, hunting and fishery	681	559	555	541	599	510	578
Mining, industry, craft and building and civil engineering	857	1,490	1,533	1,847	1,788	1,696	1,518
Production and distribution of energy	404	581	512	900	857	684	777
Transport and telecommunication	68	47	83	78	98	85	70
Housing and physical planning	39	69	37	31	60	56	50
Pollution control and nature conservation	258	370	434	356	378	357	315
Health	305	316	363	410	424	324	285
Social affairs	80	131	120	89	213	107	123
Culture, mass media and leisure	121	242	197	160	163	238	182
Education	256	309	387	390	409	383	635
Working conditions	193	140	140	141	126	129	131
Economic planning and public administration	119	87	106	85	53	99	109
Earth science	98	66	77	78	77	75	78
General scientific development	8,383	8,845	10,080	9,904	10,728	11,145	11,233
Space	228	248	316	334	238	194	251
Defence	77	86	82	76	58	58	58
Total	**12,168**	**13,585**	**15,020**	**15,419**	**16,268**	**16,142**	**16,392**

Source: Danmarks Statistik (2013)

Note: Figures from the national budget. Funding from the European Union, municipalities, and other, is not included. Figures may not add up due to rounding.

purpose *housing and physical planning* was estimated at approximately DKK90 million (USD15.5 million). The reliability of these reports may be debatable. For example, the Department of Comparative Literature at the University of Copenhagen estimate 10 per cent of its overall research to be within the area of housing and physical planning. Conversely, the Centre for Indoor Environment at the Technical University of Denmark (established by the world-leading professor in indoor climate P.O. Fanger) provided 0 per cent R&D within housing and physical planning, although indoor climate in buildings can be considered as belonging to the building/housing area. Instead, research is classified under themes such as *disease control*

Table 6.2 R&D funding sources for core public institutions (million DKK), 1999

	AAU Building technology	*DTU BYG-DTU*	*KARCH*	*AAA*	*SBi*	*DTI Building technology*
Basic public funding	16.1	29	24	12.2	32.2	6.6
Private foundations	1.2	1	0.1	0.7	0.3	
Public programmes	3.4	2	0.1	3.4	3.8	
Public sources	3.4	7	2.8	1.3	20.2	6.6
Private firms	0	2	0	0	2	39
International sources	0.2	3	0.1	0	0.9	2.8
R&D expenditure total	**24.3**	**44**	**27**	**17.5**	**59.4**	**55**

Sources: Special survey from database on research statistics by Analyseinstitut for Forskning (2000); annual reports from institutions. Figures may not add up due to rounding.

and prevention. It is therefore reasonable to assume that some research relevant to building/housing is categorised under other research purposes and vice versa.

The amount of public R&D by other public research institutions was estimated at around DKK100–150 million (USD17–26 million) after a careful examination of their research profiles, project portfolio and research publications. To this should be added R&D expenditures at the core public institutions of approximately DKK175 million (USD30 million). Thus, the total annual public R&D expenditure within the building/housing resource area was estimated at some DKK275–325 million (some USD52 million).

Out of a total public R&D budget of DKK8.93 billion (USD1.5 billion) in 2002, the building-related R&D budget amounts to some 3 per cent of the total public R&D expenditures, whereas the turnover in private companies of the resource area amounts to some 20 per cent of GDP. Thus, there is a disproportionate share of public R&D funding for building-related R&D compared to the economic importance of the building industry.

Private R&D funding

The funding of R&D for private companies within the resource area building/housing is distributed through different sources of financing (Table 6.3). R&D activities carried out by the majority of private companies, within building/housing, are financed by the companies themselves. However, there is also a substantial external financing of about 20 per cent coming from the public sector, other private enterprises and foreign sources. In 1998, the level of external financing for building/housing was 31 per cent (Analyseinstitut for Forskning, 2000). Even though the significant fluctuations from year to year call for caution, the trend is clearly defined. The resource area building/housing finances a substantially smaller part of its R&D through internal funding compared to most other resource areas.

Table 6.3 R&D funding sources for private companies (million DKK), 1999, Denmark

Source of financing	*R&D expenditure (million DKK)*	*Distribution (%)*
Public funding	128	7
Own funding	1,446	80
Other private companies	113	6
Other Danish funding	10	1
Foreign funding	101	6
Sum	**1,799**	**100**

Source: Analyseinstitut for Forskning (2000, 2001c)

Note: More recent data for the resource area are not available. Figures may not add up due to rounding.

Setting the R&D expenditure of a resource area in relation to its turnover may be a better measure of its *research intensity* (Table 6.4). There are very pronounced differences between the resource areas. Resource area building/housing is placed in the middle field, but is still far from the Barcelona targets set by the European Union.

R&D expenditure in private companies is divided into the four main business sectors covered by the resource area building/housing: primary industry, manufacturing industry, support industry and service industry. For confidentiality reasons, primary industries and manufacturing industries are merged as well as support and service industries (Table 6.5). The reporting distinguishes between two types of cost: (i) R&D performed in own business, which refers to all R&D activities of the company, including R&D that may have been executed for other companies; and (ii) purchased R&D services performed for the company by third parties.

According to this special survey, private companies within building/housing invested about DKK1.8 billion (USD310 million) on R&D in 1999.

Table 6.4 R&D expenditure in relation to turnover (million DKK), 1999, Denmark

Resource area	*R&D expenditure (million DKK)*	*Turnover (million DKK)*	*R&D share of turnover (%)*
Medico/health	6,172	71,343	8.65
IT/communication	5,046	281,974	1.79
Other business	2,843	246,539	1.15
Building/housing	1,799	329,028	0.55
Energy/environment	902	177,683	0.51
Transport	641	224,357	0.29
Food	1,128	488,882	0.23
Furniture/clothing	143	94,060	0.15

Source: Danmarks Statistik (2001); special survey from Analyseinstitut for Forskning

Table 6.5 R&D expenditure within building/housing, 1999, Denmark

	Own conducted (million DKK)	*Purchased (million DKK)*	*Sum (million DKK)*
Primary and manufacturing industries	811	–[¥]	–[¥]
Support and service industries	830	–[¥]	–[¥]
Resource area in total	**1,641**	**157**	**1,799**

Source: Special survey by Analyseinstitut for Forskning (2000)

Note: [¥]For confidentiality reasons, each cell requires a minimum of six respondents and no single respondent may account for more than 60 per cent of the total expenditure in the cell

The R&D expenditures are fairly evenly divided between the primary and manufacturing industries, on the one hand, and the support and service industries, on the other. As both the level and distribution of R&D expenditure was somewhat unexpected, a more in-depth analysis was conducted. The first analysis was related to corporate reporting on research statistics, which included companies' own estimates of the proportion of total R&D expenditure spent on research in 14 different research areas (Table 6.6).

Table 6.6 Private companies' R&D distributed on research areas, 1999, Denmark

Research area	*Internal R&D expenditure (million DKK)*
K1: Materials science	216
K2: Building and civil engineering technology	233
K3: Health	–[¥]
K4: Genetic engineering	–[¥]
K5: Biotechnology	–[¥]
K6: Food	–[¥]
K7: Energy	194
K8: Environment	192
K9: Geriatrics/assistive technology	5
K10: Defence	–[¥]
K11: Management, organisation, learning	19
K12: Software, integrated	59
K13: Software, standalone	116
K14: Hardware	91

Source: Special survey by Analyseinstitut for Forskning (2000)

Note: [¥]For confidentiality reasons, each cell requires a minimum of six respondents and no single respondent may account for more than 60 per cent of the total expenditure in the cell

The statistics are subject to two important conditions: First, the research areas are subject to interpretation and non-exhaustive, which implies that not all companies would be able to report their R&D spending in some of the areas listed. Second, the listed areas are not mutually exclusive. Therefore expenses reported in one area, for example health, can also be reported in, for example, genetic engineering. Thus, the expenditures on various research areas cannot simply be added as there may be overlaps (Analyseinstitut for Forskning, 2001c). In addition, the calculation of the private companies' R&D is not directly comparable with the calculation of the public research R&D, because the categories used are different.

However, the overview can be used to note that companies within the resource area building/housing indicate that they only conduct R&D in building and civil engineering accounting for approximately DKK233 million (USD44 million). Although building and civil engineering technology is only a subset of the building/housing resource area, it is an important and vital part of that resource area. It does not seem reasonable that building and civil engineering technology would be such a small proportion of the total R&D of DKK1.8 billion (USD310 million) in the building/housing area. Thus, the statistics indicate that a significant fraction of the R&D included is not relevant to the building/housing area.

Looking at the primary and manufacturing industries in detail, it has, unfortunately, not been possible to carry out a more in-depth investigation regarding the inventory. R&D within primary industries is relatively modest, while R&D in the manufacturing sector holds the main part of primary and manufacturing industries' R&D worth DKK811 million (USD140 million). Companies are grouped according to their *Nomenclature des Activités Économiques dans la Communauté Européenne* code (NACE, Nomenclature of Economic Activities in the European Community), a European industry classification system similar to the Standard Industry Classification (SIC) and North American Industry Classification System (NAICS) for classifying business activities. There is a risk that R&D expenditures included in building/housing should be attributed to other resource areas, especially energy/environment and other industries. Conversely, companies in other resource areas might assign their R&D costs to building/housing. However, it is not possible to firmly judge the reliability of the reporting. Consequently, the total R&D expenditure for primary and manufacturing industries of around DKK811 million (USD 140 million) is attributed to building/housing.

A second analysis took a closer look at the support and service industries, which provided a more nuanced picture (Table 6.7). R&D expenditures by consulting engineers are a very important part of the support and service industries' R&D expenditure. By contrast, the architectural firms and contractors' R&D spending is relatively small. The approved technological service institutes' share of R&D expenditure is confidential, but may be assumed to be relatively large.

Table 6.7 R&D expenditure in support and service industries, 1999, Denmark

	Internal R&D expenditure (million DKK)	*External R&D expenditure (million DKK)*	*Own financing (million DKK)*	*External financing (million DKK)*
Approved technological service institutes	–[¥]	–[¥]	–[¥]	–[¥]
Consulting engineers	459	30	328	160
Architects	38	3	18	24
Contractors	38	5	42	0
Total	**830**	–[¥]	–[¥]	–[¥]

Source: Analyseinstitut for Forskning (2001c); special survey by Analyseinstitut for Forskning (2000)

Note: [¥]For confidentiality reasons, a minimum of six respondents is required and no single respondent may account for more than 60 per cent of the total expenditure in each cell

However, there may be good reasons to be sceptical of the reported R&D expenditure, especially for the consulting engineers, because reporting from this type of businesses also contains significant R&D costs assigned to other resource areas. Only a very small part of R&D, less than 10 per cent, within consulting engineering firms is related to the research area of building and civil engineering technology (Table 6.8). The rest is distributed to other areas of research. Some of which, of course, may also be relevant for building/housing. Thus, it would hardly be reasonable to maintain that internal R&D expenditure in support and service is about DKK830 million (USD144 million).

Conversely, a number of companies outside the building/housing area perform R&D in the research area of building and civil engineering technology. The total private R&D in this research field amounts to

Table 6.8 R&D expenditure in service industries, 1999, Denmark

	Internal R&D expenditure (million DKK)
Approved technological service institutes	–[¥]
Consulting engineers	39
Architects	10
Research area *building and civil engineering technology*, total	352

Source: Special survey by Analyseinstitut for Forskning (2000)

Note: [¥]For confidentiality reasons, a minimum of six respondents is required and no single respondent may account for more than 60 per cent of the total expenditure in each cell approximately DKK352 million (USD61 million), while companies in the building/housing area itself only spend around DKK233 million (USD40 million) in building and civil engineering (Table 6.6). Thus, the statistical summaries of R&D expenditure by companies within the resource area building/housing ignores almost one third of the total research within the research field of building and civil engineering.

In conclusion, the special survey of private companies' R&D expenditures within the building/housing area reported a total R&D of DKK1.8 billion (USD310 million). Based only on a narrow definition of the research field *building and civil engineering technology*, a second estimate of private R&D expenditures would rather suggest DKK350 million (USD60 million). A more reasonable estimate would include the R&D efforts in primary and manufacturing industries plus half of the R&D efforts in support and service industries. Thus, the total private R&D expenditure within the building/housing area was estimated at approximately DKK1.2 billion (USD207 million).

The estimates and statistics of private companies' R&D must be treated with caution. Part of the explanation must be sought in the R&D statistics being based on the companies' primary business sector. The company's entire R&D expenditure is attributed to this main sector, although companies such as the consulting engineers operate in several different markets with different products and services. Another explanation could be that the sample is not representative of the building/housing resource area. Finally, the weights used to scale the data from the sample to the entire study population may lead to a systematic bias; for example, see Analyseinstitut for Forskning (2001c) for a detailed description and discussion of the methodology behind the research statistics.

Action plan for Danish building R&D

Recommendations from the Task Force for Building Research

This task force was composed of six members from the private foundation Realdania (Chairman), a leading architectural firm C.F. Møllers Tegnestue, the National Association of Housing Associations, and researchers from Copenhagen Business School, SBi and Aalborg University. The task force also comprised four public servants from the Danish Agency of Enterprise and Housing, Ministry of Science, Technology and Development, and the Danish Energy Agency. The task force was assisted by a secretariat of five at the Danish Agency of Enterprise and Housing, which in turn was assisted by two external experts including the author of this chapter.

The mandate of the Task Force on Building Research in Denmark included three main tasks to undertake (Haugbølle & Clausen, 2002):

- mapping of: (i) content and scope of existing public Danish building research, including interactions with private building research; (ii) the

institutional, organisational and financial issues regarding public building research in Denmark; and (iii) the organisational structure of the building knowledge system based on existing reports, analyses and other sources, and updating these reports;
- evaluating: (i) research needs in relation to the vision for building in the future; (ii) Danish building research in an international context; (iii) interaction between producers, providers and users of building knowledge; and (iv) how the public building research can better serve as a catalyst for private R&D investments;
- proposing: (i) prioritisation of public building research, (ii) increased interaction between public and private investment in building research, including initiatives to strengthen the incentives for firms to develop new building knowledge; (iii) initiatives to improve the dialogue between research and users of building knowledge; and (iv) reorganisation of building knowledge infrastructure.

In September 2002 the task force published its report (Udvalget vedrørende byggeforskning i Danmark, 2002). The task force pointed out that a transition of the building industry into a knowledge-driven society would raise strong requirements for the scope and quality of building research. The task force concluded that the starting point was somewhat different and identified three challenges (Udvalget vedrørende byggeforskning i Danmark, 2002):

- The amount of Danish building/housing research was lagging behind compared to other OECD countries, and the level seemed to be decreasing;
- Building/housing research did not match the current challenges faced by the industry, because the publicly funded research in recent years had been directed towards other aims;
- The building/housing knowledge system was perceived as confusing by the industry and characterised by too many independent actors.

The task force believed that these challenges could significantly be attributed to building research being driven by public support, and that orientation towards the real users therefore had been inadequate. Thus, the task force (Udvalget vedrørende byggeforskning i Danmark, 2002) proposed a strategy based on stronger demand orientation, where building research could contribute to the development of the building industry by:

- supporting building innovation and change effectively;
- contributing dynamically to the companies' ability to competently seek knowledge;
- providing businesses and authorities with relevant high-quality knowledge;

- creating the conditions for more active involvement of international knowledge.

The task force recommended a long-term strategic framework resting on four main elements (Udvalget vedrørende byggeforskning i Danmark, 2002; emphasis in the original):

- ***Increase funding*** with DKK120 million (USD21 million) annually for building research to reach the OECD level. The majority of the funding was to be provided from the building clients through the establishment of a building innovation fund.
- ***A 10-year national action plan*** for building research in 2003–2012. The action plan was suggested to be established through an agreement between the Government, major public and private clients and relevant client groups along with the firms and organisations of the building industry. The task force stressed the importance of giving the action plan a timeframe in order to ensure an effective implementation of the results of building research in the industry. The action plan was to be realised through the establishment of a number of innovation consortiums within the most significant development areas.
- A ***re-orientation*** of building research towards research areas of importance for the transformation of the building industry, not least organisation, management, learning and collaboration along with development of large-scale components and systems.
- A systematic effort towards ***improved learning*** in building along with coordination of knowledge dissemination. Learning and dissemination should be part of all research projects and innovation consortiums.

Recent developments

Ten years after the publication of the action plan, it is worth reviewing the recommendations of the action plan in relation to building-related R&D. Each of the four elements in the roadmap will be addressed.

First, the recommendation to *increase funding* by DKK120 million (USD21 million) annually has far from materialised although the private foundation Realdania has made some significant contributions. The chairman of the task force was also chairman of the board of the private foundation Realdania. The foundation has pursued its own proactive R&D strategy over the past 10 years, which included spending some DKK277 million (USD48 million) in the period 2003–2010 on research activities alone (DAMVAD A/S, 2011). Realdania has initiated six research centres, a range of standalone projects, spearhead projects and other. Half of this expenditure (DKK145 million, equivalent to USD25 million) has been spent on the establishment of six research centres within urban spaces, strategic urban research, housing and welfare, construction management, facility management and indoor environment. However, the innovation fund based on a percentage of the budget of new building projects never came into existence because of lack of support from, among others, the public building clients and disagreement with the private foundation Realdania on how to organise, manage and finance the innovation fund.

The ambition to increase the number of researchers and university research programmes through, for example, PhD programmes has not been achieved. As described previously, the organisational setup of universities and Governmental research institutes has changed due to the national university reform. These organisational changes have not in themselves provided additional funding, but some extra resources have been allocated to building/housing research due to the general funding schemes related to the establishment of new educational programmes on design and architecture, construction management and informatics, to name some.

Second, the ambition to draft a *10-year national action plan* was never met although attempts to create a common platform for the industry were made about 5 years later through the establishment of the Koordinations- og Initiativgruppen for viden i byggeriet (KIG, Coordination and Innovation Group for Knowledge in Building) embracing all major actors in the industry. A series of meetings were held by KIG where inspiration was sought from, among others, Svenska Byggbranschens Utvecklingsfond (SBUF, the Development Fund of the Swedish Construction Industry) and the Danish food industry where a year-long collaborative effort had led to the establishment of an innovation fund. In 2009 an outlook report was prepared by the head of the building technology department at the Technical University of Denmark on behalf of KIG. The report described four focus areas: sustainability and energy; economy and innovation; safety and health; and functionality and experience (Koordinations- og Initiativgruppen for viden i byggeriet, 2009).

The Task Force on Building Research had proposed that a core element in the action plan should be the establishment of innovation consortiums within strategic areas. Although at least three innovation consortiums have been established over the years, they have hardly been the direct result of the roadmap. The first initiative was Digital Construction, a development programme initiated in 2000 and officially concluded 10 years later although a follow-up programme has been established for the period 2011–2014. The second initiative is an ongoing innovation consortia on sustainable industrialised building (InnoByg) established in 2010 for a 4-year period. The third initiative is a knowledge centre on energy savings established in 2008 for a 4-year period. This was established as a knowledge centre, and not as an innovation consortium. None of these three initiatives was directly anticipated in the roadmap as the roadmap only addressed innovation consortia in rather vague terms.

The third element was to *re-orient the building research* through the transformation of basic funding through increased competition and new types of performance-based contract. The conversion of basic funds did realise for the national building research institute, but not for any other research institutes. Despite strong warnings from the national building research institute, their budget was reduced by DKK5 million (USD0.9 million) in 2003 and again in 2011 by an additional DKK5.5 million. The funds have been redistributed in calls on ad hoc themes, which have changed from year to year. Thus, a strong *short-termism* has been the result along with a reduced gearing of the funding available.

Fourth, the ambition to *improve learning* was based on a range of initiatives related to the establishment of a central knowledge centre (*one-stop shopping*), improved dissemination to educational institutions, mandatory dissemination plans for research projects, improved knowledge dissemination from the two national building defects funds, and the strengthening of networks. Especially the creation of *one-stop shopping* was considered important, which subsequently led to the initiation of two competing consortiums that developed two different strategies for the establishment of a central knowledge centre. As time would eventually tell, only smaller changes have been observed in the ways knowledge dissemination takes place partly because of the financial independence and organisational affiliation of the individual knowledge brokers.

Conclusion

This chapter has (i) described the organisational setup of Danish building and housing research; (ii) quantified the Danish public and private building/housing R&D investments; and (iii) provided an update on recent developments in the Danish building/housing R&D environment.

First, building/housing R&D activities are spread over many different actors, but most of the activities are concentrated on five core research and educational institutions with building/housing research as their primary activity.

Second, the level of R&D expenditures at the core public institutions in 2001 is estimated at around DKK175 million (USD30 million) plus DKK100–150 million (USD17–26 million) at other public research institutions, totalling approximately DKK275–325 million (USD48–56 million). The R&D expenditures in private companies within the building/housing resource area were estimated at DKK1.2 billion (USD207 million) corresponding to the R&D efforts in primary and manufacturing industries, and half of the calculated R&D efforts in support and service industries. This estimate is marked by significant uncertainty due to the methodology applied in collating these statistics. Although the exact figures may be disputable, private R&D investments primarily take place in the manufacturing industry.

Third, the Task Force on Building Research published a research roadmap for building-related R&D in 2002. The roadmap suggested a range of actions within four key areas: (i) increase R&D funding by DKK120 million (USD21 million) annually; (ii) issue a 10-year national action plan; (iii) re-orient research priorities; and (iv) improve learning. Although some actions have been realised, several of the initiatives proposed in the R&D roadmap have not been implemented or would most likely have been implemented anyway. Consequently, the R&D roadmap seems to have had little direct impact.

Acknowledgements

The original study of building-related R&D from which this data derives was financed by the Danish Enterprise and Construction Authority.

References

Analyseinstitut for Forskning (2000) *Erhvervslivets forskning og udviklingsarbejde. Forskningsstatistik 1998* (*R&D in private companies. Research statistics 1998*), Århus: Analyseinstitut for Forskning.

Analyseinstitut for Forskning (2001a) *Offentligt forskningsbudget 2001. Forskningsstatistik* (*Public R&D budget 2001. Research statistics*), Århus: Analyseinstitut for Forskning.

Analyseinstitut for Forskning (2001b) *Forskning og udvikling i den offentlige sektor. Forskningsstatistik 1999* (*R&D in the public sector. Research statistics 1999*), Århus: Analyseinstitut for Forskning.

Analyseinstitut for Forskning (2001c) *Erhvervslivets forskning og udviklingsarbejde. Forskningsstatistik 1999* (*R&D in private companies. Research statistics 1999*), Århus: Analyseinstitut for Forskning.

Bang, H. (1997) *Byggesektoren og teknologisk service* (*The building sector and technological service*), Copenhagen: Boligministeriet.

Boligministeriet (1993) *Byggesektorens forskningsaktiviteter* (*R&D activities in the building sector*), Copenhagen: Boligministeriet.

By- og Boligministeriet & Erhvervsministeriet (2000) *Byggeriets fremtid – fra tradition til innovation* (*The future of building – from tradition to innovation*), Copenhagen: By- og Boligministeriet & Erhvervsministeriet.

Christoffersen, A.K. & Bertelsen, S. (1990) *Byggesektorens F&U. Forskning og udvikling i byggesektoren. Situationen ved 80'ernes slutning* (*The building sector's R&D. Research and development in the building sector. Status at the end of the 1980s*), Copenhagen: Foreningen af Rådgivende Ingeniører.

DAMVAD A/S (2011) *Evaluering og strategigrundlag for realdanias forskningsindsats. Vejen mod realdania forskning 2.0* (*Evaluation and strategies for Realdania's research initiatives. The road towards Realdania research 2.0*), Copenhagen: DAMVAD A/S.

Danmarks Statistik (2001) *Ressourceområdestatistik 1999. Statistiske efterretninger. Generel erhvervsstatistik 2001* (*Resource area statistics 1999. Statistical report. General business statistics 2001*), Copenhagen: Danmarks Statistik.

Danmarks Statistik (2013) *FOUBUD1*, Copenhagen: Danmarks Statistik. Available at: www.statistikbanken.dk (accessed 1 May 2013).

Det Offentlige Forskningsudvalg for Byer og Byggeri (2000) *Forskning i byer og byggeri* (*Research on cities and building*), Copenhagen: By- og Boligministeriet & Forskningsministeriet.

Dræbye, T. (1997) *Teknologisk byggeviden. Videnbrug, videnformidling og videnproduktion – en kortlægning og vurdering* (*Technological building knowledge. Use, dissemination and production of knowledge – a survey and assessment*), Copenhagen: Byggeriets Udviklingsråd.

Erhvervsfremme Styrelsen (1993) *Bygge/Bolig – en erhvervsøkonomisk analyse* (*Building/Housing – a business analysis*), Copenhagen: Erhvervsfremme Styrelsen.

Fairclough, Sir J. (2002) *Rethinking construction innovation and research: A review of government R&D policies and practices*, London: Department of Trade and Industry.

Haugbølle, K. & Clausen, L. (2002) *Kortlægning af bygge/boligforskningen i Danmark* (*Survey of building/housing research in Denmark*), Hørsholm: Statens Byggeforskningsinstitut.

Koordinations- og initiativgruppen for viden i byggeriet (2009) *Nytænkning i byggeriet. Forskning skaber værdi* (*Innovation in building: Research generates value*), Copenhagen: Erhvervs- og Byggestyrelsen.

Udvalget vedrørende byggeforskning i Danmark (2002) *Byggeriet i vidensamfundet – Analyse og anbefalinger fra udvalget vedrørende byggeforskning i Danmark* (*Building in the knowledge society – Analysis and recommendations from the task force for building research in Denmark*), Copenhagen: Erhvervs- og Boligstyrelsen.

7 Finland – R&D functions in real estate and the construction industry

Suvi Nenonen, Miimu Airaksinen and Terttu Vainio

Introduction

The built environment is constructed and shaped to respond to the social, cultural and economic needs of people; so we can reside, work, worship, move, consume and enjoy. The number of new building developments depends to a large extent on the social and economic circumstances, following the needs of the economy and the occupants. Technologies to produce and operate the built environment are mainly cross-sectorial implementations of the achievements of all sectors of the economy; this technological diffusion depends on socioeconomic conditions. The development of the future building stock will vary depending on social and economic scenarios. However, one characteristic of the future built environment already in development is the increased use of embedded sensor and monitoring technologies, which will allow smart technologies to help everyday living.

Research, development and innovation policy in Finland

Companies seek competitive success through new products or by renewing their current line, by way of innovations mostly through research and development (R&D). The innovations are usually based on a group of people having different competence areas. Innovations help companies to succeed in business, build up their competitiveness and enhance productivity. This facilitates high wages and new job creation. Innovations boost productivity in the national economy and make a high standard of living and well-being possible. For these reasons, society uses a variety of measures, many of which enhance competitiveness, to motivate companies to engage in innovation activity. Finland has a particularly innovation-driven economy.

Innovations do not always succeed as expected in the markets, thus there is always an economic risk associated to these activities. For this reason, companies do not invest in innovation activities as much as society would wish. Therefore, the public sector tries to motivate the private sector to engage in innovation activities through various measures such as public research funding and other incentives for private R&D activities.

In addition to financing, many factors within the business environment influence the willingness of enterprises to innovate. In Finland, these include legislation, access to international markets and the functioning of the internal market of the European Union (EU). Such incentive schemes, and the institutions planning and implementing them, constitute the national innovation system. Finland's innovation system has been ranked among the best in the world (European Commission, 2011).

In Finland, the Ministry of Employment and the Economy is responsible for most decisions on innovation policy, which forms the base on which the innovation system develops. The development of the innovation system itself is coordinated by the Research and Innovation Council, led by the Prime Minister.

In the public sector, a number of measures are in place to encourage engagement in innovation activity. Since the public sector does not produce commercial goods, the goals of its innovation activities differ from those of the private sector. For example, among other tasks, the public sector seeks to provide citizens with more useful public services. However, public sector innovation activities also aim to improve productivity.

The European and world economies are undergoing profound changes. In particular, Asia's emerging economies are challenging the innovation systems of Finland and other EU countries. As business becomes more global, companies are influenced by the attractiveness of national innovation systems when choosing the location of their activity hubs. The Finnish innovation system must therefore be continuously developed on a national basis, so that Finland can offer excellent conditions in this respect.

This chapter aims to describe the national innovation system, its focus areas and activities in the construction industry and real estate sector, and to provide an overview of successful projects. The methodology is literature review and analysis of evaluations of innovation policies.

The construction and real estate industry in Finland

Birth of the construction and real estate industry

Due to changes in the national monetary rules, the Finish construction market experienced an expansive development during the decades of 1980 and 1990 as it had never been seen before. Residential construction was heavily developed due to the sudden deregulation of the financial market, which, in turn, released the backlog in housing demand created by former regulations (Koskela, *et al.*, 1992). High demand for industrial goods created construction needs, and low confidence in stock investments channelled investment assets into non-residential construction as well. The brisk residential and non-residential construction market jointly generated over-demand, which was reflected in rising construction costs, prices and, subsequently, inflation.

Finland's GDP collapsed during the period from 1991 to 1993. The turn in the economy signalled a rapid change in demand for construction. For a brief period, privately financed housing construction was at a virtual standstill. As a consequence, two of Finland's four leading construction firms folded (Vainio, 2008).

Industrial companies participating in the construction process joined under the leadership of the Finnish Funding Agency for Technology and Innovation (Tekes) to agree on a common strategy for the future (Matilainen, *et al.*, 1994). The stated future vision was for an internationally operating construction industry. The main market area was defined as the Baltic Sea region and the EU as well as other special global sectors. Even though the breakup of the Soviet Union brought a temporary collapse of export activities, Finland's strategic position as a neighbour and established trading partner of Russia was not to be underestimated. Internationalisation was seen as two way: imports were to be expanded as well as exports, increasing competition in what was viewed as a largely closed construction market. Instead of supply, the mode of operation was to be changed from production oriented to customer and service oriented.

This strategy required a change in operating practices and the adoption of new forms of cooperation. To achieve these objectives, Tekes launched several research programmes within the construction industry in the mid-1990s. The programmes sought new cooperation models and developed new technologies for the industry. The cooperation-building process took place phase by phase. During the first stage, in the first half of 1990, cooperation goals were set for the various parties in the construction process, namely, the construction, construction products and construction design sectors. During the second stage, led during the second half of 1990, customers were also included in the cooperation and the Finnish construction and real estate industry concept was thereby established (Vainio, 2010).

The construction and real estate industry today

The stimulus and input provided by these research-based initiatives has carried through to the present day, with participant companies having remained in collaboration for nearly two decades (see Figure 7.1).

The total value of the Finnish national wealth was EUR675 billion (USD900 billion) in 2011 (Statistics Finland, 2012b). As shown in Figure 7.2, 75 per cent is tied to buildings, roads and networks. The infrastructure is maintained, repaired and renovated annually by a sum equivalent to about 6 per cent of its value. In 2011 a total of EUR44 billion (USD57 billion) was invested in the built environment; 60 per cent in maintenance of existing buildings, constructions, traffic arteries, and networks. Investment due to new construction and replacement, which is sensitive to economic fluctuations, represented only 40 per cent (Statistics Finland, 2012a).

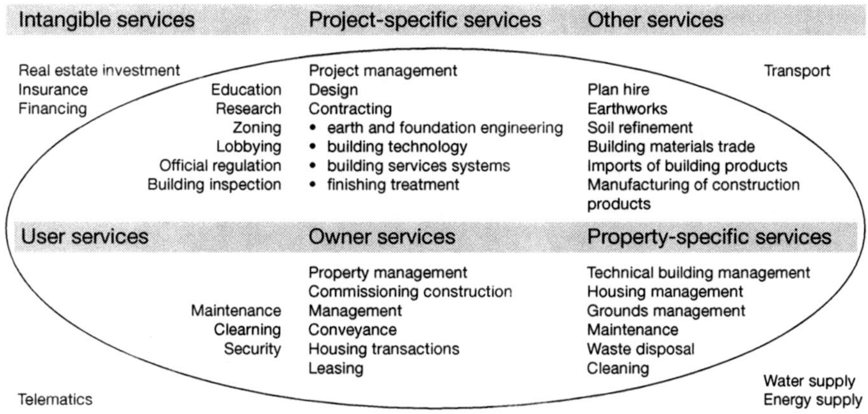

Figure 7.1 Construction and real estate industry in Finland

Source: Vainio, *et al.* (2006)

The ratio of the stable to sensitive production of the real estate and construction industry changes with the national economy. The entire sector employs over 500,000 people, which is more than one-fifth of the Finnish workforce and accounts for a quarter of the national gross domestic production (GDP). Its significance varies according to particular sectors of the national economy. The majority of total housing costs are in one way or another related to all sectors. For example, construction, repair

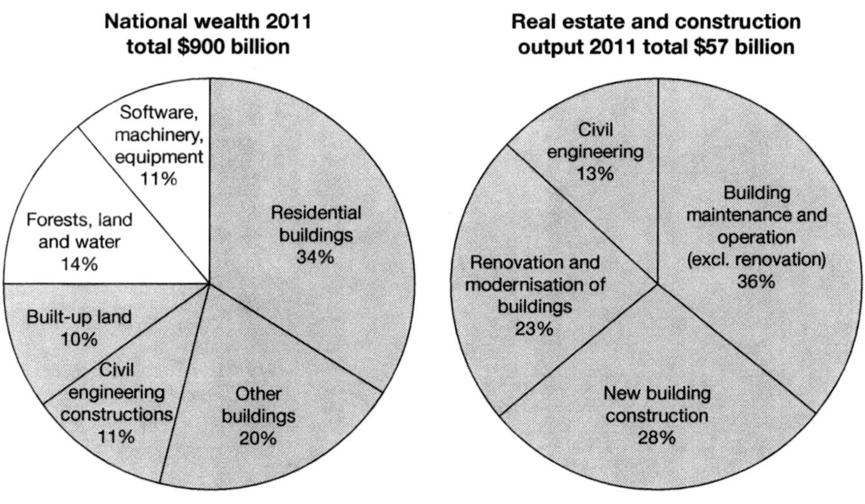

Figure 7.2 National wealth in Finland (left); real estate and construction output in Finland (right)

Source: Finland (2012a). Vainio (2012)

and maintenance of spaces are a significant item of expenditure in the education sector.

The volume of real estate and construction R&D is about USD450 million, 70 per cent financed by the private sector and 30 per cent by the public sector. The most active R&D actor has been the building product industry, mainly the building services industry. Recently, the planning sector has also invested in information technology (IT) application development such as building information modelling (BIM) (see Figure 7.3).

In general, the construction and real estate industry pay far too little attention to R&D and reserve only limited funds for development projects. This is partly due to the business structure and partly due to company conservatism. Additionally, many companies are too small and do not have access to funds for research and development. More importantly, the allocation of research projects in a company might be less effective to produce tangible outcomes than if the same effort were directed towards addressing the issue of how companies can work together and share data for good results and practices.

Collaboration requires partners to meet and share viewpoints. This applies not only to companies in the same area, but also to all organisations that contribute to satisfying customer needs and are part of the value chain. Organisations such as the Technical Research Centre of Finland (VTT), universities of applied sciences, the Confederation of Finnish Construction Industries, and the Finnish Funding Agency for Technology and Innovation should have a facilitator role in linking research to construction processes and should function as arbitrators in disputes over ownership of consortium products and inventions.

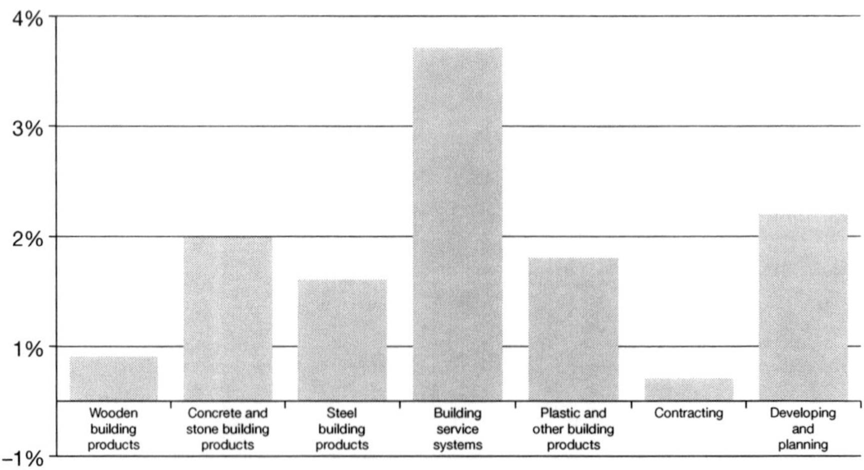

Figure 7.3 Investment in R&D in different industries inside the construction sector, Finland

Source: Vainio (2003)

Common challenges for the construction and real estate industry

The greatest challenge for the industrial society is to adapt its production and economic activities to what nature can sustain. By joining the Kyoto Protocol, Finland has committed to limiting hazardous emissions into the atmosphere. The construction and real estate industry must also take this goal seriously, as it affects over 40 per cent of the consumption of primary energy in Finland and it produces one-third of the national CO_2 emissions (European Commission, 2008).

The construction and real estate industry can use tools such as energy efficient heating, production and transportation, use of renewable energy sources, and cogeneration systems. In 2012 new energy regulations issued by the Ministry of the Environment for buildings took effect. The new regulations aim to reduce 20 per cent of the energy consumption of new buildings and created a demand for renovation activities during 2013 (Finnish Ministry of Environment, 2013).

The environmental impact of a building is the sum of the production process of construction materials, the land use related to the building as well as the energy needed for heating, air conditioning, supplying water and running the equipment. The public sector has invested in the reduction of detrimental environmental impacts of construction through programmes such as the *Finnish Government Programme for Ecologically Sustainable Construction*, approved in 1998 which was among the first of its kind in Europe (Working Group for Sustainable Construction, 2001).

A good living environment is healthy, safe, pleasant, stimulating, aesthetic and nature oriented. Communities must also be competitive in order to serve as bases supporting business and industry. A living environment that meets the needs of its residents should be linked to an eco-efficient community structure, with short distances between amenities so people can carry out their daily activities with ease, and which is also pleasant to live in. To this end, the often dispersed community structure of Finnish urban areas can be integrated by small-scale complementary building, creating additional benefits through a more independent elderly population.

Furthermore, the eco-awareness of consumers has increased significantly, creating a demanding client group also in the construction and real estate industry. Eco-efficiency is becoming one criterion for quality construction alongside safety, healthiness, pleasantness, and durability.

People spend 90 per cent of their time indoors, making indoor air quality a significant contributor to their wellbeing and productivity. Indoor air quality is mainly determined by decisions made during the construction process and is influenced by measures taken during operations. For example, indoor air quality is deteriorated by mould growth due to moisture damage, radon and other soil gases, defective ventilation, incorrect temperature for human comfort as well as emissions from building materials. For instance, according to estimates (Reijula, *et al.*, 2012; Vainio, *et al.*, 2002), the number of moisture- damaged dwellings and educational institutions discovered every year is such that repairs would cost as much as the annual renovation budget for the complete building stock. However, only a fraction of the moisture damages revealed by needs assessments are repaired.

The ageing gross area of approximately 550 million square metres of building stock, 78,000 km of roads, 28,000 km of streets, 100,000 km of fresh water pipes, 45,000 km sewers, and other infrastructure require care and certain technical repairs become timely as a building ages. In fact, the critical age of most structures ranges between 30 years (buildings) and 50 years (infrastructure sections). A large number of buildings and infrastructure parts have already reached or are about to reach this point.

The impact of the cyclical nature of new construction investments can be alleviated by timing needed repairs with periods of slow new construction. Currently, the implemented renovations do not meet the repair needs, instead, they are delayed and only carried out when the damage is evident, increasing the cost of renovation activities. Pre-emptive repair work and repairs carried out at an earlier phase of the life of infrastructure would improve living conditions and work productivity.

The generalised development of information and communication technology (ICT) has made it one of the key factors in the construction process. Information networking of projects also networks enterprises into closer cooperation, excluding those companies that do not adopt the new technology as part of their business concept. Today, organisations are able to change information on a product model basis.

In Finland, a key development in this area was the *Vera Technology Programme*, which has allowed the continued translation of research outputs into practice through an open access model. Future areas of research are the utilisation of design and production information in both infrastructure and real estate, and maintenance management. The focus of the programme is currently shifting from information exchange between design and construction to information utilisation over the entire lifecycle.

Finnish public innovation investors

Tekes, Finnish funding agency for technology and innovation

The model

Tekes supports high-quality research that generates significant commercial potential for business development while also promoting better competitiveness and welfare for society at large. Tekes funding is allocated to individual projects, many of them with global impact through leading research.

In practice, this criterion for funding allocation translates into the need for collaboration between public research institutions and private industry.

Funded research projects use the construction industry as a source for data as well as wisdom and steering for the research process, so the research outputs are relevant to the academic community and have practical implementation.

Based on the priorities outlined in its strategy, Tekes uses technology programmes to allocate funding, networking efforts and expert services to areas that are important to both business and society. Approximately half of the total funding allocated is granted to companies, universities and research institutes through technology programmes that consist of research projects and services that support business operations, such as shared visions, seminars, training programmes and international visits (Hyytinen, *et al.*, 2012).

Based on an international evaluation conducted in 2012 by van der Veen, *et al.*, Tekes is among the world's leading innovation agencies, contributing to increasing research intensity, cooperation between companies, and creating new infrastructure knowledge in Finland. These activities have helped build a strong knowledge base and competences that have increased the international competitiveness of Finnish enterprises, particularly in building energy consumption reduction technology as a form of user-driven innovations (van der Veen, *et al.*, 2012).

Past Tekes' construction and real estate industry programmes

The main goal of previous real estate and construction business technology programmes funded by Tekes was to identify Finnish real estate and construction clusters and to initiate a proper technology programme for each subcluster. The objective was to launch and reinforce research, development and innovation (RDI) activities in these fields, to increase the size of the technology programmes and to promote company-driven RDI activities. The cluster strategy was carried out through five technology programmes during the period of 1997–2007 (Table 7.1) (Rajakallio, *et al.*, 2009).

As a result, the number of new foreign investors in property markets almost quadrupled between 2002 and 2007, moving from seven new foreign

Table 7.1 The first Tekes R&D programmes of real estate and construction industry

Programme	*Period*	*Description*
ProBuild	1997–2001	Developing construction process
Rembrand	1999–2003	User-oriented real estate business
Infra	2001–2005	Civil engineering – construction and services
Cube	2002–2006	Technology programme for building services
Sara	2003–2007	Value-network oriented construction

Source: Rajakallio, *et al.* (2009)

investors to 26, although this number decreased again in 2008. This shows to what degree the markets develop with the support of R&D activities (Steinbock, 2009).

According to the evaluation carried out by van der Veen, *et al.* (2012), the RDI *knowhow* and vision of the industry has increased through Tekes' programmes, by creating prerequisites for market-led product development in the industry. The programmes have generated new, more holistic mindset of business operations, in which value networks are taken into consideration rather than the value chain. Tekes' activities have led to a significant shift in R&D in the Finnish construction industry from initiative oriented to target oriented.

Tekes' cluster programme has also had positive effects on other development areas of the industry business environment, such as the creation of the industry glossary and concepts, and development of procurement practices and network management frameworks. As Rajakallio, *et al.* (2009) showed, international cooperation in R&D projects has also increased due to Tekes' programmes. However, the development of industry trade practices and research culture had as a precondition the creation of a code of conduct for this line of business (Rajakallio, *et al.*, 2009).

Although the strategy of Tekes included the creation of an *end-user-oriented* culture, the development of the housing sector and the creation of a competitive advantage through lifecycle expertise, the cluster programme has had a less significant effect on these areas.

Recent programmes

Tekes published a new strategy in 2011, which includes intelligent and smart environments among the areas of focus. This refers to the development of housing, work and leisure environments into functional, comfortable and energy-efficient entities. Priorities on this area are: smart energy systems and sustainable material economy, safety and security of the living environment that make good use of digital systems as well as user-oriented products, services and processes (Tekes, 2011).

The latest programmes are in line with Tekes' strategy and are summarised as follows:

- The *Sustainable Community programme* (2007–2012): aims to generate renewable business activities in designing, constructing and maintaining sustainable and energy-efficient areas and buildings. One core theme of the programme is a noticeable improvement in the energy efficiency of buildings and communities as well as the promotion of the use of renewable energy sources.
- The *Spaces and Places programme* (2008–2012): the target group includes service sector players, information and communication

technology companies and the construction and real estate industry. The programme encourages the participants to cooperate across sector boundaries. The development of new business models for developing, producing and maintaining the relevant environments is focused on the user of the space.

- The *Built Environment programme* (2009–2014): increases the usability and serviceability of the built environment by developing real estate and construction practices. The programme focuses on renovation and refurbishment.

Roadmap for built environment research and development

The Roadmap Group project was collaboratively conducted by VTT, Aalto University, the Finnish Institute of Occupational Health and the Finland Futures Research Centre of the University of Turku. This group evaluated areas of emphasis and future prospects of renovation, infrastructure construction and wellbeing in Finland.

The Roadmap is intended to assist the management of Tekes' *Built Environment Programme*. The main future challenges of the real estate and construction industry are summarised in the following list:

- understanding properties and buildings as service platforms;
- need for new business models and earnings logics;
- public sector innovation challenge;
- urban living and emphasis on public and shared spaces;
- multi-purpose use of facilities;
- emphasis on local services, local production and telecommuting;
- adaptability, flexibility and individuality;
- emphasis on lifestyles, leisure times and wellbeing;
- real-time infrastructural monitoring;
- use of wood in new constructions;
- energy efficiency, renovation and development of new services;
- export as an important goal for the construction industry;
- change management in construction.

The Roadmap and recommendations for the Built Environment Programme in 2010–2014 are:

- Turning the real estate and construction industry into a driver for Finnish energy and climate policy;
- Developing cross-departmental cooperation in public administration and legislative reform;
- Innovating new services for housing and combining parallel sectors;
- Increasing participation and creating open innovation platforms;
- Conducting larger projects.

The Roadmap highlights the need to understand properties and buildings as service platforms. This statement is a relevant guideline not only for traditional technically oriented businesses but also for new possible actors. The role of change management is also emphasised. The vision towards 2020 is to exploit the user perspective as the main driver for businesses development (Nenonen, 2011).

Strategic Centre for Science, Technology and Innovation in Construction (RYM Ltd)

RYM Ltd was founded in 2009 as the Strategic Centre for Science, Technology and Innovation (SCSTI) of Built Environment in Finland. RYM Ltd is a completely new business concept in the Finnish innovation system. It can be defined as a venture for intellectual capital. The knowledge venture company RYM Ltd collects financing agreements during the first phase. The second phase marks the start of research programmes as *funds*. Finally, the third phase presents the programmes as work packages and tasks as *spinoffs*.

RYM Ltd is engaged in the production of forecasting information, acquisition and development of research funding, research programmes in selected spearhead areas, and *living labs* as test platforms for research results. The aim is to have a networked international operation, close cooperation with other sectors' SCSTI and open, multi-channel communication. The creation of a SCSTI in the real estate and construction industry (RYM Ltd) is a unique undertaking. The research strategy of RYM Ltd identifies four areas of thematic focus that will initially dominate the research agenda: (i) energy efficiency; (ii) processes and frameworks of practice; (iii) competitive social infrastructure; and (iv) user-oriented spaces. Table 7.2 presents the ongoing programmes starting after 2010.

In December 2012 RYM Ltd had 53 shareholders. RYM Ltd directs funds and expertise from both private companies and public financiers of innovation in areas of research that are deemed as most significant for international competitiveness. Joint strategic *leading-edge* research will generate superior world-class expertise over the entire lifespan of the built environment. This offers real estate and construction industry companies, and research institutes and universities, a novel way of engaging in close, well-planned and ambitious cooperation to develop and bolster *cutting-edge* expertise. The evaluation of the SCSTI was conducted in 2012. Investigations (Lähteenmäki-Smith, *et al.* 2013) indicated that RYM Ltd programmes enable true company–research cooperation that goes beyond meeting discussion and seminar talks. It also forces companies to make long-term commitments and to introduce long-term actions. Requirements for significant resources and efforts by companies are seen as positive and to enforce companies' strategic role.

Table 7.2 Programmes of Strategic Centre for Science, Technology and Innovation in the real estate and construction industry, Finland

Programme	*Time*	*Goal*	*Participants and Volume*
Built environment process reengineering	2010–2013	To create totally new user-centred procedures and business models for the real estate, construction and infra structure sectors by product model-based data management over the entire lifecycle	43 participants: 37 companies and 6 research institutes. Budget of EUR21 million (USD27 million)
Indoor environment programme	2011–2014	To find solutions that promote productivity, pleasantness and health of space users in an ecologically sustainable manner	36 participants: 26 companies and 10 research institutes. Budget of EUR20 million (USD25.7 million)
Energising urban ecosystems programme	2012–2015	To create an internationally recognised and multidisciplinary hub of excellence for urban development in Finland	11 participants: 9 companies and 2 research institutes. Budget of EUR20 million (USD25.7 million)

Source: Rajakallio, *et al.* (2009)

Other actors

The Academy of Finland provides funding for scientific research of the highest quality. The academy now provides funding for housing research. Due to the social significance and impact of living and housing, this sector must be understood to a greater depth than the building construction and settlement of large communities. Changes in family and age structures, in lifestyles and the workplace as well as energy issues and the challenges of sustainable development have created new areas of focus within the living and housing sector. This multi-disciplinary programme approaches housing from the residents' point of view as an entity that comprises environmental issues ranging from sustainable development to land use, logistics and services, and consumer issues ranging from cultural needs to health issues.

This research concerns a range of social and physical spaces as well as virtual environments. Requirements of accessibility, sustainability and versatility have gained increasing prominence in relation to living and housing. Further research is also needed in the fields of land use, other environmental aspects of housing as well as housing renovation and repairs. However, despite their prominence and relevance, housing issues have received only modest basic research attention in Finland when compared to many other European countries.

Between 1995 and 1998, the Academy of Finland carried out the *EKORA Research Programme for Ecological Construction*, and, between 1998 and 2001, the *Urban Studies Research Programme* in a joint effort with other funding bodies. Funding was also directed towards the *Spaces of Nature and Culture Research Programme* from 2001 to 2004.

Basic research on housing has been largely neglected and the research field is widely dispersed. Therefore, there is an urgent need for such research programmes, which can have an important consolidating and coordinating role. Ideally, the research programme will inspire new kinds of research project, research team and research consortium. The research programme will also promote the international networking of housing research (Academy of Finland, 2011).

Another important actor, which has been active in research activities concerning sustainable development, is Sitra, the Finnish Innovation Fund. This organisation has the duty to promote stable and balanced development in Finland, and the growth of its economy and its international competitiveness and cooperation. For example, they carried out the *Energy Programme* with several practical projects that proved that the use of energy by communities and people, as well as their greenhouse emissions, could be significantly reduced. This programme can help Finland return to the top most energy-efficient countries through good examples and influencing regulations. Energy efficiency of the built environment plays a crucial role in the reduction of emissions. In addition to energy efficiency, people's lifestyle is an important factor. Therefore, the programme aims to create the preconditions and structures that enable people to make *low-carbon* choices (Sitra, 2012).

Conclusions

R&D in the Finnish construction industry has undergone a transformation from a single product and a single research discipline to a wider scope and cross-discipline research area. The general trend in the construction industry is to act as integrators of products/services to benefit the industry and its end user. The user-oriented approach is described in Figure 7.4.

The R&D efforts (1990–2000) are summarised in Figure 7.5 through a collection of keywords that illustrate the shift in research from building techniques to more holistic and systemic research agendas.

The social dimension at a large scale was lagging behind until the year 2000, albeit hidden under many research topics. Nevertheless, the social dimension is currently seen as an essential theme in the construction industry due to the importance of users and their engagement in the process and end-product/service design.

Climate change and lack of resources have been well-known research topics since the 1970's oil crisis, including energy efficiency and energy saving.

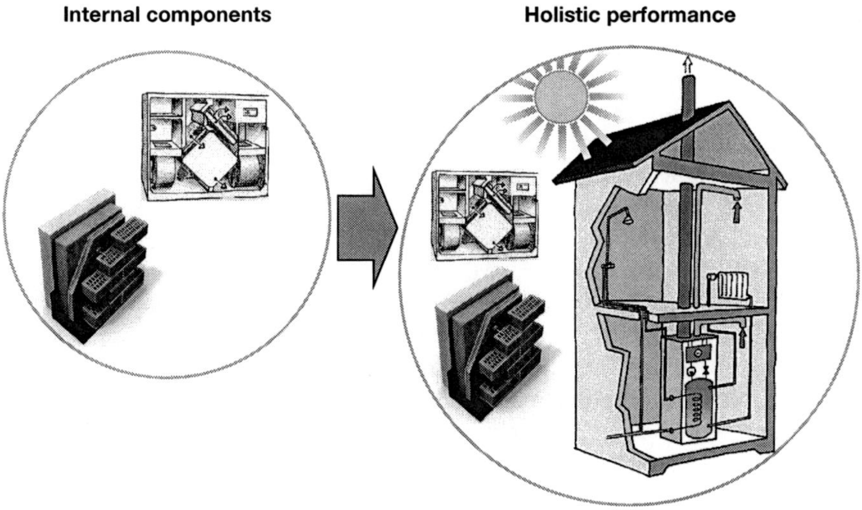

Figure 7.4 Principle of fundamental change from individual components to the holistic performance of buildings, focusing also on people's well being through research in the construction industry

The new topics of low carbon, low-energy and zero-energy buildings highlight a more holistic approach in which buildings are seen as a part of energy systems and sustainable development. The second R&D wave in wood construction is also booming today.

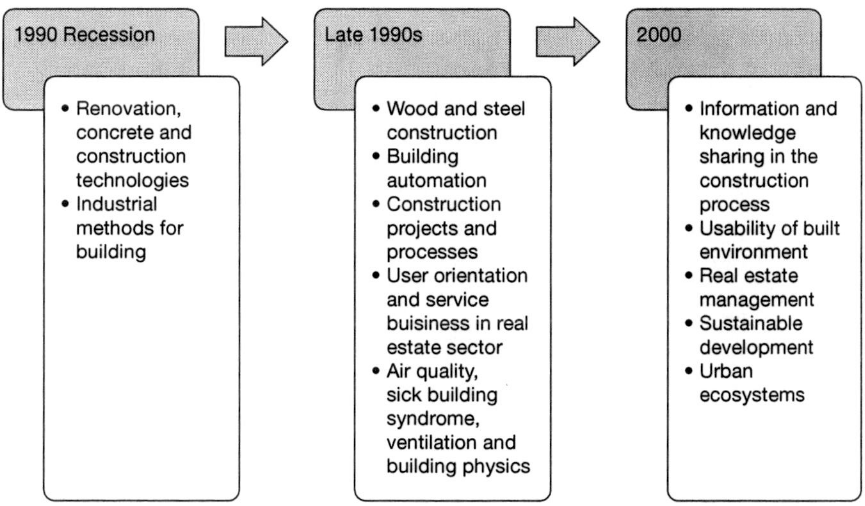

Figure 7.5 Research agenda changes, 1990–2000, Finland

Kajander, *et al.* (2012) conducted a study into *challenges for sustainability innovations* in the real estate and construction industry. Key challenges identified include: the complexity of the industry value network; team building; R&D intensity; and commercialisation management. In addition, their findings suggest that sustainability business innovation in this industry is constrained by project business orientation in the real estate and construction industry, fund raising and internationalisation issues, and lack of regulation and standards.

Interdisciplinary approaches in research have been increasingly acknowledged in the construction and real estate industries. However, methods of interdisciplinary collaboration have not been systematically analysed yet. An interdisciplinary field crosses traditional boundaries between academic disciplines. The investigations in interdisciplinary research in general indicate that the disciplinary boundaries are most thoroughly transcended when members of disparate fields develop a common language that facilitates a shared conceptual framework. This level of collaboration has the greatest potential for originality. However, it is seldom observed due to the challenges associated to the development of a common language.

A challenge in interdisciplinary research is to manage both the collaboration within academia and with the industry. Furthermore, the interpretation of the research outcomes might hinder its impact due to the articulation and interpretation of findings from a mono-disciplinary perspective by the industry.

Furthermore, the research process varies depending on the approach taken. It is always demanding to collaborate with representatives from a variety of disciplines due to the obvious risk of being considered weak or fragmented due to lack of academic rigor. One major challenge is therefore to design the research in such a way that a common language can be found. The common goal has to be shared but the ways to achieve it can be designed differently depending on the approach to the research: multi-, inter- or trans-disciplinary (Nenonen and Lindahl, 2012).

Infrastructure R&D in Finland is well established. The challenge is to translate this innovation into the field of practice, which traditionally is not research oriented. New initiatives that support the leading role of companies through innovation policy can create new ways to apply research into practice and to generate collaboration between academia and industry. The construction and real estate industry has traditionally been a disjointed industry. However, joining resources through different initiatives has promoted new collaboration.

References

Academy of Finland (2011) *Research knows no boundaries – Annual report 2011*. Academy of Finland. Available at: www.aka.fi/Tiedostot/Tiedostot/Akatemia_vk_2011_taitto_EN%20%282%29.pdf (accessed 5 January 2013).

European Commission (2008) *EU energy and transport in figures*, Statistical pocket book 2007/2008, Brussels: Directorate-General for Energy and Transport, European Communities.

European Commission (2011) *Innovation union scoreboard*, Brussels: European Union.

Finnish Ministry of Environment (2013) *Permits related to land use and building*, Finnish Ministry of Environment. Available at: www.ymparisto.fi/default.asp?node=19665&lan=en (accessed 5 January 2013).

Hyytinen, K., *et al.* (2012) *Funder, activator, networker, investor . . . Exploring roles of Tekes in fuelling Finnish innovation*, Tekes Review 2012/289, Helsinki: Tekes.

Kajander, J.K., Sivunen, M., Vimpari, J., Pulkka, L. & Junnila, S. (2012) 'Market value of sustainability business innovations in the construction sector', *Building Research & Information*, 40(6): 665–678.

Koskela, E., Loikkanen, H.A. & Virén, M. (1992) 'House prices, household saving and financial market liberalization in Finland', *European Economic Review*, 36(2–3): 549–558.

Lähteenmäki-Smith, K., *et al.* (2013) *Licence to SHOK? External evaluation of the strategic centres for science, technology and innovation*, Helsinki: Ministry of Employment and the Economy, Enterprise and Innovation Department.

Matilainen, J., Pajakkala, P. & Lehtinen, E. (1994) *Yhteistyössä innovaatioita uusille markkinoille, rakennusklusterin kilpailukyky* (*Innovations for new markets through cooperation – competitiveness of the construction cluster*), Helsinki: Research Institute of the Finnish Economy (ETLA).

Nenonen, S. (ed.) (2011) *Rakennetun ympäristön roadmap* (*Roadmap of built environment*), Tekes. Available at: www.tekes.fi/fi/document/.../rakennetun_ympariston_roadmap.pdf (accessed 7 November 2012).

Nenonen, S. & Lindahl, G. (2012) 'Interdisciplinary approaches, obstacles and possibilities – Experiences from facilities management research', in *Proceedings of EFMC*, paper presented at European Facility Management Conference, Copenhagen, 23–25 May.

Rajakallio, K., *et al.* (2009) 'Klusteria rakentamassa, kiinteistö- ja rakennusklu steriohjelmien arviointi' ('Cluster building, real estate and construction cluster evaluation of the programmes'), in *Ohjelmaraportti 7/2009 (Programme Report 7/2009)*, Espoo: Tekes.

Reijula, K., *et al.* (2012) *Home ja kosteusongelmat rakennuksissa* (*Damp and mould damages in buildings*), Helsinki: Parliament of Finland.

Sitra (2012) *Board report and financial statements 2011*, Helsinki: Sitra. Available at: www.sitra.fi/julkaisut/Toimintakertomus/2011/Sitra_Boardreport2011.pdf (accessed 5 January 2013).

Statistics Finland (2012a) 'Construction in national accounts 2011', in *Construction and Housing Yearbook 2012*, Helsinki: Statistics Finland.

Statistics Finland (2012b) *Net capital 2011*, Annual national accounts, Helsinki: Statistics Finland. Available at: www.stat.fi/til/kan_en.html (accessed 19 February 2013).

Steinbock, D. (2009) *The vital cluster: Globalization, urbanization, and Finland's real estate and construction cluster*, Helsinki: RAKLI (Finnish Association of Building Owners) and Rakennustieto oy/Building Information Limited.

Tekes (2011) *Strategy*. Available at: www.tekes.fi/en/tekes/strategy/ (accessed 30 September 2013).

Vainio, T. (2003) *Rakennusalan tutkimus ja kehitys* (*Construction sector's research and development activities*), VTT. Available at: www.vtt.fi/inf/julkaisut/muut/2003/rakennusalan_t%26k_2003.ppt (accessed 19 February 2013).

Vainio, T. (2008) *Kohti yksilöllisempää – Asuntotuotannon laatumuutokset 1990–2005* (*Towards individuality – Quality changes in housing production 1990–2005*), Espoo: VTT.

Vainio, T. (2010) 'Constructed environment or constructing environment – Is construction involved, or is it a part of change?', paper presented at the CIB World Building Congress, Salford, 10–13 May.

Vainio, T. (2012) *Rakentamisen yhteiskunnalliset vaikutukset* (*Societal impacts of the construction branch*), Helsinki: Rakennusteollisuus RT ry. Available at: www.rakennusteollisuus.fi/RT/Tilastot/Rakentamisen+yhteiskunnalliset+vaikutukset/ (accessed 1 October 2013).

Vainio, T., Jaakkonen, L., Nippala, E. & Lehtinen, E. (2002) *Korjausrakentaminen 2000–2010* (*Renovation 2000–2010*), VTT Research Notes 2154, Espoo: VTT.

Vainio, T., Nippala, E., Kauranen, H. & Pajakkala, P. (2006) *Kiinteistö- ja rakennusalojen tuottavuus. Esitutkimus* (*Productivity development at construction branch*), Espoo: VTT.

Van der Veen, G., *et al.* (2012) *Evaluation of Tekes, final report*, Helsinki: Ministry of Employment and the Economy Publications.

Working Group for Sustainable Construction (2001) *Competitiveness of the construction industry, agenda for sustainable construction in Europe*, Brussels: European Technical Contractors Committee for the Construction Industry.

8 France – role of national RDI policy in the construction industry

Frédéric Bougrain

Introduction

In 2008 France spent EUR41.7 billion (USD32.4 billion) on research and development (R&D). When expressed in terms of gross domestic product (GDP), it represented 2.14 per cent. It accounted for about 3.8 per cent of world R&D expenditure (OST, 2010). For many years, France has been characterised as a country in which state intervention was needed to guarantee the wealth and the strength of the economy. Most of the funds dedicated to R&D focused on large programmes limited to specific sectors such as nuclear, space, aeronautics, telecommunications and defence. This public support was mainly monopolised by large companies. Small and medium-sized enterprises (SMEs) as they are defined by the European Commission[1] were frequently ignored. However, this scheme has been challenged by the growing role of the European Union and French regions in national programmes. Under the new institutional framework public expenditures are not only devoted to R&D but also to innovation. Moreover, SMEs are at the core of the innovation policy.

The French construction industry, comprising building construction, installation and finishing, and civil engineering is characterised by a large number of micro enterprises.[2] Conversely, the number of firms employing more than 250 people is limited. They represent less than 0.1 per cent of the firms of the industry whereas their contribution to production reaches about 20 per cent (CGDD, 2009). For those firms, the level of R&D expenditure appears to be lower than in any other industry. As a consequence, the industry has not been able to benefit from most large R&D programmes. However, since policies supporting the innovation capacities of SMEs do not target any specific sector, construction companies can benefit from this support as long as they are innovative. Moreover, public procurement can be a source of innovation and R&D in construction. For example, large and complex buildings involving a network of actors frequently lead participants to carry out R&D activities to circumvent technical bottlenecks encountered during the course of the project.

This chapter will first characterise the evolution of the French national policy in the domain of research and innovation. Then, it will examine the role of R&D for construction companies and the importance of public funding. Finally, the analysis will examine the role of R&D in the innovation strategy of a small sample of construction companies located in the Aquitaine region.

Characteristics of innovation and research policy in France

Ergas (1987) opposes *mission-oriented policies to diffusion-oriented policies* which aim at diffusing *technological capabilities throughout the industrial structure, thus facilitating the ongoing and mainly incremental adaptation to change*. Countries such as Germany, Switzerland and Sweden tend to adopt *diffusion-oriented policies*.

According to Ergas' terminology, France's technology policy was considered for a long time as *mission oriented*. As in the United States and the United Kingdom, technology policy was linked to objectives of national sovereignty. The focus is on radical innovation, *big science is deployed to solve big problems* and most public R&D is dedicated to the defence sector and key sectors for international strategic leadership such as nuclear energy and telecommunications. Mustar and Larédo (2002) confirmed this view and proposed four traits to describe the French research system:

- The largest fraction of the public research budget is directed towards a limited number of sectors that contribute to the objectives of national independence, such as nuclear, space, aeronautics, defence, and telecommunications, to mention but some.
- Basic research is mainly done by the Centre National de la Recherche Scientifique (CNRS, National Centre of Scientific Research). Universities mainly focus on teaching and are competing with French higher education institutions which train the elite.
- Several public research institutes carry out research for the needs of several Governmental departments.
- A limited number of large companies monopolise most public support for research. However, SMEs do not belong to this network and do not benefit from public funding.

These policies faced several limitations at the end of the 1980s. It appeared that they yielded few direct benefits and crowded out a large share of commercial R&D. Only a small number of *high-tech* firms were the recipients of public subsidies. The spinoffs were limited. Traditional sectors such as construction did not benefit from the high level of public expenditure on R&D.

During the 1990s several changes affected French national policy: large programmes received less support; partnerships between the CNRS and universities were reinforced; and the research potential of universities grew. Around 2000, there were about 14,000 researchers and research engineers at CNRS versus 45,000 teacher researchers at the universities (Mustar & Larédo, 2002). Nevertheless, it was mainly the development of support for innovation and tax credits that modified the French innovation and research policy:

- The National Agency for the Valorisation of Research (ANVAR – later called OSEO, and Bpifrance since July 2013)[3] was created to support innovative SMEs. It was organised on a regional basis. Agencies located throughout France aid small firms not only through financial assistance, but also through technological advice. The main tool was a zero-rate loan that could cover up to 50 per cent of the costs of the innovation project. The firms refunded the agency only in the case of success and without paying any interest. Large firms were usually excluded from this procedure.
- Research tax credits were introduced and contributed to the development of R&D activities in SMEs. In 1987 3,500 applied to the scheme. In 2000 applicants exceeded 6,000. Several successive changes in the way R&D expenditures are defined increased the number of beneficiaries by 147 per cent between 2000 and 2007 (Table 8.1). In 2008 the scheme was simplified and in 2010 about 18,000 firms applied for research tax credits (Froger, 2012).

Moreover, two new key players appeared in the French research landscape: the regions and Europe. With the *Decentralisation Act* (1982), the regions brought complementary resources that supplemented the investments of the State. Europe also provides supports for research and innovation through its successive *Research Framework Programmes*. The European budget has grown rapidly from EUR678 million (USD527 million, 2.4 per cent of its budget) in 1985 under the *First Framework Programme* (1984–1987) to EUR6.5 billion (USD5 billion, 5 per cent of its budget) in 2008 under the *Seventh Framework Programme* (2007–2013).

Construction, R&D and public financing

R&D and the construction industry

The economic literature shows a positive relationship between firm size and the likelihood of performing R&D. Several advantages are put forward to explain this relationship:

> The greater willingness of larger firms to incur risks associated with R&D, the greater access to finance by larger firms, the greater ability of larger firms to internalise R&D spillovers due to greater

Table 8.1 Research tax credits, 2000–2007, France

	Research tax credits							
	2000	*2001*	*2002*	*2003*	*2004*	*2005*	*2007*	*Evolution 2007/2002 (%)*
Declarations	6,344	6,253	5,907	5,833	6,369	7,400	9,658	+64
R&D expenditures declared	10,248	10,712	11,668	11,300	11,600	13,500	15,300	+31
Beneficiaries	3,060	2,810	2,760	2,757	4,250	5,430	6,822	+147
Volume (million EUR)	529	519	479	428	890	982	1687	+252

Source: OST (2010)

> diversification, and the greater likelihood that larger firms will possess the complementary capabilities (e.g. marketing) necessary to exploit innovations.
>
> (Cohen & Klepper, 1996)

The construction industry is characterised by a high number of very small firms. In France in 2007, among the 369,100 firms of the industry, 339,900 (92.1 per cent) employed fewer than 10 employees and contributed to 33.4 per cent of the production (CGDD, 2009). Conversely, the number of firms employing more than 250 people is limited. They were about 300 firms in 2007, representing less than 0.1 per cent of the firms in the industry. Their contribution to production reached 20.2 per cent. This particularity and the nature of the innovation process within this industry explain why R&D expenditures are much lower than in other manufacturing sectors (Table 8.2).

Most innovations developed by contractors are made at the job site in the course of a construction project and they do not require a high amount of financial resources. Innovations are mainly project specific and they barely require R&D. Innovations are *ad hoc responses to problems encountered in the course of a construction project that an innovating builder was engaged in. They were emphatically not 'R&D projects' in any formal sense* (Slaughter, 1993). When they do some R&D, contractors usually collaborate with materials and/or equipment suppliers on specific projects. The resources dedicated to this activity are much smaller than those of their suppliers.

This situation was also typified by Pavitt (1984), who was the first to bring forward the existence of *sectoral patterns* of technical change. He categorised general contractors as *supplier-dominated firms*. Firms from this category devote few resources to R&D. They focus their innovative activities on processes. *Most innovations come from suppliers of equipment and materials, although in some cases large customers and Government-financed research and extension services also make a contribution.*

These elements were confirmed by the fourth Community Innovation Survey (SESSI, 2006a)[4], which, for the first time, integrated firms from the French construction industry. One of the aims of this large study was to characterise innovation within the industry: 39.4 per cent of the French construction firms that were surveyed declared that they were innovative between 2002 and 2004. These figures are lower than for manufacturing and service industries (46.2 per cent). Table 8.3 indicates that internal R&D is the main input for firms employing more than 250 people while smaller firms follow a different path. They rely either on training or on the acquisition of equipment. Thus focusing too much on formal R&D could be misleading since very small firms that are dominant in the industry innovate through different paths. Patel

Table 8.2 Internal R&D expenditures of the private sector, 2002–2007, France

	Internal R&D expenditures of the private sector					
	2002 (billion €)	*2007 (billion €)*	*Evolution 2007/2002 (%)*	*2002 (billion €)*	*2007 (billion €)*	*Evolution 2007/2002 (%)*
Manufacturing sector	**18.0**	**20.2**	**+ 12**	**82.3**	**81.5**	**– 1**
Aeronautics and space construction	2.3	2.7	+15	10.7	10.8	+ 1
Electric equipment	4.6	4.3	– 6	21.0	17.4	– 18
Pharmaceutical industry	2.8	3.5	+ 27	12.8	14.3	+ 12
Equipment goods	1.7	2.2	+ 27	8.0	9.0	+ 12
Transport	3.2	3.7	+ 15	14.9	15.1	+ 1
Chemistry	2.2	2.4	+ 10	10.0	9.7	– 3
Natural resources intensive industry	0.6	0.7	+ 11	2.7	2.7	– 2
Labour intensive industry	0.5	0.6	+ 29	2.3	2.6	+ 14
Non-manufacturing sector	**3.9**	**4.5**	**+ 18**	**17.7**	**18.3**	**+ 4**
Primary sector and energy	0.9	0.9	+ 9	3.9	3.8	– 4
Agro-food industry	0.5	0.5	+ 5	2.3	2.1	– 7
Construction[5]	0.1	0.1	+ 9	0.4	0.4	– 4
Services in transport and telecommunication	2.4	3.0	+ 24	11.1	12.1	+ 9
Total	**21.8**	**24.8**	+ 13	100	100	–

Source: OST (2010)

Table 8.3 Input for innovation between 2002 and 2004 within innovative contractors as a percentage, France

	Internal R&D	*External R&D*	*Acquisition of equipment*	*Acquisition of external knowledge*	*Training*
10–49	45.4	16.9	53.8	11.7	48.4
50–249	48.3	16.8	57.3	23.8	68.9
250+	69.1	31.0	47.3	22.6	53.6
Total	46.6	17.4	54.0	13.6	51.2

Source: SESSI (2006b)

and Pavitt (1994) who focused on *low-tech* industries drew a similar conclusion:

> We have shown that when we treat technical change as synonymous with R&D activities in science-based industries, we are in danger of neglecting up to nearly 40% of what is going on in technical change, especially in non-electrical machinery and in small firms.
>
> (Patel & Pavitt, 1994)

Bougrain and Haudeville (2002) also indicated that R&D investment is not a satisfactory indicator to analyse SMEs' ability to innovate. Conversely, a design office appeared to be a strong source of innovation by facilitating the use of extensive information networks.

However, this lack of R&D in construction is problematic since R&D raises the *absorptive capacity* of the firms (Cohen & Levinthal, 1990). The *absorptive capacity* is defined as the ability of a firm to recognise the value of new external information, assimilate it, and apply it to commercial ends and is critical to its innovative capabilities. It is a function of firm's prior knowledge. Thus, R&D is also a way to take advantage of external knowledge and to use existing networks. Consequently, a low level of R&D can prevent construction firms from benefiting from innovations developed by suppliers or from exploiting knowledge developed by research institutions.

Public funding for R&D in construction

R&D expenditures of the construction industry are very limited compared to other sectors. In volume, it is the sector with the smallest budget dedicated to R&D. About 90 per cent of these expenditures are financed by internal resources (Table 8.4). This is confirmed by Tessier (2008), who indicated that only 16 per cent of firms from the construction industry have access to public funding, versus 36 per cent for manufacturing companies. Only large civil engineering companies operating at the international level were able to receive some public support.

Table 8.4 R&D expenditures of the private sector by source of financing, France

	Internal R&D expenditures of the private sector						*Volume (billion EUR)*
	Type of funding						
	Public funding		*Foreign country*	*Private*		*Total*	
	Defence programmes	*Other*		*Internal R&D*	*External R&D*		
Manufacturing sector	7.6	1.4	8.7	71.5	10.8	100.0	26.0
Non-manufacturing sector	1.0	4.7	2.4	77.8	14.0	100.0	5.3
Construction	//	//	0.9	89.4	5.2	100.0	0.1
Total	**6.5**	**2.0**	**7.6**	**72.5**	**11.4**	**100.0**	**31.3**

Source: OST (2010)

Public funding to the sector mainly comes from Bpifrance, the public finance company supporting innovation. However, all financial products and services developed and proposed by OSEO are not dedicated to the construction industry. Indeed, the goal of the agency is to support any innovative SME. This is not the case of the *Programme de Recherche et d'Expérimentation sur l'énergie dans le Bâtiment* (PREBAT, Research and Experimental Programme on Energy in Building) launched in 2007. PREBAT aims to reduce the energy consumption of buildings and it is specific to the construction industry. The following sections will present how *Bpifrance Innovation* supported SMEs from the construction industry and the impact of PREBAT on the construction of low-energy buildings.

The innovation aid scheme proposed by Bpifrance

Since its creation Bpifrance has developed a whole range of products and services in order to cover the needs of innovative SMEs. Bpifrance's activities are complementary with those developed by banks or venture funds. Bpifrance covers mainly three lines of activity:

- supporting innovation and technology transfer (*Bpifrance innovation*);
- financing investments and operation cycle, in partnership with banks and finance institutions (*Bpifrance financement*);
- guaranteeing bank financing and equity investments (*Bpifrance garantie*).

The financial support brought by *Bpifrance Innovation* covers part of the expenditures dedicated to research, development and innovation (RDI). Innovation activities are defined in the Oslo Manual as:

> All those scientific, technical, organizational, financial and commercial steps, including investment in new knowledge, which actually lead to, or are intended to lead to, the implementation of innovations. These activities may either be innovative in themselves, or required for the implementation of innovations.
>
> (OECD, 2005)

The acquisition of technology, industrial design, trial production, feasibility studies and market tests and launch are integrated in those activities.

In the landscape of the French innovation and technology policy, Bpifrance focuses mainly on small projects that are closed to the market (Figure 8.1). Its activities are complementary to those of the National Research Agency, which finance mainly basic research through *call for projects* and *competitiveness clusters*.[6]

Not all financial products and services developed and proposed by Bpifrance are dedicated to the construction industry. In 2009 construction

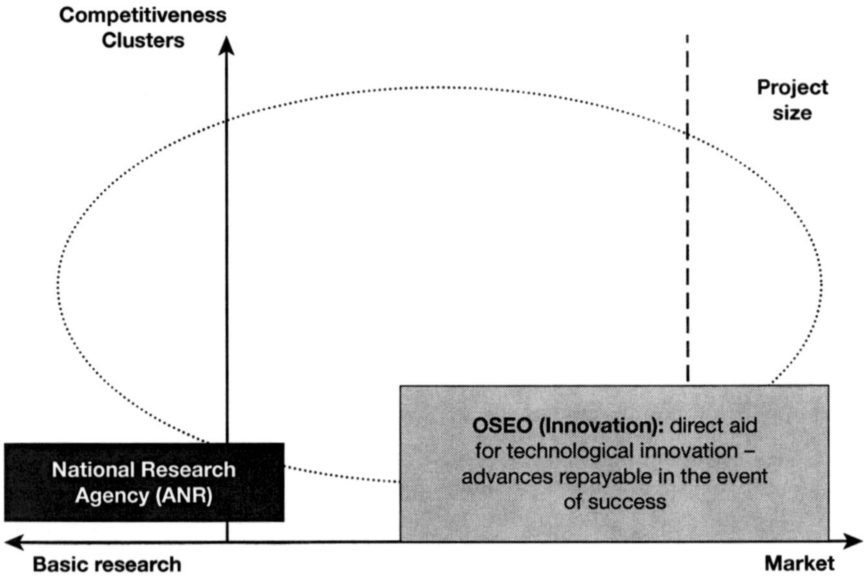

Figure 8.1 Position of Bpifrance in the French innovation and technology policies

projects represented 4.08 per cent of all projects financed by Bpifrance through the *innovation aid* scheme (OSEO, 2010a). Through the leverage effect of the innovation aid, where SMEs have to finance at least 50 per cent of their project, more public and private financing is mobilised every year.

The innovation aid contributed to the development of 131 innovative projects in the building sector (EUR10.9 million), 23 projects in civil engineering (EUR2.4 million) and 22 projects proposed by material suppliers (EUR1.35 million). However, the line between construction and other industrial sectors is frequently very thin. For example, in the energy sector among the 156 projects financed in 2009 (EUR16.6 million), 38 (EUR3.9 million) were dedicated to the building market to optimisation of heating systems and software to monitor energy consumption (OSEO, 2010b).

Bpifrance has always focused its attention on the smallest and youngest enterprises that experience greater difficulty in gaining ready access to financing:

- firms with fewer than 20 employees, also the majority in the French construction industry, received EUR6.19 million for 88 projects, accounting for half of the projects financed in 2009 (Figure 8.2);
- 33 projects (20 per cent) were developed by SMEs with less than 3 years from start-up (Figure 8.3). They received EUR2.68 million.

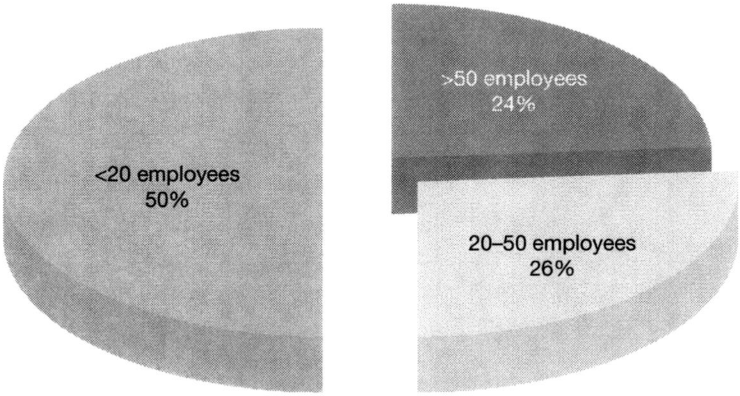

Figure 8.2 Innovation aid by size of firm, France

Source: OSEO (2010a)

Several of these projects are linked to sustainable construction, which is a source of business opportunity. New products and process are created to reduce greenhouse gas emissions and energy consumption. Examples include: photovoltaic cells, which are integrated into the envelope of the buildings; new construction systems based on products offering a smaller environmental impact; industrialisation of the construction process; and the development of building information modelling (BIM). Nevertheless, the innovation aid scheme is not frequently adapted to project-based activities and innovation developed during the course of a construction project.

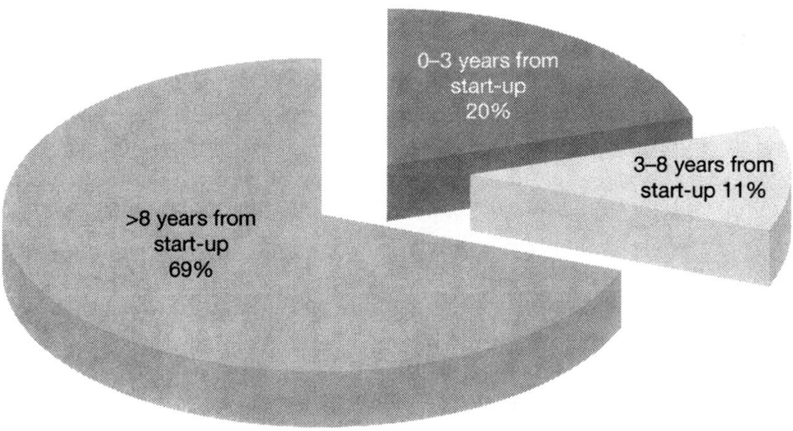

Figure 8.3 Innovation aid by time from start-up, France

Source: OSEO (2010a)

The PREBAT programme

In 2007 PREBAT was launched by the Government at the national level to spur sustainable construction. It was one of the measures taken to address climate change, to restore biodiversity and to limit environmental and health risks. The aim of this national programme that covered the period 2007–2012 was threefold (PREBAT, 2007):

- to anticipate the change of thermal regulation that would become very ambitious both for new and existing buildings;
- to find solutions to technical and economical bottlenecks;
- to develop experimental projects.

Seven priorities were defined to reach these targets:

1 To sustain and develop applied R&D: since technological breakthrough appeared necessary to reduce energy consumption and greenhouse gas emission, the development of R&D was considered as a key priority for the development of new equipment.
2 To launch several experimental low-energy building projects. The aim was to test new technologies and new systems in order to anticipate the forthcoming change of thermal regulation. Public and private clients that accepted to bear the technical risks of experimental projects would receive some support.
3 To develop actions regarding existing buildings, which will still be the major source of energy consumption in the future.
4 To consider the building as a system in order to integrate architectural, environmental and behavioural constraints.
5 To integrate the economic and financial dimensions within the research field in order to propose solutions that offer reasonable return on investment.
6 To diffuse the results of the experientations in order to make the actors working in this field aware of the innovative solutions that have been successfully implemented.
7 To set up a common reference framework at the national level, in order to favour European comparisons and the creation of networks at the European level.

The evaluations of this programme (Roux, *et al.*, 2009) revealed some shortfalls:

- Every funding institution kept its own approach;
- Large suppliers of material and equipment did not take part in this programme;
- Most research projects did not consider the building as a system. They tended to focus on specific items without examining all interconnections;

- Most research overlooked the renovation of existing buildings and focused on new construction;
- The diffusion of the results was limited.

However, the impact of PREBAT was also very positive. It contributed to the development of a strong research community gathering universities, technical centres and enterprises. It also improved the coordination among the actors belonging to this community. Finally, the main result was the development of a large number of low-energy buildings, with 1,100 new and existing buildings benefitting from this programme.

This action dedicated to low-energy buildings was financed by Agence de l'Environnement et de la Maîtrise de l'Energie (ADEME, the French Environment and Energy Management Agency) and French regional authorities. Public subsidies rose from EUR7.2 million in 2007 to EUR42.8 million in 2010. Eleven regions took part in the programme in 2007 and 20 in 2010 and 1,025 projects were supported during this period: 65 per cent concerned new buildings and 35 per cent renovation projects (Euréval, 2012).[7] By focusing on experimental projects, PREBAT reproduced the approach initiated by the *Plan Construction and Architecture* (PCA) (Campagnac, 1998). The PCA was created in 1971 to improve the quality of architecture in housing and the efficiency of the building site. However, experimental projects were not directly subsidised. The incentive for contractors who proposed innovative projects to the PCA was to avoid facing a call for tenders. This model was also criticised since the main beneficiaries of the programme were the affiliates of the major construction companies.

In the case of PREBAT, the aim was to lead the construction value chain toward low-energy buildings before the implementation of a new regulation in 2013, requiring that all new residential and office buildings have a maximum annual consumption of 50 kWh/m^2. Public and private clients who accepted either to construct low-energy buildings or to renovate buildings to drastically reduce energy consumption received financial supports. This was a stimulus to diffuse new technologies, new systems and new construction processes associated with low-energy buildings.

The subsidies delivered by ADEME and the French regions covered on average about 10 per cent of the construction costs. It represented approximately the additional investment costs linked to better energy performance.

The impact of the programme was considered as positive: 80 per cent of the 241 clients surveyed (Euréval, 2012) had never built a low-energy building before. About 75 per cent, mainly households, considered that without financial subsidy the financial equilibrium of the project would have never been reached.

However, the learning process appeared to have limitations since most clients were in charge of just one project. According to the experts who evaluated the programme, clients would have needed three consecutive projects to learn about the process of constructing/renovating low-energy buildings. In the first project, the client discovers the process and identifies the technical problems; during the second project starts to solve them; and finally, in the third project, the client monitors most technical issues. Nevertheless, public authorities, which were the main beneficiaries, representing 36 per cent of the clients, considered that they had a significant learning experience, and developed new practices. The programme was a way to raise their competencies and to make them aware of the specificities of the future regulation and of some specific technical items such as blower-door tests and implementation of new material and equipment.

Similarly the learning impact was quite important within the design team. Moreover, the programme raised the competencies of the contractors and installation and finishing companies. For example, to guarantee the energy performance of buildings, electricity and plumbing companies have to make *box out* at the design stage and they are not allowed to drill on site.

R&D and innovation: a regional case study

The context of the survey

In 2011, the Centre Scientifique et Technique du Bâtiment (CSTB, Scientific and Technical Centre for Building) and a consulting company were commissioned by the Construction Ressources Environnement Aménagement et Habitat durables (CREAHd, Construction Resources Environment and Sustainable Habitat Development); a regional institution promoting innovation in the construction industry and gathering firms, public and private research organisations and universities. The aim was to examine the innovation strategy of firms from the construction industry located in the Aquitaine region.

Four groups of firms were targeted:

- building companies
- civil engineering companies
- suppliers of material
- designers and architects.

An electronic survey based on the Oslo Manual was first sent to 3,280 firms representing the four aforementioned groups. Of the 139 firms who answered: 80 per cent employed fewer than 50 employees; 60 per cent were building companies; 50 per cent were not active in innovation during the period 2008–2011 and 6 per cent abandoned it before the implementation of an innovation. The remaining firms were innovative.

In-depth interviews were carried out with 17 innovative companies representing each group. All the firms had answered the electronic survey. Firms were questioned about: the organisation of the innovation process; funding; results of the innovation; and the role and importance of R&D within the context of sustainable construction. Telephone interviews with general managers and executives of R&D, finance and other departments, were carried out between June and September 2011. CSTB was in charge of 10 interviews. The following section will present the main findings from these 10 interviews and will focus on the role of R&D in the innovation strategy of the firms. *A* and *B* represent respectively an architect company and a design office specialised in acoustic; *C* and *D* are the suppliers; *E*, *F* and *G* are general contractors; *H*, *I* and *J* are civil engineering companies. These firms can be considered among the best within the construction industry.

The role of R&D in the innovation strategy of construction companies

- ***A and B*** were innovative but they did not conduct significant R&D. According to both firms, innovation is driven by the evolution of the regulatory environment. For example, thermal regulation is becoming more stringent, leading to the introduction of new equipment and products by suppliers. Innovations are usually developed with the partners of the building site. The suppliers are often the leaders but the innovation process is frequently coordinated by either the architect or the design office. The client that finances the project has also a key role to play. Indeed, projects that are less financially constrained are more open to opportunities for innovation.
- ***Two materials suppliers*** (*C* and *D*) belonged to large international companies with R&D facilities. *C*, employing 120 people, supplies concrete. It has three regional laboratories gathering 11 R&D engineers. They are in charge of identifying the constraints of the building site, to develop solutions and to follow the construction project. When projects are very complex, the firm receives the support of the R&D unit of the group. *D*, employing 70 people, is a quarrying company. It has no permanent budget for R&D. It develops research when its margins are better. Thus, the research activity is not continuous. However, it still benefits from the internal database managed by the parent company in order to promote exchange of best practices among affiliates.
- ***Three building companies*** (all SMEs) represented a wide range of situations:

    - *E* is a contractor with fewer than 10 employees. It has no R&D department but the manager has developed an R&D project with both public and private research laboratories in order to create a new product. It has also received the financial support of several local agencies promoting innovation. However, due to the limited capacity of this company the innovation process is very long and the SME may lack a commercial network to sell its innovation

- *F* employs 53 people. Three of them work part time in R&D activity. The rest of the time they are involved in marketing activities that are complementary to their research work. It is one way to get feedback from the markets. This research activity allowed this contractor to diversify its activity and to supply other contractors with products. In 2010, the turnover was EUR7.6 million, and about EUR3 million was connected to the commercialisation of new products. *E* and *F* can be considered as *user-innovators* who became *user-manufacturers*.[8] They regularly cooperate with a local engineering school for most of their research projects. Due to a bad experience with OSEO, now Bpifrance, they preferred to innovate by using their internal resources
- *G* is a subsidiary to one of the major contractors in France. It has no internal R&D competencies. R&D is centralised in a lab located near Paris. It can assist subsidiaries located everywhere in France. One of the main sources of innovation is the internal database gathering information on the internal innovation awards' scheme. These awards are open to all employees and concern all types of innovation. The aim is to prevent employees from *re-inventing the wheel* and to promote the exchange of best practices between companies located in remote places.

- Of the ***three civil engineering companies***, two are affiliates of large companies and each has its own R&D laboratory:

  - *H*, employing 120 people, belonged to large international companies with R&D facilities. It aims at keeping the link between the research activity and the needs of the market. One of the key issues of its lab is to follow the evolution of the regulation and to anticipate changes in order to sell new products before its competitors. *H* frequently cooperates with local engineering schools and suppliers. In the past, it also received financial support from OSEO for some products developed for building sites. Head of units are frequently the main barrier to R&D. Indeed, when the construction market has to face a crisis and when prices go down, they try to reduce R&D budgets. However, since the company has developed a culture of innovation, most budgets are not diminished in the long term.
  - *I* has no permanent R&D lab but several engineers are mobilised for complex projects such as the high-speed train. Only internal resources are used to finance research projects. Most innovations developed on site do not require R&D. Innovation also comes from suppliers of products and equipment. *I*, as *H*, considers that public R&D funding is not adapted to the research activity connected to the construction site. Indeed the lifecycle of a construction project is very short while most public support is dedicated to new products with longer lifecycles.

- *J* is a SME with fewer than 20 employees. It has no R&D activity. Nevertheless, it has recently created a design office that can be a strong source of innovation in SMEs. The innovations developed by the manager concern the organisation of the company. The aim is to enhance both the service quality and the reputation of the firm in public tenders.

Concluding remarks and future directions

The construction industry is frequently criticised for its limited capacity to invest in R&D. It is true that a basic statistical comparison between construction and all manufacturing industries shows a disadvantage for the construction industry. Nevertheless, R&D in construction is not organised as formally as in manufacturing companies. Only large companies can support a dedicated R&D department. R&D is more development and occurs at the building site.

Moreover, several activities that are at the core of the construction business are not classified as R&D by the Frascati Manual. For example, *investigations of proposed engineering projects, using existing techniques to provide additional information before deciding on implementation, is not R&D* (OECD, 2002). Similarly:

> The vast bulk of design work in an industrial area is geared towards production processes and as such is not classified as R&D, There are, however, some elements of design work which should be considered as R&D. These include plans and drawings aimed at defining procedures, technical specifications and operational features necessary to the conception, development and fabrication of new products and processes.
>
> (OECD, 2002)

Thus, the project-based nature of the industry does not fit well with the official classifications.

R&D does not have to be considered as an end. It has to be considered as a way to achieve a better performance or to develop a new production process. As was illustrated by Kline and Rosenberg (1986) with the chain-linked model, there are several paths leading to innovation. The innovation model is not linear. Several innovations are developed due to feedbacks from the market. Moreover, the flexibility of the construction process allows for adaptation and development of existing products.

Further research needs to be carried out to examine the complementary role of R&D in construction, the *absorptive capacity*, and analyse the impact of training on innovation. Indeed, training is considered among the smaller construction companies as the main input to innovation.

This knowledge could improve the efficiency of the RDI policy and would contribute to a better allocation of public funding. Up until now, French public policy has not taken into account the diversity of the construction industry. Manufacturers developing new equipment/product and large contractors were the main beneficiaries. However, most small contractors, which represent the majority of firms in the construction industry, and firms such as facility management enterprises that propose new services to the users of buildings, were neglected.

Notes

1 Small and medium-sized enterprises are defined as enterprises that employ fewer than 250 persons and whose annual turnover does not exceed EUR50 million (European Commission, 2005).
2 Micro enterprises are defined as enterprises that employ fewer than 10 persons and whose annual turnover or annual balance sheet total does not exceed EUR 2 million (European Commission, 2005).
3 Although the mission of Bpifrance is broader than that of its predecessors, the core activity dedicated to the support of innovative SMEs has not been modified for over 30 years. For example, the zero-rate loan scheme, which was first applied in 1979, still exists.
4 The fourth Community Innovation Survey (CIS4) was launched in France in 2005 and 24,675 firms employing more than 10 employees received the postal survey. The answer rate was approximately 86 per cent. Collected information concerned corporate innovative behaviour during the period 2002–2004.
5 OST defines *construction* as CGDD. The definition includes firms engaged in building construction; installation and finishing; and civil engineering. Manufacturing, architectural and engineering consulting firms and facility managers (downstream activities) are excluded.
6 Clusters are geographic concentrations of interconnected companies, specialised suppliers, service providers, research centres and educational institutions. They are present in a particular field in a region. By bringing together different actors, the aim is to develop synergies and cooperative efforts. Altogether, 71 clusters have been labelled in France. Three of them concerned construction.
7 The evaluation conducted by Euréval (2012) did not examine whether the buildings had reached the expected energy performance.
8 According to Baldwin, *et al.* (2006), user-innovators seek to develop new designs for their own personal use or (in the case of user firms) internal corporate benefit. They do not anticipate selling goods or services based on their innovations, although they may later go into business as user-manufacturers. Designing for use and testing by use are the essential characteristics of user-innovators.

References

Baldwin, C., Hiernerth, C. & von Hippel, E. (2006) 'How user innovations become commercial products: A theoretical investigation and case study', *Research Policy*, 35: 1291–1313.

Bougrain, F. & Haudeville, B. (2002) 'Innovation, collaboration and SMEs internal research capacities', *Research Policy*, 31: 735–747.

Campagnac, E. (1998) 'National system of innovation in France: Plan construction et architecture', *Building Research & Information*, 26(5): 297–301.

Cohen, W.M. & Klepper, S. (1996) 'A reprise of size and R&D', *The Economic Journal*, 106: 925–951.

Cohen, W.M. & Levinthal, D.A. (1990) 'Absorptive capacity: A new perspective on learning and innovation', *Administrative Science Quarterly*, 35: 128–152.

Commissariat Général au Développement Durable (CGDD) (2009) *Entreprises de construction: Résultats de l'EAE 2007 (Construction companies: Results of the 2007 annual enterprise survey), Chiffres et Statistiques (Figures and Statistics)*, n°58. Available at: www.statistiques.developpement-durable.gouv.fr/fileadmin/documents/Produits_editoriaux/Publications/Chiffres_et_statistiques/2009/Chiffres%20et%20stats%2058%20résultats%20EAE%20 2007%20Construction.pdf (accessed 9 April 2013).

Ergas, H. (1987) 'Does technology policy matter?', in *Technology and global industry: Companies and nations in the world economy*, Guile, B.R. and Brooks, H. (eds), Washington, DC: National Academy of Engineering.

Euréval (2012) *Evaluation du programme mis en oeuvre par l'ADEME et les régions pour le soutien et la valorisation des bâtiments exemplaires du PREBAT (Evaluation of the programme implemented by the ADEME and areas for support and development of exemplary buildings of PREBAT*), Paris: ADEME.

European Commission (2005) *The new SME definition user guide and model declaration*, Enterprise and industry publications. Available at: http://ec.europa.eu/enterprise/policies/sme/files/sme_definition/sme_user_guide_en.pdf (accessed 3 April 2013).

Froger, V. (2012) *Les bénéficiaires du crédit impôt recherche sont de plus en plus nombreux (The beneficiaries of the research tax credit are becoming more numerous)*, Les Echos. Available at: http://entrepreneur.lesechos.fr/entreprise/financement/actualites/les-beneficiaires-du-credit-impot-recherche-sont-de-plus-en-plus-nombreux-10021094.php (accessed 20 December 2012).

Kline, S. & Rosenberg, N. (1986) 'An overview of innovation', in *The positive sum strategy: Harnessing technology for economic growth*, Landau, R. & Rosenberg, N. (eds), Washington, DC: National Academy Press.

Mustar, P. & Larédo, P. (2002) 'Innovation and research policy in France (1980–2000) or the disappearance of the Colbertist state', *Research Policy*, 31: 55–72.

OECD (2002) *Frascati manual proposed standard practice for surveys on research and experimental development*, Paris: OECD.

OECD (2005) *Guidelines for collecting and interpreting innovation data*, Oslo Manual, Paris: OECD, European Commission.

OSEO (2010a) *L'innovation dans les PME en 2009 – Bâtiment-travaux publics, synthèse sectorielle (Innovation in SMEs in 2009 – Public building works, sector summary)*, Paris: OSEO.

OSEO (2010b) *L'innovation dans les PME en 2009 – Energie, synthèse sectorielle (Innovation in SMEs in 2009 – Energy Sector Summary)*, Paris: OSEO.

OST (2010) *Indicateurs de sciences et de technologies (Indicators of science and technology)*, Paris: Observatoire des Sciences et des techniques. Available at: www.obs-ost.fr/fileadmin/medias/PDF/R10_Complet.pdf (accessed 2 January 2013).

Patel, P. & Pavitt, K., (1994) 'The continuing, widespread (and neglected) importance of improvements in mechanical technologies', *Research Policy*, 23: 533–545.

Pavitt, K. (1984) 'Sectoral patterns of technical change: Towards a taxonomy and a theory', *Research Policy*, 13: 343–373.

PREBAT (2007) *Programme de recherche et d'expérimentation sur l'énergie dans le bâtiment – programme de travail 2007–2012 (Research programme and experimentation on energy in the building-work programme 2007–2012*). Available at: http://rp.urbanisme.equipement.gouv.fr/puca/edito/Prebat2007_2012.pdf (accessed 2 April 2013).

Roux, D., Gillet, A. & Roudier, J. (2009) *Evaluation du PREBAT 2005–2009*. Available at: www.prebat.net/IMG/pdf/2009_evaluation_prebat1_bis-2.pdf (accessed 3 April 2013)

SESSI (2006a) *Enquête communautaire sur l'innovation en 2004 (CIS4)* (*Community Innovation Survey 2004*), Ministère de l'Economie, des Finances et de l'Industrie. Available at: www.industrie.gouv.fr/sessi/enquetes/innov/cis4/cis4.htm (accessed 4 April 2013).

SESSI (2006b) *Enquête statistique publique, réalisation sessi – CIS4 2004* (*Public statistics survey, construction*), Ministère de l'Economie, des Finances et de l'Industrie. Available at: www.insee.fr/sessi/enquetes/innov/cis4/EH.html (accessed 5 April 2013).

Slaughter, S. (1993) 'Innovation and learning during implementation: A comparison of user and manufacturer innovations', *Research Policy*, 22: 81–95.

Tessier, L. (2008) 'La structure et les métiers de la construction guident son innovation' ('The structure and construction trades guide innovation'), *SESP en bref*, Ministère de l'Ecologie, de l'Energie, du Développement Durable et de l'Aménagement du territoire, Volume 24.

9 Germany – researching sustainability

Alexandra Staub

The national context

The German Government has sponsored building research since 1919, making it a fundamental national characteristic. In 1969, the *Arbeitsgemeinschaft für Bauforschung* (*Working Group for Building Research*) brought together for the first time all major research institutes in the country to discuss and coordinate the funding and evaluation of research efforts. Results were disseminated over the nationally funded Fraunhofer-Informationszentrum Raum und Bau (Fraunhofer IRB, Fraunhofer Information Centre for Regional Planning and Building). While the initial objectives were to reduce building costs and improve productivity in housing construction (BBSR, 2013a; Triebel, 1983), in the past few decades, the emphasis has shifted to social and ecological building practices and a healthy living climate. Research to improve building practices has helped guide policies for new building programmes and housing legislation, and has supported industries in their quest for better performance at lower cost (BBSR, 2013a).

In 1999, restructuring allowed the Bundesministerium für Verkehr, Bau und Stadtentwicklung (BMVBS, Federal Ministry of Transport, Building and Urban Development) to channel oversight of sponsored research to the Bundesamt für Bauwesen und Raumordnung (BBR, Federal Office for Building and Regional Planning). The BBR programmes allow companies, university institutes, planners and others involved in the building industry to access funds to improve the production of housing. Simultaneously, the Federal Government initiated annual competitions for building research funding based on predetermined questions that were deemed politically and socially important (BBSR, 2013a). By doing so, the Federal Government was able to expand the kind of building research being carried out. The results were disseminated through the Federally funded Fraunhofer IRB through presentations and symposia and through brochures and online documents, all of which guaranteed access to the research outcomes by a broader section of the population.

Other ministries have sponsored research directly or indirectly related to building practices, including the Federal Ministry of Education and Research through the programme *Bauen und Wohnen im 21 Jahrhundert* (*Building and living in the twenty-first century*) in 2000; the Ministry for the Environment, Nature Conservation and Nuclear Safety; and the Ministry of Economics and Technology. However, the BMVBS remains the largest sponsor, allocating over EUR12 million (USD15.4 million) to construction research programmes in 2012 (Figure 9.1).

In addition to the Federal research programmes, individual States have sponsored their own research agendas. With the exception of the programme *Stadtumbau West* (Urban Reconstruction West) (BMVBS, 2012a) coordination between national and State levels has remained limited:

- Bavaria has initiated a programme titled *Climate Programme Bavaria 2020* to optimise the reduction of CO_2 emissions (Bayerisches Staatsministerium für Umwelt, Gesundheit und Verbraucherschutz, 2007).
- Baden-Württemberg allocates almost 5 per cent of its gross national product (GNP) to research, mainly on technology development programmes, including solar and other renewable energy. This State Government has the highest investment in research and development (R&D) in Germany, as a percentage of total expenditure. The aim is to support existing industrial research institutes, to guarantee a well-functioning technology transfer system in small and medium-sized enterprises (SMEs), and to guarantee economic revenues for the region (Baden-Württemberg, 2013).
- Lower Saxony is home to major research institutes, such as ForWind; the German Wind Energy Institute; DEWI-Offshore and Certification Centre; the Institute for Solar Energy Research, the Energy Research Centre; and a District Heating Research Institute. Many of these institutes partner with universities in the State while others have consulting functions for industries (Niedersächsisches Ministerium für Umwelt, Energie und Klimaschutz, 2013a). Lower Saxony is a founding member of the Institut für Bauforschung (IFB, Institute for Building Research), which has existed since 1946, although the current institute evolved from a different organisation founded in 1920. The IFB is part of an extensive network that includes the BMVBS and the State of North Rhine Westphalia, as well as the International Council for Research and Innovation in Building and Construction (CIB). The IFB supports research on economics, as well as social, urban, architectural and environmentally equitable practices (Niedersächsisches Ministerium für Umwelt, Energie und Klimaschutz, 2013b).
- Historical preservation experts in the individual States are organised into their own organisation, the Vereinigung der Landesdenkmalpfleger in der Bundesrepublik Deutschland (Association of the Conservation Authorities in the Federal Republic of Germany), and sponsor building

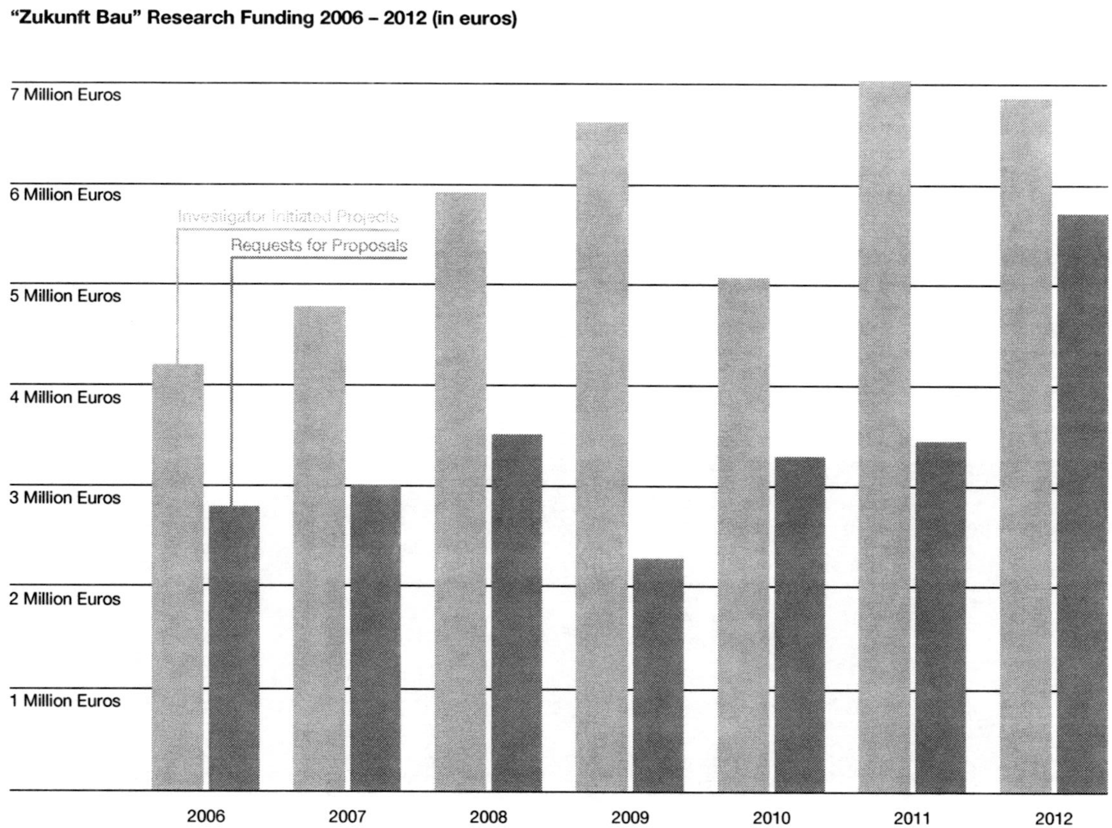

Figure 9.1 Federal expenditures for PI-determined grant proposals (*Antragsforschung*); Intermittent requests for proposals (*Auftragsforschung*), 2006–2012, Germany

Source: BBSR

research on historical buildings, often in partnership with university institutes (Bayerisches Landesamt für Denkmalpflege, 2013).

Building research does not stop at State or national borders and many programmes make use of European Union (EU) funding in addition to local sponsorship. In contrast to other European Governments, tax incentives for R&D are generally not used as a policy tool in Germany (Czarnitzki & Lopes Bento, 2011).

Overall, building-related R&D in Germany is characterised by a dense integration, although not necessarily an intense coordination, between Government ministries, Government- or university-sponsored institutes and private industries. Since most universities in Germany are Government funded, much of German R&D is thus spearheaded, or at least aided, by the public sector.

R&D investment 1990–2010

The main policy instrument for building-related R&D in Germany is direct Government subsidies, with major funding coming through the BMVBS. Projects are supervised by the subordinate Bundesinstitut für Bau-, Stadt- und Raumforschung (BBSR, Federal Institute for Research on Building, Urban Affairs and Spatial Development) within the BBR. An examination of projects sponsored by the ministry from 1990–2010 shows four main directions of research (BBSR, 2010):

- *Technical studies*, for example to improve building materials or elements.
- Studies on how to *lower building costs* through new materials or procedures.
- *Ecologically oriented projects*, such as sustainability standards, material recycling, energy saving measures, or green roof and façade systems.
- Studies that can potentially influence *socio-political policy directions*, such as: projects that focus on single-family housing with potential for suburban sprawl; studies on housing for the elderly as an answer to an ageing population; attempts to tie into international norms as a prerequisite for closer international collaboration; studies on various alternative housing types; and studies that examine the future of large scale housing estates, especially those built in Eastern Germany before 1989.

It is clear why the Federal Government has an interest in sponsoring such enquiries, as they deal with issues that are linked to longer term policy directions, both nationally and internationally. In this vein, the BBSR has partnered with the European Spatial Planning Observation Network

(ESPON) and INTERREG, which foster transnational cooperation in spatial development (BBSR, 2013b, 2013c).

Future building

Since 2006 the Federal Government has financed building research through its programme *Zukunft Bau* (*Future Building*), supervised by the BBR. The aim of this programme is to increase German competitiveness on the European construction market and to reduce technical, cultural and organisational deficits in the building process (BBSR, 2013d). There are two funding mechanisms: twice yearly grant application cycles (*Antragsforschung*); and intermittent requests for proposals (*Auftragsforschung*). The twice-yearly application cycles are aimed at institutions and industries and fund-applied building research in one of seven categories:

- energy efficiency and renewable energy in buildings as well as calculation tools
- new concepts and prototypes for energy saving building, zero- or plus-energy house concepts
- new materials and technologies
- sustainable buildings and building quality
- demographic changes
- regulations and the awarding of contracts
- modernisation of existing buildings.

Researcher-initiated projects

Institutes applying for funding may engage further institutes for specialised subcategories of a given project, and grants obtained through the *Antragsforschung* mechanism may be used to seek co-financing with partners in the EU. In general, projects are to be completed within 2 years. Grant funds may not be used to realise building projects or develop industry products, and applicants are expected to contribute additional funding of their own or have additional funding from third parties. A panel of experts appointed by the BMVBS evaluates the grant applications and ranks them, after which available funds are disbursed. Each funded project is accompanied by a small group of advisors appointed by the BMVBS (BMVBS, 2012b).

Research projects defined by the BMVBS (*Auftragsforschung*) fall into five clusters:

- value-added chains in the building industry
- sustainable building and building quality
- surrounding conditions

- current challenges and new markets
- energy-efficient and climate-appropriate building.

Current *Zukunft Bau* projects as well as those of older programmes are announced on the BBSR website (BBSR, 2013e; Forschungsinitiative Zukunft Bau, 2013).

Other R&D mechanisms

Additionally, since 2011 the BBR has maintained a programme to support the construction of *plus-energy houses* (BMVBS, 2012c). The programme provides financial support on a variety of levels, outlined in more detail below.

Although much of German R&D has remained at the scale of individual buildings, the shaping of urban environments as an ecological, economic and social concern has become a second, and long-term, priority. The German Government has recognised that national culture is affected not only by what is built, but also by the decision-making process leading up to construction. While the idea of using sponsored research and publically funded building programmes to shape cultural perceptions has often not been expressed as such by the initiators of these research agendas, many of the programmes have dealt with issues beyond just providing housing or urban facilities for the population. Programmes have considered the needs of an ageing population, young families with children, citizens desiring to become more involved in the decision-making process, and cities that have lost their economic base and thus need restructuring (BBSR, 2013f).

In addition, studies have examined German policies compared to those of EU partners. For the sponsors, the questions of *how* cities should be structured, *how* transportation should be arranged and *what forms* of housing should be built for the population, have become questions of socio-cultural management as much as a charge of practical implementation.

Despite this, there remains a striking absence in Federal or State R&D sponsorship for urban projects attempting to address social issues more concretely, for example communities that attempt to provide alternatives to single-family suburban housing and reliance on the automobile. *Car-free* communities, such as Freiburg-Vauban or parts of Munich-Riem, may be seen as prototypes, but have been funded largely through grassroots efforts or municipal Governments (Wohnen ohne Auto, 2013) (see Figure 9.2).

Vauban's planning, for example, began in late 1994 as a seven-person citizens' initiative. Calling themselves Forum Vauban, each member contributed DM14,000 (German marks, EUR7,158 or USD9,204) toward the goal of realising a new type of community. Two months later, Forum Vauban had over 60 members. The urban design competition for the new quarter was sponsored by the City of Freiburg, which decided in 1995 to involve local citizens, led by the Forum Vauban group, in the planning

Figure 9.2 Vauban, a car-free community

process. Forum Vauban began receiving DM40,000 (EUR20,452 or USD26,299) annually from the mayor's office and, in 1995, DM30,000 (EUR15,339 or USD19,724) for an advertising campaign, as well as DM160,000 (EUR81,807 or USD105,196) from the Deutsche Bundesstiftung Umwelt, a foundation maintained by the Federal Government to foster environmental projects. With this funding, Forum Vauban was able to hire several employees to oversee the project. Plans for the quarter were finalised in 1997.

Construction of individual residential buildings was started in 1998, financed by groups of individuals (*Baugruppen*) or cooperatives (*Baugenossenschaften*). Vauban, whose spatial organisation encompasses extensive pedestrian zones, bike paths and a sophisticated layering of public, semi-private and private green spaces in addition to its zero- or plus-energy buildings, was chosen as a *Best Practices* example by the UN Conference Habitat II in Istanbul in 1996 because of the exemplary cooperative planning process between the municipality and its citizens. This raised the project's international visibility immensely and, in 1997, Vauban received a total of EUR1.4 million (USD1.8 million) from the EU *LIFE-Programme* (European Commission, 2013). These funds assured personnel costs would be covered, and financed several ecological projects such as solar collectors and a car-sharing service (Forum Vauban, 2013a, 2013b).

Despite the clear success of such projects attempting to redefine urban structures, and despite some indirect funding efforts, the German

Government has not made such research one of its priorities. A start may have been made with a community of *plus-energy* homes commercially developed by the firm WeberHaus as part of its Övolution line. Built in close proximity to the Vauban project, the homes are the result of Federally sponsored R&D and present an overall community structure as part of their energy-conscious offering (Plusenergiehaus, 2013).

Case studies

Baukultur*: the culture of building*

In 2001, the German Government formally recognised the connection between individual building programmes and their larger social implications. The result was a systematic programme to research the question of *Baukultur in Deutschland,* a term that may loosely be translated as an examination of culture and building practices, both with regards to housing and urban design (Kähler, 2001). In 2004 the term was made part of the Federal building code, with *Baukultur* becoming a mandatory point of consideration when developing any sort of master plan. Part of the ensuing policy assessments encompassed results from the programme *Experimenteller Wohnungs- und Städtebau* (*Experimental Housing and Urban Design*), abbreviated ExWoSt, which as early as 1988 had begun to consider innovative solutions to housing and urban design questions through their potential as socio-political forces (BBSR, 2013g; Bundesministerium für Wirtschaft und Technologie, 2013). One example is the ExWoSt research project *3stadt2*, which between 2002 and 2003 sponsored five model projects in attempts to link community groups with public–private partnerships.

The *3stadt2* projects focused on developing *instruments*, namely, cooperative planning processes, to link Government, citizens, and investors as the three main participants in urban development. The new methods had to be both quantitative and transparent (BBR, 2002). The cooperative method was chosen because:

> Classical planning processes are often characterised through expensive and time-consuming decision-making processes . . . The reason for this is first of all a lack of cooperation between various administrative offices that continues into the political arena, and often ends with citizen dissatisfaction.
>
> (BBR, 2002)

The programme therefore aimed to find not only new methods, but to avoid costly debates and debacles, as well as to improve urban design overall.

Methods explored included town hall-style meetings to increase user participation in the planning process, and workshops that teamed up investors, municipal representatives and members of local citizen groups. Although results could not be quantified, accompanying studies found that compared to conventional procedures, cooperative planning processes accelerated the urban design process, encouraged a broad political consensus as well as greater public acceptance, improved urban design quality, allowed early recognition of potential conflicts, encouraged creative processes and unconventional methods and increased the planners' motivation (BBR, 2004a, 2004b).

The *3stadt2* projects were as much a search for a new culture of process tools and forums as they were a quest for specific design and planning results. The extensive information and documentation of all projects in print and on the internet may be seen as part of the information package that is altering the way citizens perceive changes in the built environment. The importance of steering both the means and the end is reflected in the terminology used, as studies refer to *instruments* that are to be applied to shape both process and product. With its focus on *Baukultur*, Germany has attempted to make an invisible concept more visible.

Planning for climate change

Beginning in 2007, the German Government initiated a series of studies relating to climate change, spatial planning and energy concepts, spanning scales ranging from the regional to the individual built object and with a wide range of economic and technological implications. An initial conference organised by the BMVBS in 2007 in Berlin defined the scope and nature of the research questions and allowed the Government to solicit input for the *Modelvorhaben der Raumordnung* (MORO) project *Raumentwicklungsstrategien zum Klimawandel* (*Spatial Strategies for Climate Change*). This two-phased planning study took place in 2009–2013 to develop and test climate change adaptation strategies in eight regions encompassing different climate types. A preliminary study, the results of which were published in 2008, prepared the request for proposals for the eight regions (BMVBS/BBR, 2008b). A year later the investigators published a *blueprint*, which aimed to inform policy makers about the types of process that might be required for effective planning in light of climate changes (BMVBS/BBSR, 2009). In 2010 a final paper summarised the specific impacts on various regional typologies to be expected from climate change, listed existing political and administrative *instruments* and recommended further ways in which the Government could steer development (BMVBS/BBR, 2010).

The stated goals of this first stage focused almost exclusively on developing *concepts*, *networks* and *spatial instruments* (BBSR, 2013h). This was essentially an attempt to bring public awareness into play while organising and developing administrative infrastructure to tackle a looming problem. Further studies would focus on improved urban planning and building construction in order to circumvent technical and physical problems associated with climate change.

By 2008, a study on how climate change has affected building practices in Germany came to the conclusion that most measures had focused on energy efficiency to reduce CO_2 emissions or flood protection, with too little effort put into the effects of climate change on construction types and building practices themselves (BMVBS/BBR, 2008a). Funding for this study was provided by BMVBS, with the principal investigators (PIs) coming from the Institut Wohnen und Umwelt (IWU), a research institute under the joint auspices of the State of Hesse and the City of Darmstadt. At this time, the Government saw climate change as a broad technical and social problem. Building codes had to reflect certain extreme weather conditions to make buildings more impervious to damage as well as changing temperature and precipitation conditions, and thermal comfort, for example through the use of new insulation materials or different glass types.

The 2008 study on building practices began by defining the current state of knowledge through a literature review, interviews with experts on building materials and building damage and examination of industry meeting protocols and insurance reports. The authors acknowledged that with respect to climate change and building practices, Germany was still in the early stages of R&D, with Great Britain and Switzerland noted as countries that were more advanced in the field. There followed an examination of specific problems arising from different climate change factors: heatwaves, downpours, high winds, hail, winters with more humidity, summers with longer sun exposure, and other factors such as changes in snow load or water table. Buildings were analysed with regards to portions above ground, subterranean portions, and heating, ventilation and air conditioning (HVAC) systems, with a further analysis examining statics, construction, performance over time, functionality and requirements for protection from heat, moisture and fire. This was followed by a roadmap for future research.

The study, while remaining broad in its scope, provided detailed scenarios and possible approaches to solving climate-induced problems. For example, in order to minimise the effects of higher summer temperatures, the authors named adequate insulation, shading devices, building orientation, use of thermal massing elements, reduction of interior sources of heat, the use of special glazing and heat exchange systems as prerequisites for avoiding reliance on air conditioning, especially in housing.

These passive methods are certainly not new, yet the authors also pointed out the need for research on building materials with enhanced performance and resilience, as well as new construction methods. Increased extreme rainfall and the ensuing flood risk were discussed in terms of roof drainage, backwater in stormwater drainage systems, surface drainage, flooding, changes in the water table and landslides. Here, the authors listed potential problems as opportunities for new construction approaches. More general phenomena, such as increasingly damp and humid winters, were linked to the possibility of a rise in problems regarding building materials, especially wood, while an increase in wet snow, as opposed to the much lighter powder snow, was noted as calling for structural examinations. In all these areas, linking existing material and construction knowledge to specific needs calculated from climate change data served to develop priorities and point out knowledge deficits and thus opportunities for new R&D.

One difficulty was seen not so much in planning new buildings, but in protecting older ones, especially those with historical significance. Wooden and half-timbered buildings were seen as especially endangered. Since the effects of climate change cannot be determined with absolute certainty, the authors suggested constructing buildings so that they could be regularly upgraded, in order to avoid costs for potentially unnecessary measures.

In addition to research on buildings, and their materials and construction, the BMVBS has sponsored research on how the effects of climate change can be accommodated through urban design measures. In 2011, a preliminary study presented an integrated *Communal Climate Change Strategy and Action Set* that was tested in several model communities (BMVBS/BBR, 2011). The focus here was on mitigation of and adaptation to the effects of climate change, as well as the integration of other urban planning demands, with a special emphasis on typologies of new urban developments. This study examined long-term urban strategies for specified problems with respect to conflicts and congruencies between climate change demands and existent urban models, taking into account physical as well as administrative solutions.

The concern for creating buildings whose building physics will stand the test of time comes from a cultural belief that buildings should be long lasting, thus also the concern for adequate urban design. German industrial norms and construction standards are high, and the centralised approach to studying effects of climate change, as well as the focus on *instruments* and procedures, indicates the Government's desire to identify best practices and convert them into new industrial norms and building codes. It also reveals a desire to be proactive, and to work through alternative scenarios in order to find the most technically appropriate and cost-effective measures for an identified problem.

The broad series of sponsored studies, from those on regional planning practices to those examining individual building components, show an awareness of how pervasive the problem of climate change will be, and that only a holistic approach can fully deal with the problems to be expected. Thus, this research is far from complete, but may be considered the start of a larger agenda. While some of the conferences and publications have elaborated on the need for unspecified political action and have thus been rather administrative in nature, the publications written by experts in the field have developed valuable information and a clear programme for further R&D in the years to come.

Houses as power plants

Energy-conscious construction is legislated in Germany. From 1977 to 1995 heating efficiency requirements for new buildings were regulated through a series of *Wärmeschutzverordnungen* (literally, *Heat Insulation Ordinances*). In 2002 the name of the governing laws was changed to *Energiesparverordnungen* (*Energy Saving Ordinances*), based on European Union Directives on the Energy Performance of Buildings (EurLex, 2010). To help enforce these laws, in January 2009, a law was passed under which owners who want to sell or lease their house or apartment must make an *energy passport* available to prospective buyers and tenants, in order to help them estimate heating expenses before concluding the sale or lease. Currently, owners constructing new buildings are required to cover part of their energy demands through renewable energy sources and it is probably only a matter of time before the Government will require new houses to cover *all* of their energy needs in this way (BMVBS, 2011). To pave the way, the German Government has sponsored research on buildings that produce more energy than they use for heating, cooling and running appliances. Called *plus-energy houses*, the projects are an example of how German R&D has been coupled with national economic interests.

The project has taken a three-pronged approach:

- Individual owners constructing new houses have benefited from a programme that in 2011 disbursed a total of EUR1.2 million (Forschungsinitiative Zukunft Bau, 2013).
- In a second approach, established prefabricated housing companies such as Bien-Zenker, Weber, Schwörer and Huf have developed *plus-energy* homes that are commercially available, as well as larger units with up to 100 housing units (BBSR, 2013i) (see Figure 9.3). An accompanying Fraunhofer-Institute for Building Physics programme evaluates the houses in use (Fraunhofer Institut Bauphysik, 2001). The Federal Government has provided research funds through the CO_2 certificate trade and a tax on nuclear fuel elements. Overall, EUR70,000 were made available for the monitoring of energy use, as well as a 20 per cent bonus for innovative technologies (BBSR, 2013i). The prefabricated housing companies are able to provide an extensive fabrication and distribution system, with the houses displayed as model homes in a prefab home park near Cologne, or as individual prototypes across Germany. The houses make use of elements such as innovative façades, high-efficiency lighting, energy management, a low-temperature heating system, renewable energies, electrical storage capacity and docking stations for electric cars (BMVBS, 2013a).
- A third approach has been the Federal Government's own series of *plus-energy* houses, developed through university institutes and other public organisations. These houses strive to be aesthetically superior

Figure 9.3 HUF green [r]evolution plus-energy house

Source: HUF Haus

> while reducing energy use through optimising both materials and processes. Two of the sponsored houses, developed at the Technical University at Darmstadt, won the Solar Decathlon competition in Washington, DC, in 2007 and 2009 respectively (BMVBS, 2013b) (see Figure 9.4). The 2007 house was displayed in six German cities from 2009–2011 before finding a final location in Dortmund.

In 2010 the BMVBS sponsored a further *plus-energy* house competition for teams of university institutes and commercial architecture offices. The winning entry, titled *Effizienzhaus Plus mit Elektromobilität* (*Efficiency-plus-house with electromobility*) was based on using the house's surplus energy to power two electric cars and an electric scooter (BMVBS, 2012d, 2013c) (see Figure 9.5). All parts of the house were to be recyclable, and materials and functioning of the house were optimised so that the energy required for its use as a home would be minimised. The house was displayed at a central location in Berlin before a four-person test family moved in for 15 months. The side of the house facing the street contained the parking spot for the cars as well as a series of monitors and displays that served to inform, and perhaps inspire, a watching public, while a blog maintained by the test family gave personal accounts of their experiences with the house and its vehicles.

Figure 9.4 Solar Decathlon 2007 winning entry by the Technical University Darmstad
Source: BMVBS

Figure 9.5 Plus-Energy house with electromobility, back of house with family spaces
Source: Ulrich Schwarz

The marketing-like publicity surrounding the upscale plus-energy house, combined with the living experiences documented by a family many Germans could identify with, have served to promote energy-conscious building as a path for the future. The Government's choice to select *electro-mobility* as the winning concept of the 2010 competition ties into national economic interests in two ways: Germany is positioning itself as a producer of increasingly advanced building systems while the Government continues to support Germany's automotive industry into a post-oil era. The focus on energy and technology demonstrates that the German Government has chosen to develop the country's identity towards that of a *high-end, high-tech* powerhouse.

R&D impact: the effects of Government sponsorship

Federal Government R&D, as well as the more indirect financial support of research through the expertise and technical support of publicly funded institutes and universities, has established an infrastructure of invested players who reinforce Government-defined policy aims, such as energy efficiency. This also contributes to a viable export market. This network has also established a framework for international collaborations at the EU level, which has paid out on several levels as outlined below.

EU funding

Czarnitzki and Lopes Bento (2011) found that innovation input in the economy was higher when research was funded through EU grants and/or national grants, than when such funding was absent. Firms often find that the high R&D costs for uncertain outcomes make finding funding through private channels difficult, or that R&D knowledge financed by one firm can be used by another, leading to a disincentive to invest in such research. Czarnitzki and Lopes Bento point out that Germany is one of the few Organisation for Economic Cooperation and Development (OECD) countries that in recent history has *not* cut annual budget provisions for R&D (Czarnitzki & Lopes Bento, 2011).

Industry/university collaboration

The relationship between industries and universities is also worth examining. University-based researchers continue to be funded by and conduct research for industries; Hottenrott and Thorwarth (2010) call it *one of the main channels through which knowledge and technology are transferred from [academia] to the private sector* (Hottenrott & Thorwarth, 2010). From a private sector standpoint, this collaboration is entirely positive, although from the academic standpoint the results are mixed; publication output of professors who were sponsored by industry decreased both quantitatively and qualitatively in subsequent years, although patent citations increased (Hottenrott & Thorwarth, 2010). This indicates, not surprisingly, that industry-sponsored studies focus on applied, as opposed to pure, research, and that if Governments wish to advance overall scientific innovation, including basic research, the funding must come from another source.

In a follow-up study to examine workflows, Hottenrott and Lawson (2012) showed that industry–academia partnerships lead to a greater likelihood that university researchers will source their ideas from industrial partners, especially if the partners were larger firms (Hottenrott & Lawson, 2012). This brings up the question of leadership once again. Both firms and Governments may be assumed to have an interest in defining research directions. However, industry interests are above all economic, while Governments may have other motives such as overall employment rates or national prestige.

Energy efficiency

Many of Germany's R&D efforts have dealt with energy efficiency. On an economic level, this has paid out. Rennings and Rammer (2009) report that when firms with *energy- and resource- efficiency innovations* (EREIs) are compared to firms without such innovations, the EREI firms are more productive (higher revenue) and more cost efficient, since greater energy and material efficiency is associated with lower costs per unit. The same study found that such firms also tend to focus on improved process quality, leading to better product quality. While this study deals with individual firms, the mechanisms studied parallel those used at larger scales, such as those of the case studies described above. The authors define environmental innovations as *new or modified processes, techniques, practices, systems and products which make it possible to avoid or reduce environmental damage*, and analyse the role of both *technology-push* and *market-pull* factors. The former includes infrastructure measures or subsidies that promote R&D, while the latter includes corporate image, increasing energy prices and consumer preferences for environmentally friendly products, which in Germany may be heightened through publicity measures surrounding projects such as the plus-energy houses. Regulation to spur technical innovations may also encourage R&D, especially where market demand is initially weak.

Mennel and Sturm (2008) agree with this last assessment in a study that examines the need for State regulation versus reliance on market factors in achieving greater energy efficiency and an equitable resource distribution. The primary criterion they use for evaluating energy efficiency measures is cost efficiency: when a goal can be achieved with the least possible costs for society. Using this criterion, Mennel and Sturm list political measures such as informational campaigns, CO_2 certificates, and specific CO_2 and energy taxes that reduce reliance on fossil fuels as effective measures. Mennel and Sturn deem *(tradable) white certificates* as both bureaucratic and one sided for their reliance on reducing energy demand, and compulsory standards as less cost efficient and thus less effective means to increase energy efficiency.

Achtnicht (2010) finds a good degree of consumer awareness already existing. In a study of German homeowners' preferences in heating and insulation technologies where cost and environmental benefits of energy-saving measures were explained, he concludes that homeowners are aware of their responsibility and are willing to contribute to climate protection, although uncertainties and lack of access to information can hinder the use of existing technologies. Achtnicht suggests that being able to trace consumer preferences helps to fine-tune policy instruments and that information policies, including an official certification system for energy advising, are more effective than stricter standards in increasing energy efficiency in homes. In this, the findings suggest that the German Government's spending on informational campaigns is an effective investment, although quantifying this statement remains difficult.

Overall, increased consumer awareness of products that improve energy efficiency and heighten building durability or occupant comfort can be expected to lead to increased market demand. While the German Government has established funding mechanisms to encourage basic research, applied research in the area of building components and products has led to increased opportunities for companies in this sector. At the same time, the wide network of private industry and university or other academic institutes coupled with an efficient system of information dissemination assures a high standard of research, awareness of cost efficiency, and ready access to research results.

Future directions

For centuries, Germany's construction industry has been organised into trades and guilds, a factor that has often been criticised as stifling innovation. The current focus on *high-tech* products, coupled with increased prefabrication of building elements and construction methods based on environmental demands, has allowed the building industry to accelerate into the twenty-first century while retaining its basic cultural identity. Essentially, Germany's strong technology and engineering sector has been merged with the building sector to redefine the construction industry as a *high-tech trailblazer*. The focus on new energy systems and innovations in the building envelope are testimony to this. This approach has allowed the building industry to become more competitive internationally, an aim clearly stated by the Federal Government.

In the past decade, Germany has followed a two-pronged approach. First, the Federal and State Governments have worked to shift public awareness through informational campaigns and a focus on *building culture* and energy reform. Second, the Governments have more conventionally supported R&D in the building sector especially as it relates to energy efficiency, renewable energy systems or materials and processes that will increase user comfort in the face of climate changes. At the same time, the Federal Government has undertaken studies that compare German policies to others in the EU, in part to fulfil the aim of keeping the German construction industry internationally competitive. The aim is thus to expand market demand, while undertaking a rigorous programme of R&D to provide adequate innovation in supply. Because of the well-established cooperation between the public and private sectors, the Federal and, to an extent, State Governments have taken on a strong leadership role in defining R&D. This has allowed certain industries, such as those in the area of renewable energies, to flourish.

Future research might examine how to optimise collaboration and information transfer between technology firms and the building trade and construction industry. The Federal Government's plus-energy house with integrated electrical vehicles is one example of such partnerships. Innovations must be perceived as convenient and cost effective to find consumer acceptance, and a focus on elements that work together efficiently and seamlessly will make such products more attractive to the consumer. A holistic planning approach must also take into account lifestyle choices and various user groups to allow organisations involved with the building industry to maximise both business and policy impact.

References

Achtnicht, M. (2010) 'Do environmental benefits matter? A choice experiment among house owners in Germany', FCN Working Papers 27/2010, E.ON Energy Research Center, Future Energy Consumer Needs and Behavior (FCN).

Baden-Württemberg (2013) *Forschung: Hochtechnologie-standort nummer 1* (*Research: High-tech location number 1*). Available at: www.baden-wuerttemberg.de/de/bw-gestalten/erfolgreiches-baden-wuerttemberg/forschung/ (accessed 9 April 2013).

Bayerisches Landsamt für Denkmalpflege (2013) *Weitere information zur bauforschung* (*For more information on building research*). Available at: www.blfd.bayern.de/bau-und_kunstdenkmalpflege/stabsstelle/bauforschung/weitere_informationen/index.php (accessed 9 April 2013).

Bayerisches Staatsministerium für Umwelt, Gesundheit und Verbraucherschutz (2007) *Klimaprogramm Bayern 2020* (*Bavarian climate programme 2020*). Available at: www.bayern.de/Anlage2093555/Klimaprogramm%20Bayern%202020.pdf; English translation available at: www.stmug.bayern.de/umwelt/klimaschutz/klimaprogramm/doc/klimaprogramm2020_en.pdf (accessed 9 April 2013).

BBR (2002) *ExWoSt – Informationen 3stadt2 – Neue kooperationsformen in der stadtentwicklung* (*ExWoSt – Information 3stadt2 – New forms of cooperation in urban development*), no. 24/1, Bonn: Bundesamt für Bauwesen und Raumordnung (BBR, Federal Office for Building and Regional Planning). Available at: www.bbsr.bund.de/cln_032/nn_23550/BBSR/DE/Veroeffentlichungen/BMVBS/ExWoSt/24_29/exwost24.html (accessed 9 April 2013).

BBR (2004a) *ExWoSt – Informationen 3stadt2 – Neue kooperationsformen in der stadtentwicklung*, no. 24/5, Bonn: Bundesamt für Bauwesen und Raumordnung. Available at: www.bbsr.bund.de/cln_032/nn_23550/BBSR/DE/Veroeffentlichungen/BMVBS/ExWoSt/24_29/exwost24.html (accessed 9 April 2013).

BBR (2004b) *ExWoSt – Informationen 3stadt2 – Neue kooperationsformen in der stadtentwicklung*, no. 24/6, Bonn: Bundesamt für Bauwesen und Raumordnung. Available at: www.bbsr.bund.de/cln_032/nn_23550/BBSR/DE/Veroeffentlichungen/BMVBS/ExWoSt/24_29/exwost24.html (accessed 9 April 2013).

BBSR (2010) *2009 Antragsliste*, Unpublished document, Bonn: Bundesinstitut für Bau-, Stadt- und Raumforschung im Bundesamt für Bauwesen und Raumordnung (Federal Institute for Research on Building Urban Affairs and Spatial Development in the Federal Office for Building and Regional Planning).

BBSR (2013a) *Bauforschung* (*Building research*). Available at: www.bbsr.bund.de/nn_21268/BBSR/DE/FP/FoerderungBauforschung/foerderungbauforschung_node.html?_nnn=true (accessed 9 April 2013).

BBSR (2013b) *ESPON*. Available at: www.bbsr.bund.de/cln_032/nn_21268/BBSR/DE/FP/ESPON/espon_node.html?_nnn=true (accessed 9 April 2013).

BBSR (2013c) *INTERREG*. Available at: www.bbsr.bund.de/cln_032/nn_21268/BBSR/DE/FP/INTERREG/interreg_node.html?_nnn=true (accessed 9 April 2013).

BBSR (2013d) *Zukunft bau* (*Future construction*). Available at: www.bbsr.bund.de/cln_032/nn_21268/BBSR/DE/FP/ZB/zukunftbau_node.html?_nnn=true (accessed 9 April 2013).

BBSR (2013e) *Ausschreibungen forschungsprojekte* (*Tenders research projects*). Available at: www.bbsr.bund.de/cln_032/nn_112742/BBSR/DE/Aktuell/Forschungs projekte/forschungsprojekte_node.html?_nnn=true (accessed 9 April 2013).

BBSR (2013f) *Programme*. Available at: www.bbsr.bund.de/cln_032/nn_112742/BBSR/DE/FP/forschungsprogramme_node.html?_nnn=true (accessed 9 April 2013).

BBSR (2013g) *ExWoSt*. Available at: www.bbsr.bund.de/cln_032/nn_21686/BBSR/DE/FP/ExWoSt/exwost_node.html?_nnn=true (accessed 9 April 2013).

BBSR (2013h) *MORO*. Available at: www.bbsr.bund.de/cln_032/nn_23558/BBSR/DE/FP/MORO/Forschungsfelder/2009/RaumKlima/Phase1/01_Start1.html (accessed 9 April 2013).

BBSR (2013i) 'Deutsche bauforschung' ('German construction research'). Email (5 April 2013).

BMVBS (2011) *Wege zum effizienzhaus-plus* (*Ways to efficiency-plus-house*), Berlin: Bundesministerium für Verkehr, Bau und Stadtentwicklung (BMVS) (Federal Ministry of Transport, Building and Urban Development). Available at: www.forschungsinitiative.de/PDF/Effizienzhaus-Plus_Barrierefrei2.pdf (accessed 9 April 2013).

BMVBS (2012a) *Stadtumbau west: Evaluierung des bund-länder-programms* (*Urban redevelopment west: Evaluation of the federal–state programme*), April 2012, Berlin: Bundesministerium für Verkehr, Bau und Stadtentwicklung. Available at: www.bbsr.bund.de/cln_032/nn_627458/BBSR/DE/Veroeffentlichungen/BMVBS/

Sonderveroeffentlichungen/2012/StadtumbauWestEvaluierung.html (accessed 9 April 2013).

BMVBS (2012b) *Förderrichtlinie zukunft bau* (*Funding guidelines future construction*). Available at: www.bmvbs.de/SharedDocs/DE/Artikel/B/forschungs initiative-zukunft-bau-foerderrichtlinie-zuwendung-forschungsvorhaben. html?nn=75494 (accessed 9 April 2013).

BMVBS (2012c) *Richtlinie über die vergabe von zuwendungen für modellprojekte im effizienzhaus plus-standard* (*Directive on the award of grants for model projects efficiency-plus-house standard*). Available at: http://m.bmvbs.de/SharedDocs/DE/Artikel/B/forschungsinitiative-zukunft-bau-foerderrichtlinie-modelle-ehp-standard.html?nn=36330 (accessed 9 April 2013).

BMVBS (2012d) *Effiziensbaus plus mit elektromobilität: technische informationen und details, Berlin: Bundesministerium für Verkehr* (*Efficiency-plus-house with electric mobility: Technical information and booking details. Berlin: Federal Ministry of Transport*), Bau und Stadtentwicklung. Available at: www.bmvbs.de/cae/servlet/contentblob/78972/publicationFile/51909/effizienzhausplus_elektromobil_de_aufl2.pdf (accessed 9 April 2013).

BMVBS (2013a) *Modellvorhaben netzwerk* (*Model project network*). Available at: www.bmvbs.de/DE/EffizienzhausPlus/Modellvorhaben/Netzwerk/effizienzhaus-plus-neubauten_node.html (accessed 9 April 2013).

BMVBS (2013b) *Plus-energie-haus: Bauen für die zukunft 2007-2011* (*Plus-energy-house: Building for the future 2007–2011*). Available at: www.bmvbs.de/SharedDocs/DE/Artikel/B/plus-energie-haus-bauen-fuer-die-zukunft.html (accessed 9 April 2013).BMVBS (2013c) *Effizienzhaus-plus* (*Efficiency-plus-house*). Available at: www.bmvbs.de/DE/EffizienzhausPlus/effizienzhaus-plus_node.html (accessed 9 April 2013).

BMVBS/BBR (2008a) *Folgen des Klimawandels: Gebäude und Baupraxis in Deutschland* (*Consequences of climate change: Building and construction practice in Germany*), BBR-Online-Publication 10/2008, Bonn: Bundesamt für Bauwesen und Raumordnung. Available at: www.bbsr.bund.de/nn_187722/BBSR/DE/Veroeffentlichungen/BBSROnline/2008/ON102008.html (accessed 9 April 2013).

BMVBS/BBR (2008b) *Raumentwicklungsstrategien zum klimawandel – Vorstudie für modellvorhaben* (*Spatial development strategies on climate change – Study for model projects*), BBR-Online-Publication 19/2008, Bonn: Bundesamt für Bauwesen und Raumordnung. Available at: www.bbsr.bund.de/cln_032/nn_23582/BBSR/DE/Veroeffentlichungen/BBSROnline/2008/ON192008.html (accessed 9 April 2013).

BMVBS/BBSR (2009) *Entwurf eines regionalen handlungs- und aktionsrahmens klimaanpassung ('Blaupause')* (*Draft regional framework for action climate action and adaptation* (*'blueprint'*), BBSR-Online-Publication 17/2009, Bonn: Bundesamt für Bauwesen und Raumordnung. Available at: www.bbsr.bund.de/cln_032/nn_23582/BBSR/DE/Veroeffentlichungen/BBSROnline/2009/ON 172009.html (accessed 9 April 2013).

BMVBS/BBR (2010) *Klimawandel als Handlungsfeld der Raumordnung: Ergebnisse der Vorstudie zu den Modelvorhaben Raumentwicklungsstrategien zum Klimawandel* (*Climate change as the sphere of spatial planning: Results of the preliminary study for the project model spatial development strategies on climate change*), Forschungen Heft 144, Berlin: Bundesministerium für Verkehr, Bau und

Stadtentwicklung. Available at: www.bbsr.bund.de/cln_032/nn_23494/BBSR/DE/Veroeffentlichungen/BMVBS/Forschungen/2010/Heft144.html (accessed 9 April 2013).

BMVBS/BBR (2011) *Klimawandelgerechte Stadtentwicklung: Ursachen und Folgen des Klimawandels durch urbane Konzepte begegnen* (*Climate-proof urban development: Encounter causes and consequences of climate change through urban concepts*), Forschungen Heft 149, Berlin: Bundesministerium für Verkehr, Bau und Stadtentwicklung. Available at: www.bbsr.bund.de/cln_032/nn_23494/BBSR/DE/Veroeffentlichungen/BMVBS/Forschungen/2011/Heft149.html (accessed 9 April 2013).

Bundesministerium für Wirtschaft und Technologie (2013) *Richtlinien des Bundesministers für Verkehr, Bau- und Wohnungswesen für Forschungsvorhaben zur Weiterentwicklung des Wohnungs- und Städtebaues (experimenteller Wohnungs- und Städtebau) vom 2* (*Guidelines of the minister of transport, building and housing research projects for the development of the housing and urban development (experimental housing and urban development)*, November 1987. Available at: www.foerderdatenbank.de/Foerder-DB/Navigation/Foerderrecherche/suche.html?get=4aa561e46fff16fb87d819d09c769842;views;document&doc=10797&typ=RL (accessed 9 April 2013).

Czarnitzki, D. & Lopes Bento, C. (2011) 'Innovation subsidies: Does the funding source matter for innovation intensity and performance? Empirical evidence from Germany', Discussion paper no. 11-053, Mannheim: ZEW-Centre for European Economic Research. Available at: http://papers.ssrn.com/sol3/papers.cfm?abstract_id=1908764 (accessed 9 April 2013).

EurLex (2010) *Directive 2010 / 31 / EU of the European parliament*. Available at: http://eur-lex.europa.eu/LexUriServ/LexUriServ.do?uri=OJ:L:2010:153:0013:01:EN:HTML (accessed 9 April 2013).

European Commission (2013) *LIFE programme*. Available at: http://ec.europa.eu/environment/life/ (accessed 9 April 2013).

Forschungsinitiative Zukunft Bau (2013) *Forschungsinitiative Zukunft Bau* (*Research initiative for future construction*). Available at: www.forschungsinitiative.de (accessed 9 April 2013).

Forum Vauban (2013a) *Abstract*. Available at: www.vauban.de/info/abstract.html (accessed 9 April 2013).

Forum Vauban (2013b) *Geschichte Teil 5 1997–2000*. Available at: www.vauban.de/info/geschichte5.html (accessed 9 April 2013).

Fraunhofer Institut Bauphysik (2001) *Fertighäuser im Wandel. Vom Niedrigenergiehaus zum Null-heizenergiehaus* (*Prefabricated houses changing. From low energy to the zero-energy house*), Stuttgart: Fraunhofer Institut Bauphysik. Available at: www.weberhaus.de/fileadmin/weberhaus_de/redaktion/niedrigenergiehaus_nullheizenergiehaus.pdf (accessed 9 April 2013).

Hottenrott, H. & Thorwarth, S. (2010) 'Industry funding of university research and scientific productivity', Discussion paper no. 10-105, Mannheim: ZEW-Centre for European Economic Research. Available at: http://onlinelibrary.wiley.com/doi/10.1111/j.1467-6435.2011.00519.x/full (accessed 9 April 2013).

Hottenrott, H. & Lawson, C. (2012) 'Research grants, sources of ideas and the effects on academic research', Discussion paper no. 12-048, Mannheim: ZEW-Centre for European Economic Research. Available at: http://papers.ssrn.com/sol3/papers.cfm?abstract_id=2118513 (accessed 9 April 2013).

Kähler, G. (ed.) (2001) *Statusbericht Baukultur in Deutschland. Ausgangslage und Empfehlungen* (*Status report culture of building in Germany. Background and recommendations*), Bundesministerium für Verkehr, Bau- und Wohnungswesen. Available at: www.hoembergundpartner.de/downloads/Statusbericht.pdf (accessed 9 April 2013).

Mennel, T. & Sturm, B. (2008), 'Energieeffizienz – Eine neue aufgabe für staatliche regulierung?' ('Energy efficiency – A new role for government regulation?'), ZEW Discussion Paper No. 08-004, Mannheim: Zentrum für Europäische Wirtschaftsforschung GmbH (ZEW).

Niedersächsisches Ministerium für Umwelt, Energie und Klimaschutz (2013a) *Niedersächsische Institute* (*Lower Saxony Institute*). Available at: www.umwelt.niedersachsen.de/energie/nds_institute/ (accessed 9 April 2013).

Niedersächsisches Ministerium für Umwelt, Energie und Klimaschutz (2013b) *Institut für Bauforschung* (*Institute for Building*). Available at: www.umwelt.niedersachsen.de/energie/nds_institute/IFB/97585.html (accessed 9 April 2013).

Plusenergiehaus (2013) *Ein Haus für die Zukunft* (*A house for the future*). Available at: www.plusenergiehaus.de/index.php?p=home&pid=10&L=0&host=1 (accessed 9 April 2013).

Rennings, K. & Rammer, C. (2009) 'Increasing energy and resource efficiency through innovation – an explorative analysis using innovation survey data', Discussion paper no. 09-056, Mannheim: ZEW-Centre for European Economic Research. Available at: http://madoc.bib.uni-mannheim.de/madoc/volltexte/2009/2479/ (accessed 9 April 2013).

Triebel, W. (1983) *Geschichte der Bauforschung. Die Forschung für das Bau- und Wohnungswesen in Deutschland* (*History of building research. The research for construction and housing in Germany*), Hannover: Curt R. Vincentz Verlag.

Wohnen ohne Auto (2013) *Autofreie projekte in Deutschland, Europa und weltweit, Übersicht* (*Carfree projects in Germany, Europe and worldwide, Survey*). Available at: www.wohnen-ohne-auto.de/projekte (accessed 9 April 2013).

10 Hong Kong (China) – R&D funding in the construction industry

Geoffrey Qiping Shen and Jingke Hong

Background

Hong Kong, as one of the largest international finance centres and trading entities in the world, provides a natural platform for the blend of Chinese and Western cultures. After the Second World War, Hong Kong experienced a rapid development in economic and social aspects. Up until now, Hong Kong has been one of the safest, wealthiest and most rapidly developing cities with a high degree of autonomy in the world. In 2011, the gross domestic product (GDP) of Hong Kong was approximately HKD1,896 billion, with an annual average growth rate of 5.64 per cent between 1990 and 2011 (National Bureau of Statistics of China, 2012). The total population of Hong Kong is around 7 million people over an area of 1,070 sq km, giving it one of the highest population densities in the world.

R&D investments in Hong Kong

Research and development (R&D) investments, in creating an education hub and providing a bond between institutional collaborations and economic development, play an important role in the growth of Hong Kong as an international metropolis. The research sector has experienced long-term development, aiming to provide high-quality education, develop individual potential and cultivate the capability to face future challenges. To achieve these objectives, the University Grants Committee (UGC), established in 1965, aimed to provide appropriate tools and incentive mechanisms to encourage and assist institutions to improve their international competitiveness as well as to play a proactive role in strategic planning and policy making. The Research Grants Council (RGC), established by the Government of Hong Kong in 1991 as a subsidiary under the UGC, focused on meeting the academic research needs of the tertiary institutions. The Research Endowment Fund, which was mainly determined by UGC and RGC, could be divided into four parts: (i) *block grants*; (ii) *earmarked research grants*; (iii) *other funding schemes and activities*; and (iv) *other Government and private funds* (Figure 10.1).

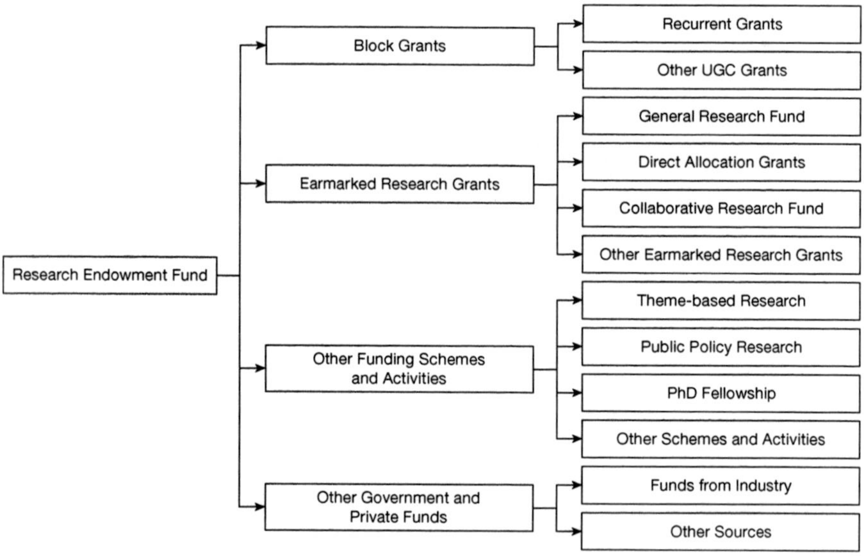

Figure 10.1 Structure of the Research Endowment Fund in Hong Kong

The earmarked research grants complement block grants from the UGC, which provides institutions with funds for infrastructure research and outlays such as researchers' salaries, laboratory costs and other overheads related to UGC- or RGC-funded research activities. The other funding schemes and activities are used to support some special research programmes such as projects with long-term goals that are strategically beneficial to the development of Hong Kong, to promote public policy research in higher education institutions or to attract the *best and brightest* students in the world.

In 2012 there were eight institutions funded by the UGC to cultivate highly trained professionals and promote economic and social development. These institutions are: City University of Hong Kong (CityU); Hong Kong Baptist University (HKBU); Lingnan University (LU); the Chinese University of Hong Kong (CUHK); the Hong Kong Institute of Education (HKIEd); the Hong Kong Polytechnic University (PolyU); the Hong Kong University of Science and Technology (HKUST); and the University of Hong Kong (HKU).

From 1997 to 2010, the research expenditure of UGC-funded institutions has increased steadily, from HKD3.8 billion (USD490 million) in 1997 to HKD6.9 billion (USD900 million) in 2010, and the average growth rate was 4.7 per cent. During the same period, the ratio of research expenditure to total expenditure of UGC-funded institutions increased by 7.74 per cent, from 21.98 per cent to 29.72 per cent (UGC, 2012f, 2012g).

Figure 10.2 shows the subventions from UGC and its relative proportion as a percentage of Hong Kong's GDP from 2002 to 2011. In 2011 the subvention from UGC was HKD16.3 billion (USD2 billion), which has grown

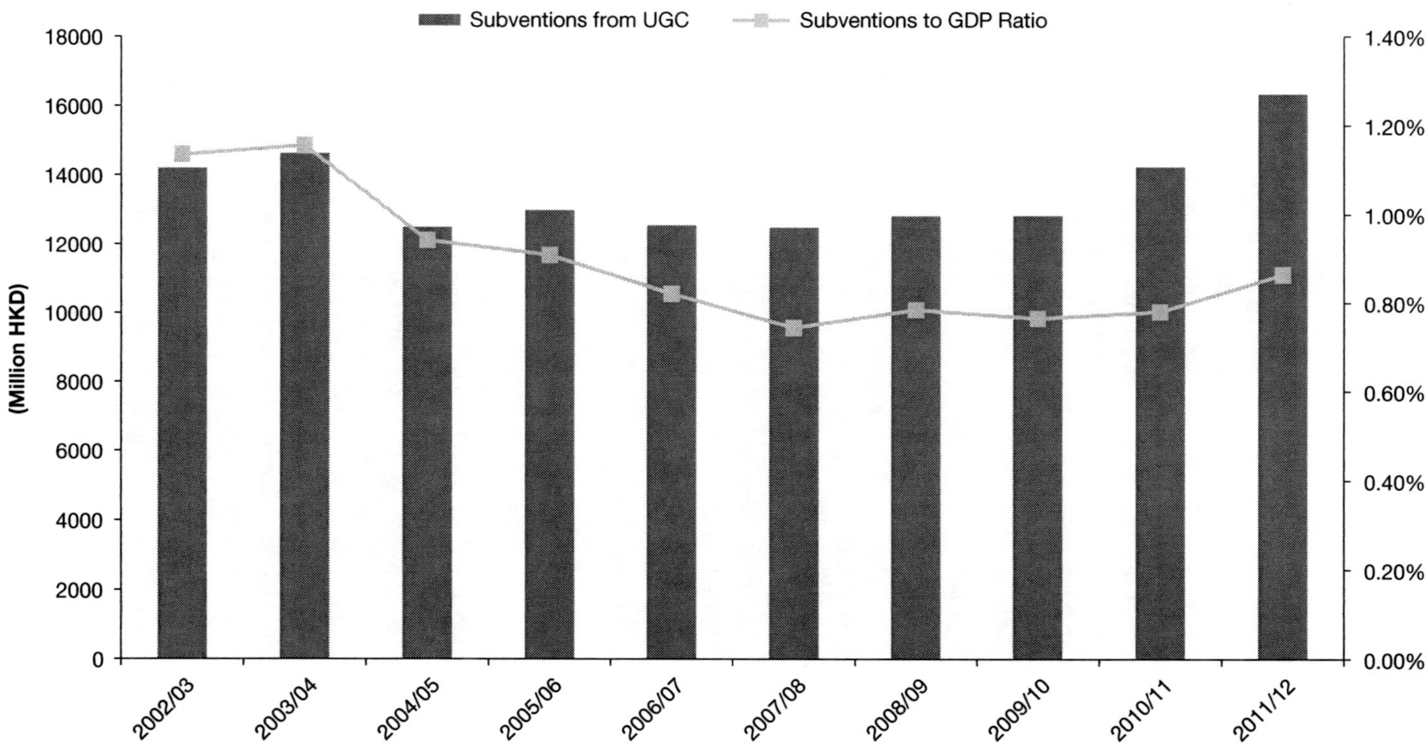

Figure 10.2 Subventions from UGC and its ratio to GDP, 2002–2011, Hong Kong

Source: University Grants Committee (2012h, 2012i)

by 14.8 per cent when compared with 2002. Nevertheless, it is important to note that the share of subventions in Hong Kong's GDP actually declined in the past 10 years, from 1.13 per cent in 2002/03 to 0.86 per cent in 2011/12, and fell to its lowest in 2007/08 with a proportion of 0.74 per cent.

From the two figures above, it is clear that the UGC-funded institutions gradually increased their investments and expenditures on research in order to enhance their competitiveness and performance in academic achievements, especially considering the increased importance of these grants to each institution's future development and the highly competitive environment they are now in. By contrast, from the point of view of the whole regional economy, the emphasis on regional research development displays the opposite trend: continued decrease of the relative proportion in regional GDP.

Earmarked research grants

The primary source of funding available to a UGC-funded institution to support research and other professional and academic activity is the *institutional recurrent* or *block grant*. However, funding for research projects is also provided through other channels, including the earmarked research grants administered by the RGC (UGC, 2012a) which are the main source of funding for academic research in the local higher education sector. They are allocated on a competitive basis.

In 2012, funding was allocated to five broad categories: *direct allocation grants*; *General Research Fund*; *Collaborative Research Fund*; *joint research schemes*; and *research postgraduate students conference/seminar grants*.

Direct allocation grants

Direct allocation grants are allocated directly to the institutions to support small projects costing less than HKD200,000 each (USD26,000), carried out by new junior faculty members on a competitive basis. According to *RGC Annual Reports* from 1994–2011, the level of funding for the direct allocation grants was relatively constant for this period. Especially after 2004, the funding level was approximately HKD65 million (USD8.4 million) per year over 8 years. The distribution of the amount of direct allocation grants between eight UGC-funded institutions was also fairly constant, so that although the allocated proportion varied from approximately 2.8 per cent to 22.9 per cent, the share granted to each institute has been relatively constant with only small fluctuations during the past 17 years. From 1994 to 2011, the institutions receiving direct allocation grants were ranked as follows: the University of Hong Kong, 22.9 per cent; the Chinese University of Hong Kong, 22.1 per cent; the Hong Kong Polytechnic University, 16.0 per cent; City University of Hong Kong, 14.0 per cent; the Hong Kong University of Science and Technology, 11.9 per cent; Hong Kong

Baptist University, 7.2 per cent; the Hong Kong Institute of Education, 3.1 per cent; and Lingnan University, 2.8 per cent. The Hong Kong Institute of Education and Lingnan University, two comparatively younger members in UGC-funded institutions, received the least allocations with an average share of 3.1 per cent and 2.8 per cent in each of the 17 years.

General Research Fund

The aim of the General Research Fund is to supplement institutions' own research support to those who have achieved or have the potential to achieve excellence. Compared to the direct allocation grants, the General Research Fund was more useful to reflect the R&D investment in Hong Kong, its influence and competitiveness as well as amount of funding being substantially above those of other grants. It represented a sharp increase of 804.9 per cent in 2011 in relation to the funds first granted by RGC in 1991 (Figure 10.3). Over this 20-year period, the average rate of increase was 12.2 per cent. This was an achievement considering the stagnant economy and decreased emphasis on R&D.

For the distribution of amount of the General Research Fund, the institutions with the top two rankings are the same as those in the direct allocation grants, they are: the University of Hong Kong, 27.0 per cent; and the Chinese University of Hong Kong, 23.5 per cent. The Hong Kong University of Science and Technology received the third largest share, 12.0 per cent. Lingnan University showed the least competitiveness in this research funding with only 0.38 per cent.

Collaborative Research Fund

This funding scheme serves to support multi-investigator and multi-disciplinary projects. As observed in Figure 10.4, the trend of Collaborative Research Fund grants increased gradually from HKD24.8 million (USD3.2 million) in 2000 to HKD56.3 million (USD7.3 million) in 2011, representing an increase of over 126 per cent with respect to the funds granted in 2000. However, the amount of funding decreased after its first peak in 2008. Although the amount of these grants depended to some extent on how RGC allocated the proportion among different types of funding and on the quality of the applications in every year, it is considered satisfactory that the collaborations between different institutions have improved and been enhanced since 2005.

A summary of the development of competitive research funding under the support of RGC in Hong Kong using the grants mentioned above, namely, the General Research Fund and Collaborative Research Fund is shown in Figures 10.3 and 10.4. As a result, the absolute quantity of the total funds has increased, reflecting the excellent performance and development in the field of academic research in Hong Kong.

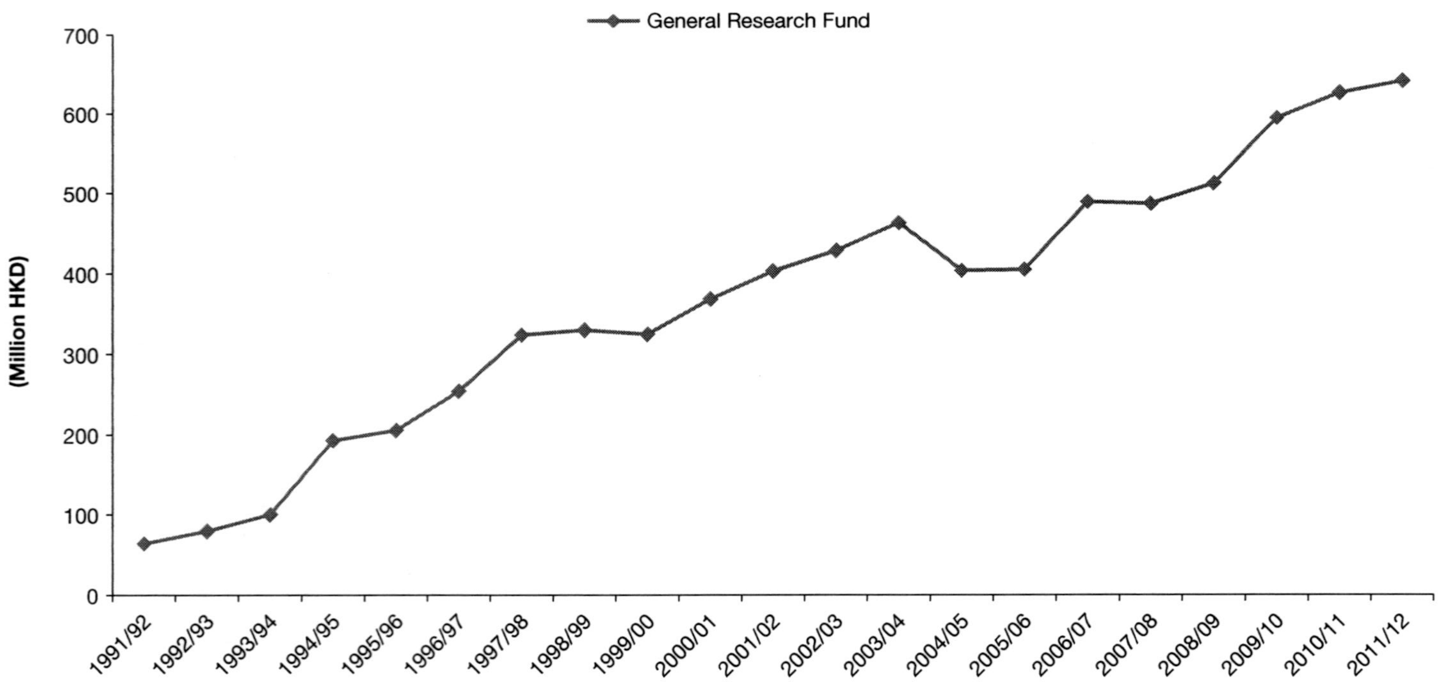

Figure 10.3 General Research Fund, 1991–2011, Hong Kong

Source: Hong Kong Research Grants Council (1994–2011)

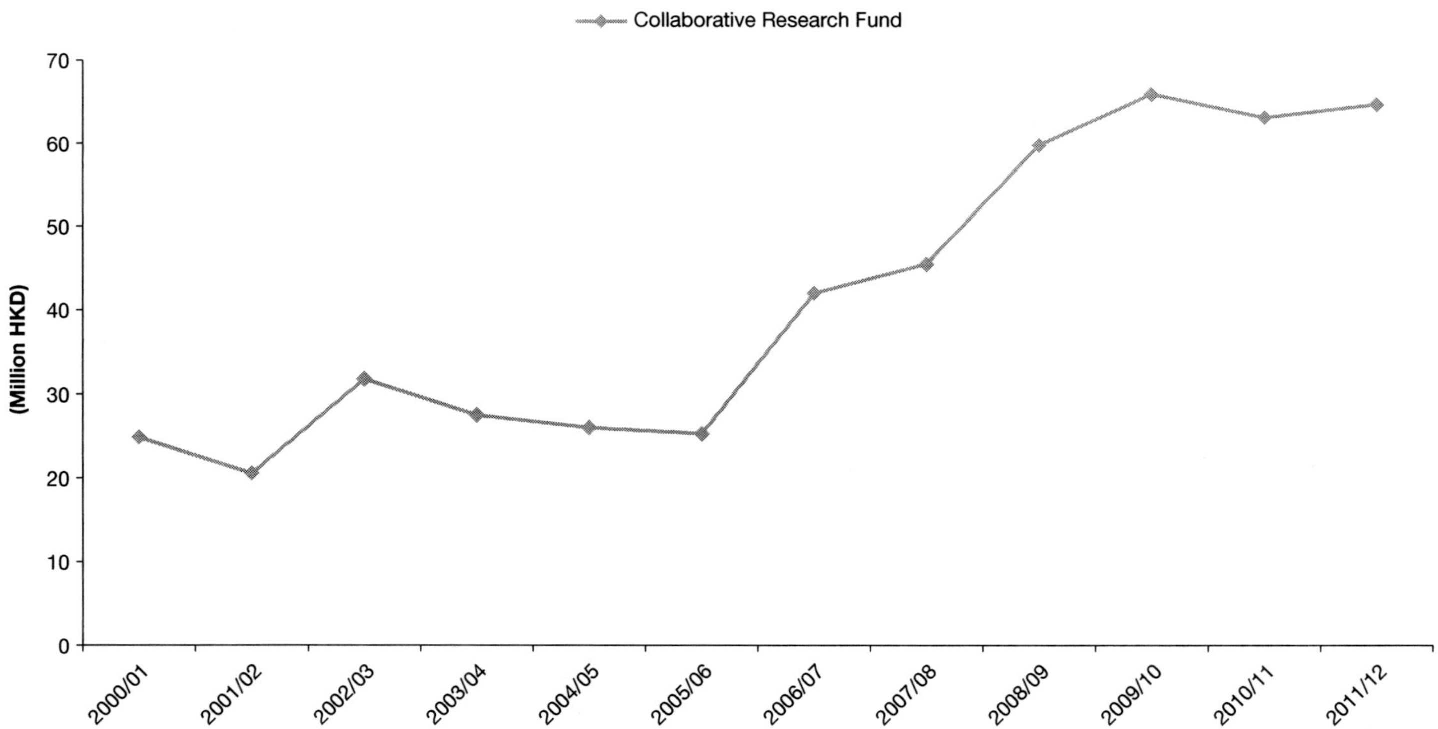

Figure 10.4 Collaborative Research Fund, 2000–2011, Hong Kong

Source: University Grants Committee (2012b)

Other funding schemes and activities

Theme-based research scheme

This scheme aims to focus academic research efforts of UGC-funded institutions on themes of strategic importance to the long-term development of Hong Kong. Beginning in 2010, three themes and 11 grand challenge topics were established within the framework of the theme-based research scheme in order to encourage research achievements in those areas. The three themes included:

- promoting good health
- developing a sustainable environment
- enhancing Hong Kong's strategic position as a regional and international business centre.

As observed in Table 10.1, apart from Lingnan University and the Hong Kong Institute of Education, the other six UGC-funded institutions have already participated in this collaboration project. Although the

Table 10.1 Theme-based research scheme in the first and second round, Hong Kong

	Project title	*Participating UGC-funded institutions*	*Funding ('000 HKD)*	*Total*
First round	The liver cancer genome project: translating genetic discoveries to clinical benefits	CUHK, HKUST, HKU	45.0	247.7
	Massively parallel sequencing of plasma nucleic acids for the molecular diagnostics of cancers	CUHK	32.0	
	Personalised medicine for cardiovascular diseases: from genomic testing and biomarkers to human pluripotent stem cell platform	CUHK, HKUST, HKU	40.0	
	Cell-based heart regeneration	CityU, CUHK, HKUST, HKU	60.0	
	Challenges in organic photovoltaics and light-emitting diodes – a concerted multi-disciplinary and multi-institutional effort	CityU, HKBU, PolyU, HKUST, HKU	57.4	
	Transforming Hong Kong's ocean container transport logistics network	CityU, HKUST, HKU	13.3	

	Project title	*Participating UGC-funded institutions*	*Funding ('000 HKD)*	*Total*
Second round	Functional analyses of how genomic variation affects personal risk for degenerative skeletal disorders	CityU, HKU	74.6	203.1
	Stem cell strategy for nervous system disorders	CUHK, PolyU, HKUST, HKU	60.8	
	Sustainable lighting technology: from devices to systems	PolyU, HKU	21.7	
	Cost-effective and eco-friendly LED system-on-a-chip (SoC)	HKUST, HKU	30.6	
	Enhancing Hong Kong's future as a leading international financial centre	CUHK, PolyU, HKU	15.4	

Source: University Grants Committee (2012d, 2012e)

amount of funding in the first round was more than that of the second round, the average amount per item was approximately equal to HKD41.3 million (USD5.3 million) and HKD40.6 million (USD5.2 million) respectively, confirming that the support from RGC is consistent in this funding series.

Public Policy Research Fund

This funding scheme was launched by the Government in 2005 to promote public policy research in higher education institutions. Public Policy Research Fund grants have been allocated through 10 rounds over the past 10 years. The trend of funds allocated is fluctuant and fell to its lowest point on the sixth round, to below HKD4 million (USD0.52 million). However, the amounts granted through this fund increased in the tenth round to a total HKD4.8 million (USD0.62 million) when compared with the funding of the first round, representing an increase of 66.7 per cent over the 10-year period (University Grants Committee, 2012c).

The distribution of funds through this scheme was completely different between rounds and universities. This depended to a large extent on the policy preferences of RGC in successive years and on the development of new academic achievements at these institutions. Given the competitive environment and uncertainties in the final results, it is important to note that the University of Hong Kong was the only institution that received this research funding in every round, demonstrating their solid academic capacity in policy research.

Other Government and private funds

Grants from industry

Due to changes in economic circumstances, the funds allocated by industry grants varied significantly between 1994 and 2011 (Figure 10.5). It peaked in 2004 at HKD55.1 million (USD7.1 million) while reducing in the subsequent years. The share of industry grants in total funding for new research projects for the period 1996–2010 was always below 5 per cent, indicating a huge potential for industrial support for the development of research funding in Hong Kong.

Innovation and Technology Fund

The Innovation and Technology Fund (ITF), which is managed by the Innovation and Technology Commission, aims to increase the added value, efficiency and competitiveness of national economic activities in Hong Kong. The special objective of ITF is to encourage and assist Hong Kong enterprises to promote their technological level and introduce innovation to their businesses. There are four programmes approved by the ITF, they are: *Innovation and Technology Support Programme*; *General Support Programme*; *University–Industry Collaboration Programme*; and *Small Entrepreneur Research Assistance Programme*. At the end of September 2012, 3,025 projects had been approved and HKD6922 million (USD890 million) allocated to different programmes. The Innovation and Technology Support Programme received HKD5869.5 million (USD760 million), the largest share of the total funds accounting for 84.8 per cent of the total share (Innovation and Technology Commission, 2012).

The distribution of approved projects among different technology areas is dominated by: electrical and electronics, 27.8 per cent; information technology, 27.6 per cent; and manufacturing technology, 15.6 per cent. Together these three areas received 71.0 per cent of all ITF funds allocation. The allocation of the remaining 29 per cent was distributed to: nanotechnology, 9.2 per cent; biotechnology, 6.2 per cent; materials science, 4.4 per cent; environmental technology, 1.9 per cent; Chinese medicine, 1.2 per cent; and others, 6.0 per cent.

Construction R&D investment

In Hong Kong, there is a subject discipline named civil engineering surveying, building and construction (CESBC) subsidised by RGC under the Panel of Engineering, which also includes three other subjects: computing science and information technology (CSIT); electrical and electronic engineering (EEE); and mechanical, production and industrial (MPI). It is important to note that all of the engineering subject disciplines have experienced sharp growth (Figure 10.6), and CESBC ranked first with an the average increase

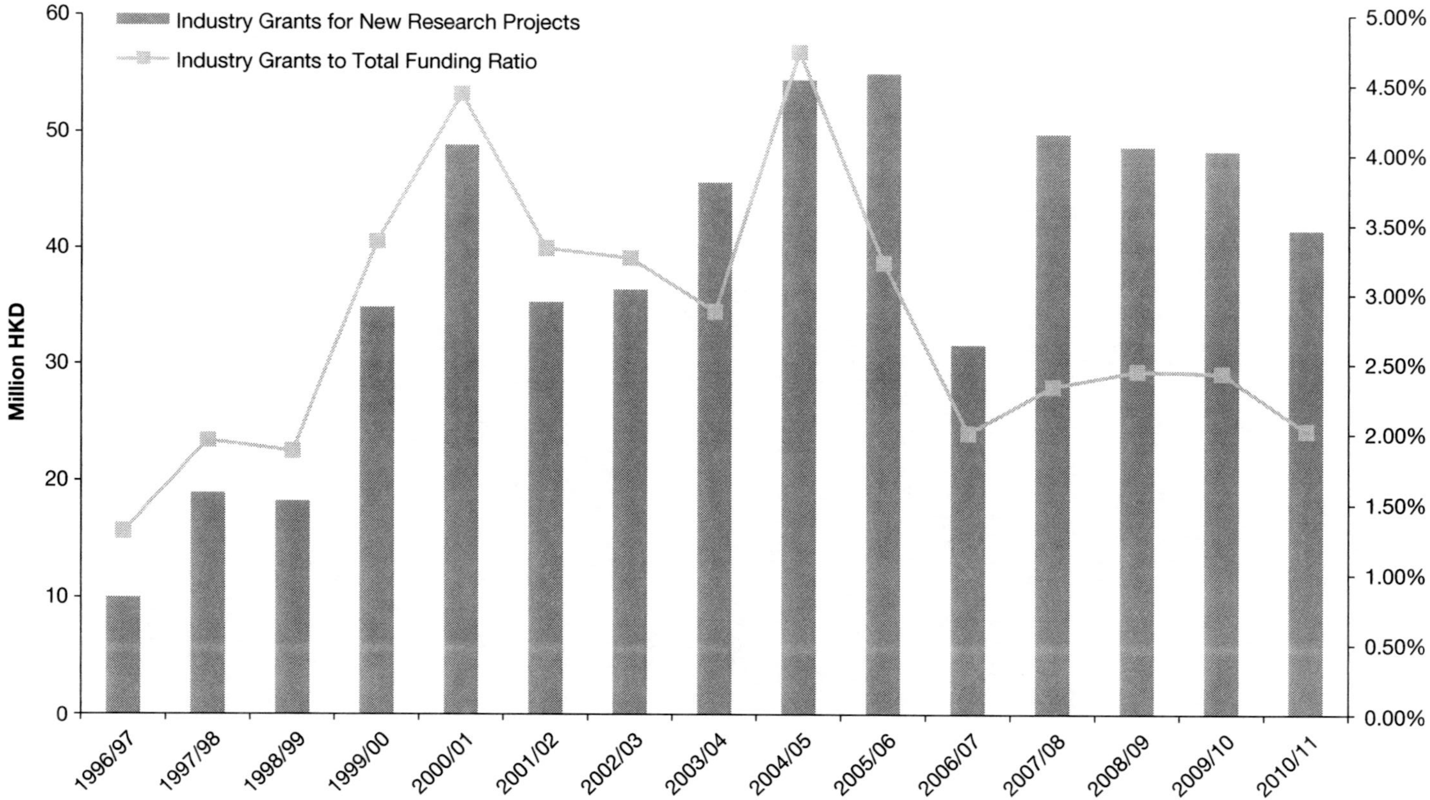

Figure 10.5 Industry grants for new research projects and their ratio to total funding, Hong Kong

Source: Hong Kong Research Grants Council (1994–2011)

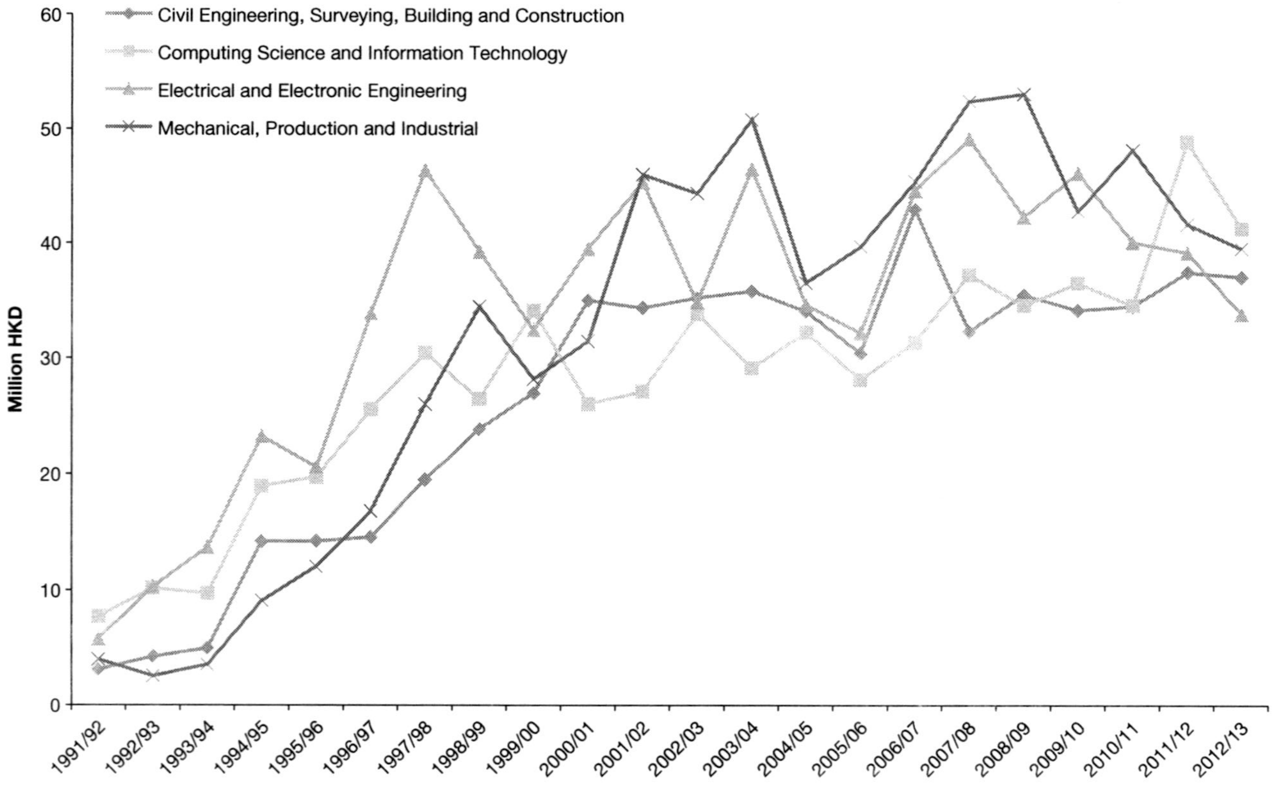

Figure 10.6 Amount of grants for each of four subject disciplines under the Panel of Engineering, 1991–2012, Hong Kong

Source: Hong Kong Research Grants Council (1994–2011)

rate of 12.2 per cent, reflecting the flourishing and rapid development of the construction research field and its competitiveness with regards to other disciplines.

The average proportion of the four disciplines in engineering remained relatively steady from 1991/92 to 2011/12, ranging from 22 per cent to 28 per cent. The proportion of CESBC grants in the total engineering funding allocation increased from 14.5 per cent in 1991/91 to 24.4 per cent in 2012/13 with an average growth rate of 2.5 over the past 22 years.

The distribution of research investment among different institutions for CESBC is shown in Figure 10.7. It is clear that the Hong Kong Polytechnic University ranked first from 1991 to 2012 according to the number of General Research Fund Scheme (GRF) and Early Career Scheme (ECS) grants received and the total grant value won, receiving consistently around or above 40 per cent of all grants allocated to these disciplines.

The above results indicate that construction R&D investment in Hong Kong has been experiencing a vigorous and flourishing period, such that both the level of research funding and the number of projects have increased under the support of the RGC. However, further potential for future development of construction research investment still exists through greater collaboration with the industry.

Case studies

The following section presents two cases studies to further analyse the impact of R&D investment in the construction industry of Hong Kong. The first case[1] presents a successful collaboration between industry and university, which was unusual in that Sun Hung Kai Properties (SHKP) complemented university laboratory and computing facilities by making available to the researchers facilities at buildings, for testing research theories. The second case[2] was a project approved by the Research Grants Council, which was the main source of funding for academic research in the local higher education sector.

Industry–university research collaboration

In 2005, SHKP provided a donation of HKD25 million (USD3.2 million) to the former Faculty of Construction and Land Use (currently the Faculty of Construction and Environment) at the Hong Kong Polytechnic University (PolyU). The aim was to enhance collaboration with the tertiary institutions and increase the competitiveness in technology innovation and construction practice, encouraging practically oriented research for the industry in general. Five separate research projects had been selected, one of which was named *Enhanced Energy Efficiency for Office Building A/C System*. This was a case based on collaboration between industry and academic personnel. The studies were not confined to university laboratories and

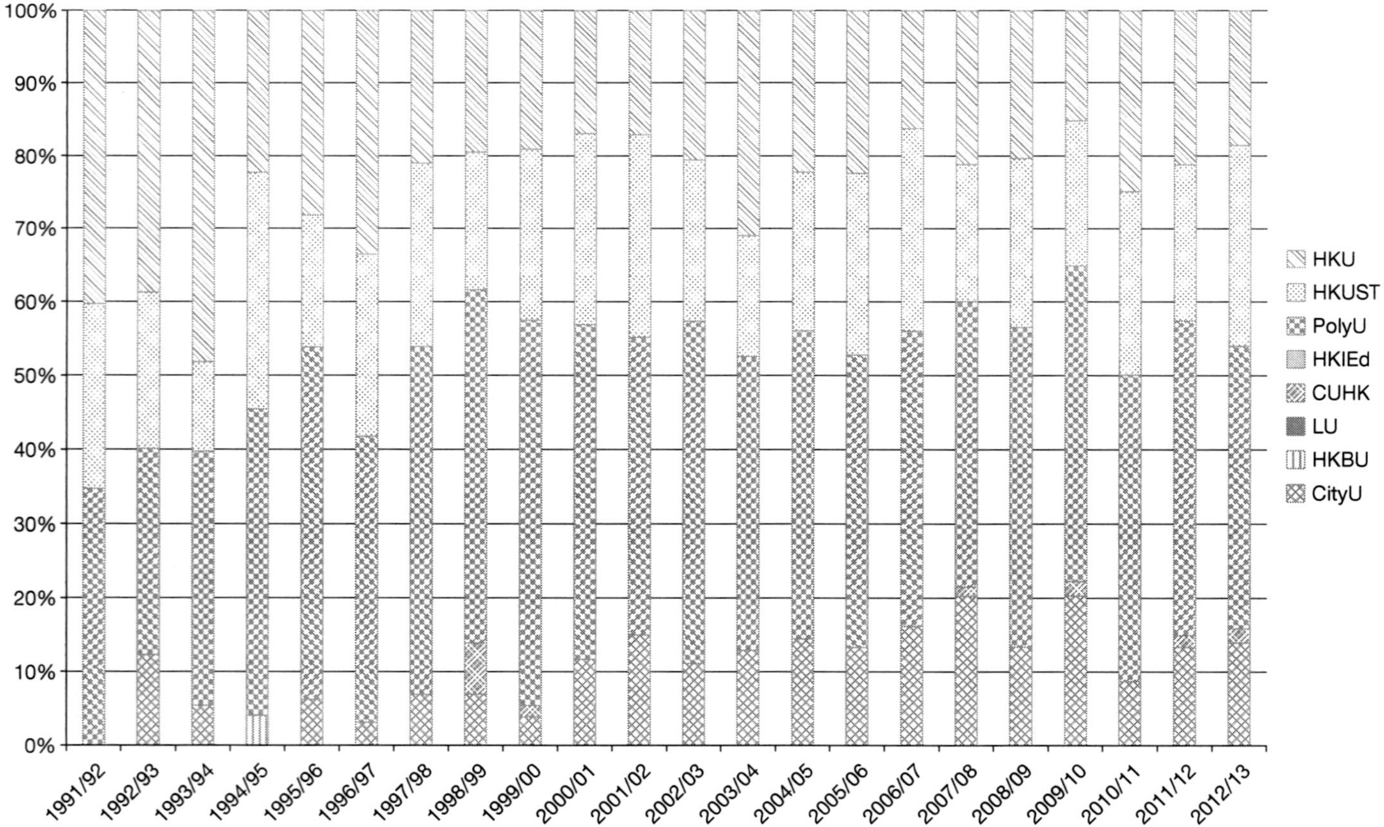

Figure 10.7 Distribution of CESBC in the eight UGC-funded institutions, 1991–2012, Hong Kong

Source: Hong Kong Research Grants Council (1994–2011)

SHKP provided actual buildings as *test beds*. In addition, SHKP technical staff were intensively involved.

The hypothesis underlying this programme was that practically oriented research moves more quickly, and is *more to the point*, when academics and practitioners choose to work together.

The PolyU research team had diagnosis and optimisation expertise in simulating building heating, ventilation, air conditioning and refrigeration (HVAC&R) systems, and optimising plant configurations and controls. Using research-based control systems and theories, the team proposed to save energy while maintaining internal comfort by using system diagnosis and optimisation throughout the building's lifecycle. This essentially lab-based expertise was applied to the International Commerce Centre (ICC) building to validate the academically derived control systems and theories and to understand the level of energy savings possible in practice.

After five years of continuous effective collaboration with the client, mechanical and electrical (M&E) consultant, contractor and others, the project had:

- saved 7 million kWh of energy per annum (approximately 18 per cent of the air conditioning chiller plant energy consumption) by optimally redesigning the plant system configuration and the system controls, a saving that is worth approximately HKD6 million (USD0.77 million) per year. This achievement derived from detailed system simulations subsequently validated on site;
- produced and transferred software for online optimal control, which interfaced with the given building automation system and communicated with monitoring instruments throughout the building;
- implemented demand controlled multi-zone ventilation, which was not initially part of the specification;
- allowed the ICC building to be upgraded to *Platinum* grade from *Gold* under Hong Kong Building Environmental Assessment Method (HKBEAM) certification through reduction of energy consumption: four credits; reducing peak demand: one credit; and use of advanced control technologies: one credit.

During the work, many detailed engineering system changes were made. Two installed pumps became redundant and although not originally provided by the design, users of the building can now exercise local ventilation control. Apart from the savings in energy costs, the project reinforces SHKP's reputation for sustainable environmental development policies and its commitment to social responsibility. SHKP staff confirmed that the ICC savings in plant energy consumption due to the advanced control technologies are of direct monetary benefit.

To summarise, the collaborative model of this research programme shows that technical progress can be rapid when both researchers and industry practitioners collaborate, giving researchers access to real *test bed laboratories*. It is important to continue to enhance this kind of collaboration between industry and academy in order to maintain the technologic status of Hong Kong and a long-term successful future.

Best practices for project briefing

Many design and construction problems in the development cycle can be tracked back to the initial project briefing process. Research has identified that the briefing phase of construction project development can be problematic (Shen, 2008). Various studies have indicated that more effort is needed to specify user requirements and expectations. Professor Shen and his team, from the Faculty of Construction and Environment of the Hong Kong Polytechnic University, have used findings from the Research Grants Council to identify key problem areas in a series of five studies. Additionally, they have created tools that assist clients and professionals in the building industry responsible for the design, construction and management of projects.

This work has resulted in two guidebooks on value briefing (Yu, *et al.*, 2006). One has been written for professional and technical staff, whereas the other is designed for executives of relevant companies in the development process. The guides can be used as a checklist for those who are new to facilitative briefing, and as reference documents for those who will lead and manage the briefing process. The research created benchmarks for best practice in this area and provided guidelines and directions for both researchers and practitioners.

The research explored the potential of a best practice framework to systematically identify and precisely represent client requirements, demonstrating that value management is a good framework for such a purpose.

A PhD student was successfully trained over the duration of this research project. The resultant thesis has received a number of accolades including the Innovation Award from the CIB Student Chapter of the Hong Kong Polytechnic University. In terms of the achievements from this research, numerous papers have been published in refereed journals and international conferences, with one receiving a Highly Commended Paper Award from the Emerald Literati Network.

The future of R&D investment in Hong Kong

R&D investment in Hong Kong can be summarised as follows:

- Government-oriented investment subsidised by UGC and RGC dominates research funding.
- Research expenditure has increased in each UGC-funded institution according to its financial report, while the total share of subventions from UGC experienced a decrease as a percentage of GDP.

- Funds from industry and other private sources are still limited and present a small fraction of the total funding.

Based on these characteristics, a set of strategies and actions to improve existing research funding systems should be designed in order to achieve better performance and higher efficiency. A renovated strategy would lead to higher quality and more efficient research.

As the case study shows, under the guidance of Government policies, collaboration between public institutions and private companies should be enhanced to provide financial support for institutional research as well as for theoretical bases for company practice.

The share of expenditure of educational funding as a fraction of regional GDP is a comprehensive indicator that reflects the support of a nation or region to public education. The proportion of educational funding from developed countries is stable at around 5 per cent annually. However, R&D investment from the UGC in Hong Kong accounts for only 0.89 per cent of regional GDP, illustrating the large potential for future development.

It has been suggested that the current funding system may be changed to incorporate more performance-oriented allocation, meaning that the funds allocated will be linked to the research outputs of each institution. Such suggestions and strategies could reform the R&D investment system in Hong Kong, provide solid and well-rounded foundation for the continuous development of the higher education sector to achieve greater impact and recognition, and serve as a source of innovation for the community as well as facilitate the sustainable development of this sector to meet the increasing demands of our society.

Notes

1 This case study is based on an executive report for construction industry leaders issued in 2012 but not publicly available.
2 This case study is based on Shen (2008).

References

Hong Kong Research Grants Council (1994–2011) *Annual report 1994–2011*, Hong Kong: Research Grants Council.

Innovation and Technology Commission (2012) *The innovation and technology fund statistics of approved projects*. Available at: www.itf.gov.hk/l-eng/ Stat View101.asp (accessed 16 November 2012).

National Bureau of Statistics of China (2012) *China statistic yearbook 2012*, Beijing: China Statistics Press.

Shen, Q.P. (2008) 'Best practice for project briefing', *Newsletter of the Research Grants Council of Hong Kong, China*. Available at www.ugc.edu.hk/rgc/ rgcnews14/west/06.htm (accessed 20 December 2012).

University Grants Committee (2012a) *Research funding*. Available at: www.ugc.edu.hk/eng/rgc/fund/fund.htm (accessed 15 November 2012).

University Grants Committee (2012b) *The RGC report of collaborative research fund from 2000–2011*. Available at: www.ugc.edu.hk/eng/rgc/result/other/other.htm (accessed 15 November 2012).

University Grants Committee (2012c) *The RGC report of public policy research – list of projects funded from 1st–10th round*. Available at: www.ugc.edu.hk/eng/rgc/result/other/other.htm (accessed 16 November 2012).

University Grants Committee (2012d) *The RGC report of theme-based research scheme – first round exercise funding result*. Available at: www.ugc.edu.hk/eng/rgc/theme/results/trs1.htm (accessed 15 November 2012).

University Grants Committee (2012e) *The RGC report of theme-based research scheme – second round exercise funding result*. Available at: www.ugc.edu.hk/eng/rgc/theme/results/trs2.htm (accessed 15 November 2012).

University Grants Committee (2012f) *The UGC report of expenditure of UGC-funded institutions as a whole 1997–2003*. Available at: http://cdcf.ugc.edu.hk/cdcf/searchStatisticReport.do;jsessionid=0DF784FE5A4ED445095F4E74FFB63CDA# (accessed 11 December 2012).

University Grants Committee (2012g) *The UGC report of expenditure of UGC-funded institutions as a whole 2004–2010*. Available at: http://cdcf.ugc.edu.hk/cdcf/searchStatisticReport.do;jsessionid=0DF784FE5A4ED445095F4E74FFB63CDA# (accessed 11 December 2012).

University Grants Committee (2012h) *The UGC report of key statistics on UGC-funded institutions from 2002–2006*. Available at: http://cdcf.ugc.edu.hk/cdcf/searchStatisticReport.do;jsessionid=0DF784FE5A4ED445095F4E74FFB63CDA# (accessed 13 November 2012).

University Grants Committee (2012i) *The UGC report of key statistics on UGC-funded institutions from 2006–2011*. Available at: http://cdcf.ugc.edu.hk/cdcf/searchStatisticReport.do;jsessionid=0DF784FE5A4ED445095F4E74FFB63CDA# (accessed 13 November 2012).

Yu, T.W., Shen, Q.P., Kelly, J. & Hunter K. (2006) *A how-to guide to value briefing*, Hong Kong: Hong Kong Polytechnic University.

11 India – R&D scenario in the construction industry

Arun Kashikar

The national context

India is poised for a massive upturn in economic and social growth. It is also on the path to becoming a technology-driven superpower in the coming years. For India to derive maximum growth through science and technology (S&T), its S&T fundamentals have to be strong and excel. Investments in research, until recently, lagged behind that of China, the European Union and the USA. The Indian Government has made concerted efforts to drive investments in S&T and this is reflected in the *XI Five Year Plan (2007–12)* (Delloitte, 2011). This plan includes ambitious programmes and covers:

- substantial increase of support to basic research
- enlarging the pool of scientific manpower
- strengthening S&T infrastructure
- implementing selected national flagship programmes that have a direct bearing on the technological competitiveness
- establishing globally competitive research facilities and centres of excellence.

The plan includes a fourfold increase for education in relation to the previous plan. It also targets growth of R&D as a share of gross domestic product (GDP) from 0.9 per cent in (2011) to 1.2 per cent by 2012. The Indian innovation system is still in its *nascent stage*: 75–80 per cent of the domestic R&D is undertaken by the public sector, 20–25 per cent by the private sector and close to 3 per cent by universities (Delloitte, 2011).

R&D projects present the particular challenge of often having significant uncertainties in regards to the return on investment. Investment in R&D may not always produce immediate returns. The Indian Government motivates R&D activities by creating an environment that offers:

- growth of a knowledge-based economy through effective fiscal policies, incentivising investments in R&D, tax and other benefits
- enhanced mobility of population and increased talent pool

- cooperation and interaction between academic institutions and industry
- protection of intellectual property rights (IPRs) with the right balance between protection and mobility of knowledge (Delloitte, 2011).

The present level of investment in India for R&D in the S&T sector is 0.88 per cent of GDP, of which 74 per cent is contributed by public sector. According to the Government, this is *similar to the public investment in most other countries*, in the range of 0.7 to 1.0 per cent. However, the private investment in R&D as percentage of GDP in India is 0.23 per cent, which is less than many developed and emerging countries in the world. The Government aims to increase the total R&D expenditure from the current 0.88 per cent to 2 per cent of GDP by the end of XII plan period (Press Information Bureau, 2013).

R&D institutions in India

In the Indian construction industry, several institutions, some with a significant heritage, continue to undertake significant developmental work both across the broader industry and in specific niche areas. The following provides an overview of some of these institutions.

Construction Industry Development Council (CIDC)

CIDC was established jointly by the Planning Commission of India and India's construction industry in 1996, with the main objective of advising the Government on policy formulation related to the construction industry. The council facilitates interaction and networking with international organisations to promote emerging technologies and best practices, in addition to developmental work such as: standardising construction contracts and procedures; training at the skilled worker and construction management levels; devising mechanisms for workers' welfare; and creating an environment that ensures equality of opportunity for all Indian contractors (CIDC, 2013).

Central Building Research Institute (CBRI)

The CBRI, established in 1947, has had the responsibility of *generating, cultivating and promoting building science and technology in the service of the country* (CBRI, 2013). The institute assists the building construction and building material industries to find time- and cost-effective solutions to the problems of materials; *rural and urban housing; energy conservation; efficiency; fire hazards; structural and foundation problems; and disaster mitigation*. Information technology (IT) also serves to disseminate the results of research to the broader community and transfers the developed technologies to the industry for further commercialisation (CBRI, 2013).

The following technologies were developed by CBRI and are commercially used:

- Clay fly ash bricks: achieve 30 per cent saving in fuel in firing compared to traditional clay brick technology and are suitable for manual and semi-mechanised production;
- Coir-cement board: panels of coir and cement conforming to BS 5669 – Part 4 (1989) and ISO 8335 (1987) have low thermal conductivity, and better sound insulation and fire resistance. This is a cost-effective alternative to timber, particle boards and fibre boards, suitable for walling, door panelling, windows, partitions and false ceilings;
- EPS door shutters: developed as substitute for wood, environment friendly, conforms to ITADS: 3-1992 IS:4020-1994. These shutters have excellent sound and thermal insulation and are lighter than wooden door shutters;
- Phosphogypsum: CBRI developed the process plant to produce phosphogypsum complying with the Indian standard IS:12679-1989. This is a substitute for high purity natural gypsum. Phosphogypsum is then used in various applications such as *rapid wall systems* used for load-bearing panel of prefabricated houses, making of gypsum plaster and in cement manufacture.

Many more technologies such as a sand lime brick plants, gypsum calcination plants, lime kiln pollution control devices, vertical shaft lime kiln, and gypsum binder are developed and commercialised by CBRI, which have led to 93 Indian and two international patents.

Building Materials and Technology Promotion Council (BMTPC)

BMTPC, established in 1990, has the objective to bridge the gap between researchers and industry to promote large-scale application of new building materials and technologies. The main function of BMTPC is to help the industry to up-scale proven technologies, to enhance their widespread use and to assist commercial production. It also facilitates systematic dissemination of appropriate technologies for the benefit of the construction industry and different sections of the population (BMTPC, 2013). This Council strives to:

- promote development, production, standardisation and large-scale application of cost-effective innovative building materials and construction technologies in the housing and building sector;
- promote new waste-based building materials and components through technical support and encourage entrepreneurs to set up production units in urban and rural regions;
- develop and promote methodologies and technologies for natural disaster mitigation and management, and retrofitting/reconstruction of

buildings including disaster-resistant design and planning practices in human settlements;

- provide S&T services to professionals, construction agencies and entrepreneurs in selection, evaluation, up-scaling, design engineering, skill upgrade, and market technology transfer, from research to practice, in the area of building materials and construction.

Institute of Steel Development & Growth (INSDAG)

INSDAG was established with the objective to promote steel construction in India. This institute is a not-for-profit organisation with more than 600 members comprising organisations, institutions, associations and professionals. The institute primarily works towards the development of technologies for steel usage and marketing of steel in the construction industry. Some of its roles are:

- creating awareness among potential users about affordability and benefits of steel
- providing a prompt advisory service on materials, construction practices, interpretation of codes of practice and creating an environment for better usage of steel by acquiring and disseminating knowledge about best practices
- upgrading the skills of the work force by organising refresher courses/training programmes and offering better technical knowhow, design aids and teaching aids
- communicating the benefits of steel compared to other competitive materials (INSDAG, 2013).

Central Institute of Plastics Engineering & Technology (CIPET)

This institute is located in Chennai, Tamil Nadu, and is devoted to training and technical services for the plastics and allied industries (CIPET, 2013). R&D projects undertaken here include the development of: mechanically and thermally stable biodegradable plastic composites; high-performance thermoplastics and thermosetting nano-composites; and polyolefin-based nano-composites.

National Council for Cement and Building Materials (NCB)

This council was established in 1962 as the Cement Research Institute of India and re-designated as National Council for Cement and Building Materials in April 1985. NCB is an apex body dedicated to continuous research, technology development and transfer, education and industrial services for the cement and allied building material industries. The council carries out research, technology development and transfer, education and industrial services (NCB, 2013).

Central Road Research Institute (CRRI)

CRRI was established in 1948, with an objective to carry out R&D in roads and transportation-related areas. The institute provides technical and consulting services to various user organisations in India and other countries. Since 1962, this institute has also organised national and international training programmes, including continuing education courses to disseminate R&D findings (CSIR, 2012). R&D areas undertaken by CRRI include:

- traffic engineering and transportation planning
- pavement engineering and materials including flexible pavement, rigid pavement and pavement evaluation
- geotechnical engineering including ground improvement and landslide investigations
- road development planning and management including GIS-based network planning and master plan for rural roads
- bridges and structures including bridge design, investigation, instrumentation and rehabilitation measures (CSIR, 2012).

Research Designs and Standards Organisation (RDSO)

RDSO was established in 1957, by integrating the Central Standards Office (CSO) and the Railway Testing and Research Centre (RTRC), under the Ministry of Railways and is located at Lucknow. RDSO functions as the technical advisor to railway board/zonal railways and production units and performs the following functions:

- development of new and improved designs
- development, adoption and absorption of new technology for use on Indian railways
- development of standards for materials and products specially needed by Indian railways
- technical investigation, statutory clearances, testing and providing consultancy services
- inspection of critical and safety items of rolling stock, locomotives, signalling and telecommunication equipment and track components (RDSO, 2013).

Case studies

Government-led programme: national science and technology extension programme in innovative building materials and housing in India

India has always had an acute housing shortage, especially for the poor who can hardly afford the cost of conventional building materials

(CBRI, 2009). However, technology has been available for the production of alternative building materials, which could be used to build low-cost houses in rural and semi-rural areas. The limiting factor has been the lack of effective mechanisms to transfer the existing technology to all sectors of the population. In response to this issue, the Government of India launched the *5-Year Innovative Building Material and Housing Action Plan* in 1990.

Responsibility for the plan was given to the CBRI and several national agencies. Two main tools were used to enable local poor people to acquire the skills and knowledge to set up a production unit for low-cost residential construction materials and then use them to build residential dwellings in their local area:

- Trainers were trained throughout the country to assist the locals, especially women and unskilled youth, to develop skills and knowledge in production and construction.
- Decentralised production facilities and housing construction demonstrations were set up to show how to produce and use the materials.

The deployment of the plan led to: *employment generation, especially for the poor and women, decentralisation of production, and development of human resource* (CBRI, 2009).

An effort was made to establish the demonstration units at sites where the beneficiaries had funds to start their own production units or houses. However, units were also established at sites sponsored by collaborating organisations and agencies. By incorporating a training programme and decentralising production 30,000 locals were trained and over 6.3 million person days of employment generated. In addition, the projects gave first-hand experience of building low-cost housing to thousands of masons, builders, social workers, contractors, engineers and architects. It also *brought about a new awareness among the people about low cost housing in the country* (CBRI, 2009).

The action plan (1990–1995) was designed to *involve mass participation to disseminate the innovative technology as widely as possible among the rural and semi-urban population, to achieve a high multiplier effect* (CBRI, 2009). The action plan had the following key features:

- priority to the rural sector; greater than 50 per cent
- benefits for castes, tribes, women and the poor
- direct participation of the beneficiaries
- appropriate structural changes for income distribution to the beneficiaries
- shift from *large, capital-intensive and urban-based to small-scale, rural-based industry with focus on employment and production for consumption by the masses.*

The aim was to have mass participation by setting up 100 demonstration units and six integrated training programmes annually. This was accompanied by awareness campaigns through mass media, get-togethers and exhibitions. Over the five-year programme, a total of 34,000 demonstration units, costing INR860 million (USD16 million), ranging from a small production unit or hut to a cluster of as many as 2,000 dwellings were set up in rural or semi-rural areas across 18 states of India. The savings achieved in the demonstration units was INR80 million (USD1.5 million), and less use of scarce building material such as cement and steel (CBRI, 2013).

Although it is difficult to assess the total impact of the S&T extension programme on the poor, it is clear that the impact goes beyond building materials and housing.

Industry-led programme: R&D in Tata Housing

Scarcity of labour and high demand for affordable housing also encourages the private corporations and developers to engage in R&D, leading to reduction in manual labour, construction cost and time. Tata Housing is one of the major organisations undertaking R&D to develop affordable housing throughout India (Tata Housing, 2013).

One of the initiatives undertaken by Tata Housing is the use of reinforced concrete hollow block masonry construction. This is an improvement introduced over conventional load-bearing wall construction, which was used in the last decade but is now largely replaced by reinforced cement concrete (RCC) framed construction throughout India.

Conventional load-bearing construction was replaced due to two issues: limited strength of bricks depending on local clay available; and the limited earthquake resistance of the buildings constructed using bricks with conventional methods. Recent advancement has made it possible to manufacture high-strength concrete blocks, which can replace conventional clay bricks. This enabled Tata Housing to keep the wall thickness to 200 mm even for seven storey buildings. Introduction of reinforcement and ductile detailing of building made it one of the best systems for earthquake-resistant structures (IS 4326, 1993, Code of Practice).

This technology has many advantages, namely:

- Customer acceptance is not an issue, as it looks and feels like the more conventional technology.
- There is a significant cost saving in the order of 15 per cent, in addition to time saving when compared to conventional construction methods.
- Specialised labour is not required as it is an adaptation of the conventional load-bearing structure used in India for many years.

Fifty five-storey buildings are being constructed using this technology at Tata Housing's Shubh Griha project located at Ahmedabad. Similarly, the use of *lost in place shutters* (*Plasswall*) for walls reduces construction time and saves cost of formwork in addition to saving cost for plaster and improving finish quality. The building constructed using this technology has shear wall slab system. Six row houses at Tata Housing's project Shubh Griha-Boisar are constructed using this technology

Use of *bio-enzyme* in internal road construction is another innovation introduced as a result of R&D undertaken at Tata Housing. Most of the projects being developed by Tata Housing are township projects and these projects have a network of permanent internal roads. In addition to these permanent roads, temporary roads are required during construction to cater for construction vehicle movement. Normally, these construction roads get damaged during construction activity and therefore have to be completely rebuilt. Subgrade is an integral part of the road pavement structure as it provides support to the pavement from beneath. *Bio-enzyme stabilisation* is one of the innovative and eco-friendly methods that can be used for stabilising the subgrade. Bio-enzymes are liquid products of fermentation of organic matter. The Bio-enzyme alters engineering properties of soil such as the capacity to bear loads.

This technology has many advantages such as:

- saving of 15–20 per cent in construction cost
- improved performance of roads, resulting in reduced maintenance cost
- all layers of road except the riding surface can be constructed before the start of construction activity at site. This, after being used as construction road for the entire construction period, can be topped with a riding surface to convert into permanent roads for the township.

Internal roads at Shubh Griha project at Ahmedabad are constructed using this technology.

Future focus

Until recently, construction in India was carried out mainly by manual labour due to the availability of low-cost construction workers, predominantly migrants from rural India. The productivity of this model has been very low compared to international norms, due to lack of focused training attributable to the migratory nature of labour and their low education level.

India faces an increasing affordable housing shortage of approximately 18.78 million units (NBO, 2011). It is impossible to meet this demand using the manual construction techniques used so far. Furthermore, the cost of manual construction labour is increasing exponentially, mainly due to various successful government schemes that have provided livelihoods to workers in their place of origin.

This reduction of the migrant workforce, the resulting need for mechanisation, and foreign investment in India's technology industry due to the global economic recession, has triggered significant R&D and innovation activities in India. Future R&D in India will be more focused on precast and prefabrication technologies, as well as the improvement of labour productivity, as means to meet the considerable housing shortfall. About 10 precast plants are under different stages of completion in India at different locations.

The combination of significant water, fuel and energy shortages, and India's commitment to climate change initiatives, will also drive R&D towards sustainability and climate change initiatives. In fact, investment in green energy technology has recently started to grow at a fast pace. R&D in the construction industry in India is likely to see exponential growth in the coming years.

References

BMTPC (2013) *Building materials & technology promotion council*. Available at: www.bmtpc.org/ (accessed 25 March 2013).

Bureau of Indian Standards (1993) *IS 4326, Earthquake resistant design and construction of building – Code of practice*, New Delhi: Bureau of Indian Standards.

CBRI (2009) *S&T extension programme. A national S&T extension programme in innovative building materials and housing in India*, Roorkee: Central Building Research Institute. Available at: www.cbri.res.in/index.php?option=com_content&view=article&id=97&Itemid=99 (accessed 25 March 2013).

CBRI (2013) *Central building research institute*. Available at: www.cbri.res.in/ (accessed 25 March 2013).

CIDC (2013) *Construction industry development council*. Available at: www.cidc.in/new/ (accessed 25 March 2013).

CIPET (2013) *Central institute of plastics engineering & technology*. Available at: http://cipet.gov.in/ (accessed 25 March 2013).

CSIR (2012) *Central road research institute*. Available at: http://crridom.gov.in/ (accessed 25 March 2013).

Deloitte (2011) *Research & development expenditure – a concept paper*. Available at: www.deloitte.com/assets/Dcom-India/Local%20Assets/Documents/Whitepaper_on_RD_expenditure.pdf (accessed 25 March 2013).

INSDAG (2013) *Institute for steel development & growth*. Available at: www.steel-insdag.org/ (accessed 25 March 2013).

NBO (2011) *Report of the technical group on urban housing shortage (TG-12) (2012–17)*, Government of India, Ministry of Housing and Urban Poverty Alleviation, National Building Organisation. Available at: http://nbo.nic.in/Images/PDF/urban-housing-shortage.pdf (accessed 25 March 2013).

NCB (2013) *National council for cement and building materials*. Available at: www.ncbindia.com/ (accessed 25 March 2013).

Press Information Bureau (2013) *Status of scientific research in the country government of India*, New Delhi: Ministry of Science and Technology.

Available at: http://pib.nic.in/newsite/PrintRelease.aspx?relid=92924 (accessed 25 March 2013).
RDSO (2013) *Research designs & standards organisation*. Available at: www.rdso.indianrailways.gov.in/ (accessed 25 March 2013).
Tata Housing (2013) *Tata housing*. Available at: www.tatahousing.com (accessed 25 March 2013).

12 The Netherlands – innovations in the Dutch construction industry

Geert Dewulf, Emilia van Egmond and Masi Mohammadi

Introduction

Innovation is generally considered to contribute to the competitive success of firms in many different ways (Tidd & Bessant, 2009). Due to the increasingly shorter lifecycle of products and services, firms are urged to innovate at a faster pace. Additionally, construction firms are faced with the need to innovate in order to succeed in a more competitive market and to respond to the increased societal pressure to meet the demand for sustainable and cleaner industrial production, higher quality output, lower costs and higher added value. Despite its many cross-industry relations, the construction industry is a tradition-based sector and lags behind in innovativeness when compared to, for example, the manufacturing industries. Interactions throughout the construction supply chain are ad hoc, project based, and knowledge is protected due to perceived uncertainties and vulnerability to risks (Egmond & Mohammadi, 2011). This forms a barrier to learning and knowledge exchange of R&D findings and project experiences that is detrimental to innovation (Franco, *et al.*, 2004).

Public clients, such as the Dutch Highways Agency, and Government programmes, such as those focused on sustainable construction systems, start with pilot projects and experiments. The envisaged effects of these appeared limited. Evidence shows that these pilots are characterised by much freedom and few regulations as well as that these institutional factors have a negative impact on the adoption pace in the industry.

This chapter will discuss the barriers to develop a mature R&D investment climate in the Dutch construction industry. The chapter describes two different cases. The concept of strategic niche management (SNM) will be used to explain the barriers for the emergence of sustainable research and investments programmes. SNM is a concept from business management literature describing system transition that aims to develop a transition pathway towards a new sustainable system that constantly improves (Geels, *et al.*, 2010). Building on this framework, this chapter will explain how the characteristics of the Dutch construction industry and the Dutch institutional SNM impede R&D, and learning from R&D and innovative projects and programmes.

The chapter is structured as follows: first, the Dutch construction industry and R&D policy will be described. Second, the framework for analysis will be explained. Thereafter, the two cases will be discussed. One case study will focus on R&D and the barriers for product innovation, while the other case study will focus on a public–private R&D and innovation programme illustrating the institutional context and barriers for sustainable R&D investments in construction. Finally, conclusions and lessons learned for R&D investments and innovative programmes based on the Dutch cases will be presented.

The Dutch construction industry

The main features of Dutch construction are very similar to any national construction industry worldwide. Its importance to the national economy is not a subject for debate. Its contribution to gross domestic product (GDP), fixed capital formation, Government revenue and employment is generally significant. In terms of production output, the construction industry proves to be one of the largest industries.

The construction industry also contributes to a country's standard of living, by means of delivering housing, business productivity and public services such as education, transport and healthcare (OECD, 2009).

The industry involves a wide range of stakeholders, notably supplying industries such as building materials, building equipment and machinery, architecture and design, and information technology (IT) (Figure 12.1). Furthermore, construction is characterised by a client's leadership, as customers have significant bargaining power over suppliers in the value chain (Egan, 1998).

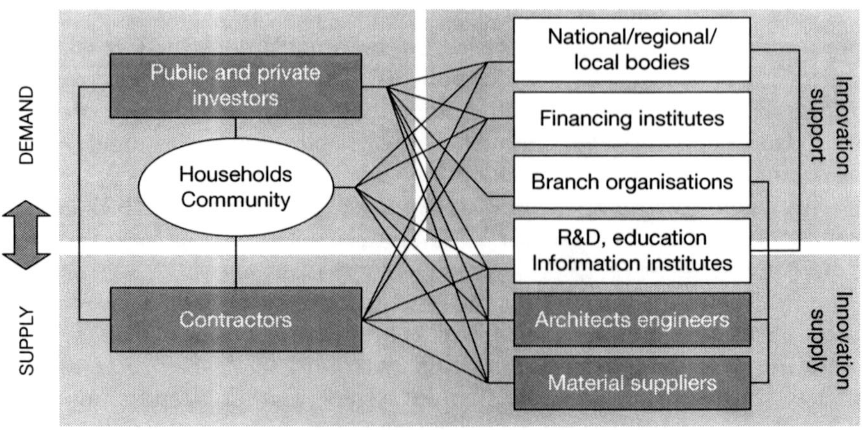

Figure 12.1 Innovation system for the construction industry, Netherlands

Source: Egmond (2007)

The recent credit crisis and economic downturn have had serious implications for the construction industry in all the countries of the European Union (EU). This can be noticed in the decline of output by 14.2 per cent between the first quarter of 2008 and the third quarter of 2009 as well as the job loss in the sector: the EU27 employment index for construction decreased 8.8 per cent between the first quarter of 2008 and the second quarter of 2009 (Eurostat, 2009).

The construction industry represented 11 per cent of the gross national product (GNP) just before this crisis. Also in terms of employment, construction-related jobs constituted 9.5 per cent of the total employment in the Netherlands (Centraal Bureau voor de Statistiek, 2011). The contribution of the Dutch construction industry to GDP decreased to 5.5 per cent in 2010, while its contribution to employment decreased to 6.5 per cent (Centraal Bureau voor de Statistiek, 2011). Despite a slight economic growth of 1.5 per cent in the second half of 2011 compared with the second quarter of 2010 and an increase in the number of jobs of 0.3 per cent, the whole economy is still experiencing the burdens of the substantial downturn and business confidence has remained low since the crisis started in 2008. Construction-based investment remains low, with about 9 per cent less building permits issued in 2011 as compared to 2010, thereby reaching its lowest level since 1953 (Centraal Bureau voor de Statistiek, 2011).

The overall economic situation challenges the production sectors in the country including construction. R&D investments could help in solving problems faced by the construction industry. R&D and innovation activities, namely, the development of new technologies, products, production processes and knowledge, take place in constant cycles because of emerging new society needs due to changes in rules and regulations, populations, technology or other developments (Egmond, 2009). It hereby has to be noted that these activities do not always result in achieving the targets of sustainable development, since the diffusion, acceptance, adoption, and implementation of these seem to be hampered. Diffusion is accomplished through human interactions and communication between members of a social system (Egmond, 2007).

Research and development investments in the Dutch construction industry

Generally, construction is considered to be not at all innovative. Based on the R&D intensity, number of people employed in R&D, and the number of highly skilled professionals employed, the sector is classified by the OECD Frascati Manual as a low-tech industry (OECD, 2002). Furthermore, in recent years there has been no significant statistical evidence of growth of labour productivity in construction. The reason given for this is the lack of standardisation in the industry, which may make productivity growth difficult to achieve (OECD, 2009). Despite the fact that new building products and materials are being developed, many publications report that the diffusion of new technologies is progressing slowly in the construction industry.

Construction is found to be deeply embedded in local laws, regulations, institutions and in long-established professional practices. These are important causes behind the underutilisation of technological opportunities from R&D investments. Conversely, the construction industry can also be perceived to be innovative due to its very creative nature, unique and seldom identical location, and client bound projects that require tailored solutions. Construction provides a fertile environment for R&D investments within projects. Nevertheless, it is a fact that its characteristics are often not well reflected on traditional statistics and R&D investment indicators such as patents, publications and R&D personnel. At the same time, there is an underestimation of productivity growth due to the highly diverse nature of this industry (OECD, 2009).

In the Netherlands, only 0.22 per cent of construction turnover is spent on R&D investments compared to the 3.6 per cent in capital intensive industries and 1.7 per cent in labour intensive industries. Additionally, the proportion of knowledge workers, both academics and R&D personnel, is rather low (de Bruijn & Maas, 2005). In general, it can be stated that Dutch R&D investments are rather modest from an international perspective, at just 1.7 per cent of GDP. This can be attributed to the economic structure in the Netherlands, with a relatively large commercial service sector and inherently limited R&D intensity (Eurostat, 2008).

R&D investments and the inclination to innovate

In general, much of the R&D investment occurring in construction tends to be incremental. Many innovations are process oriented and occur on site, mostly through problem solving on the work floor such as innovations in construction process logistics, health and safety arrangements, and labour and work planning and management. However, many of these solutions are not recorded as innovations. These innovations stem mostly from contractors and are not preceded by R&D activities. The contractors' role in innovation is limited, being responsible for only 11 per cent of all innovations, and restricted mainly to process innovation. However, the contractors' innovativeness seems to be increasing lately (Pries & Dorée, 2005). Construction firms generally are more inward looking, focusing on projects and project control, thereby attempting to improve their technology and related processes (Gann & Salter, 2000). The main motivation of contractors to innovate is to improve productivity and to secure the continuity of their business based on lowering operational costs. This is reinforced by high competition in the construction market, which is price and cost driven, with many small and medium-sized enterprises (SMEs) producing similar products with similar technology and similar materials (Gann & Salter, 2000). Pries and Dorée (2005) found that only 25 per cent of innovation was carried out *in response to specific market demands*. This is underpinned by Egmond and Mohammadi (2011) who reported that customers' demands are still seldom taken into consideration in construction.

However, construction suppliers do invest heavily in R&D (de Bruijn & Maas, 2005). The supplying industry produces 65 per cent of all registered innovations, and approximately 80 per cent of all product innovations. Major construction suppliers are in the top 20 leading R&D firms. In 2009, AKZO-Nobel invested EUR48 million (USD66 million) in R&D and had 490 R&D employees; DSM invested EUR233 million (USD323 million) and had 1,460 R&D employees; and Corus Netherlands invested EUR76 million (USD105 million) and had 487 R&D employees (Boeters, 2010). Due to the high upfront investments in machineries and equipment, supplying firms are R&D oriented. The construction industry consists of *supplier-dominated* firms (Pavitt, 1984), which is characteristic for traditional industries such as textiles and furniture. Following Pavitt (1984) such firms primarily innovate by acquisition of machinery and equipment. Construction forms an important market for specialised machinery and equipment suppliers thanks to its relative large size in economic terms (OECD, 2009).

Nevertheless, a large number of product innovations have been developed in this industry by suppliers and property owners. For example, various industrialised building systems were developed after the Second World War for reconstruction in the Netherlands such as steel-based framework systems for social housing and large-scale precast systems for housing projects. The latter gained a market share of approximately 50 per cent during the 1960s (Pries & Dorée, 2005). During the last decades of the twentieth century new materials and other building systems were developed such as stacking systems with blocks and bricks like the sand lime blocks.

A diversification in construction output has also taken place during the last decades, due to various new building concepts resulting from R&D investments, mainly from real estate developers. Lately, the increased awareness of sustainability has stimulated measures to improve the use of building materials, leading to the substitution of the traditionally used building materials by innovative *eco-materials* in more than 50 per cent of the built houses, and inclusion of products such as water-based acrylic binders, recycled PVC rainwater pipes, water-saving toilets, and water-saving showers (Klunder, 2002). A major focus in technological innovation for sustainable construction has been on residential construction (Egmond, 2010).

Yet the impact of these R&D investments by suppliers in the Netherlands remained limited, since most product developments can be characterised as innovation by addition to existing products, although they do not create a tangible difference in the traditional way of working in construction (Lichtenberg, 2009).

Despite the relatively low R&D investments by individual construction firms, innovations do take place in the industry but often as a result of a joint effort. Pries and Dorée (2005) found that 50 per cent of all innovations have been established in collaboration between firms. However, Government support is found to be an absolute condition for these efforts to improve Dutch construction and to create loyalty to sustainable construction (Egmond, 2010). This is in line with MacMillan and McGrath (1997) who state that Governments play an important and central role in supporting innovation, through, for example, building regulations and building codes.

Public R&D supporting schemes

Environmental and sustainability aspects of production processes, and more specifically construction processes, are considered of pivotal importance in Dutch policy making. Many of the current newly introduced regulations are aimed at boosting investment in eco-friendly infrastructure and the built environment. After promoting and stimulating the improvement of building practices for the benefit of sustainability since the 1990s, the Dutch Government decided to shift to a market approach. In 2004 the market *picked up* the ideas of sustainable building, while the industry considered sustainability a collective responsibility of the Government (van Hal, 2002). However, to support the recovery of a construction industry that was severely affected by the current financial crisis, the Dutch Government decided in 2011 to make EUR5.9 million (USD7.6 million) available for innovation programmes carried out in a collaborative effort between a minimum of 15 and a maximum of 35 SMEs, as an effort to boost innovation. This is in line with EU policies, which strongly focus on SMEs (Subsidies voor het MKB, 2008).

Several Government subsidies have been additionally directed at stimulating energy efficiency improvements, with approximately EUR1.2 billion during 2008–2011. At the same time, building regulations and strong building codes were set for cost efficiency over the lifetime of the measures. Energy performance certificates for buildings are mandatory and sanctions for non-compliance were reinforced in 2012. These are implemented within the framework of the EU directive on energy performance of buildings. Furthermore, through the *Energising Development Programme*, the Government supports international efforts to stimulate the adoption of higher efficiency alternatives to fuel-based lighting in off-grid communities (Pasquier & Saussay, 2012). Energy efficiency has become a key policy issue during the last few years in the Netherlands and rather ambitious energy-efficiency improvement targets have been set, such as the target of all new buildings being energy neutral by 2020. As indicated in the Dutch building code 2007, this will be done gradually: in 2011 the energy performance coefficient for residential buildings was reduced from 0.8 to 0.6 and is expected to be 0.4 by 2015. The aim is to reduce energy use by 50 per cent in all new buildings (Bowbesluit, 2007).

Other schemes are more specific for a certain sector, for example, targeted at social housing including reductions of value added tax (VAT) for new social housing and the construction of public buildings, and fiscal deductions for energy-related work (OECD, 2009). These schemes also include funding opportunities for R&D activities.

However, the bulk of Government packages are not necessarily aimed specifically at the construction industry. At present the larger schemes do not consider the construction industry itself but are focused on innovation in nine top sectors: *energy, logistics, creative industries, high-tech industries, chemical industries, horticulture, water, food and life sciences* (Human Capital in Topsectoren, 2011). Some of these categories also include key suppliers of new technologies and materials that might enhance innovation in construction. A CIOB survey indicated that innovation outside the industry indeed helps to induce innovation within construction. For example, in the UK construction industry, 11 innovations in information and communication technology (ICT) products were considered to have the greatest impact on construction innovation (Dale, 2007).

However, it remains unclear to what extent the Governments' schemes and related R&D have had an impact on innovation and industrial renewal. The Dutch construction industry has proved to have the capability to deliver difficult and innovative projects. Still, the construction industry needs to be improved in order to contribute more to the enhancement of sustainability and socioeconomic growth in the Netherlands. Manseau and Seaden (2001) stated that most of the available public policy instruments to support innovation have not been of great use to the construction industry. Barriers to innovation and industry renewal will be discussed through case studies in the following sections

Case studies

Institutional characteristics of the Dutch construction industry have a significant effect on R&D investments and their impact. The characteristics of the network of actors, including the professional practice and regulators, determine the speed of the innovation process. The technological regime, defined as the pattern made of knowledge, rules, regulations, expectations and assumptions, guides the development of innovations (Nelson & Winter, 1982).

In this chapter, the concept of SNM is used to analyse the barriers to the emergence of sustainable research and investment programmes in the Dutch construction industry. Innovations are then seen as technological niches that offer opportunities and promising results, but still are undersupplied in and by the market because of high uncertainty, high *upfront* costs or because the technical and social benefits are insufficiently valued in the market (Egmond & Mohammadi, 2011).

Only innovations that are compatible with the existing technological regime will be adapted. Regime shifts are therefore often needed to make innovations sustainable in the long run. Changes in rules and procedures but also changes in culture, habits and attitudes, are needed to improve the emergence of sustainable research and investment programmes. To illustrate this, we present two case studies. The first case is an example of a nationwide programme to enhance process innovation in the construction industry. The second case describes a programme to enhance product innovation in a major part of the construction industry: the steel structure industry.

PSIBouw revisited: programme for system and process innovation in the Dutch construction industry

During the early 2000s the business integrity of the industry has been called into disrepute as a result of a Parliamentary investigation on tendering practices. The poor image of the Dutch construction industry and an atmosphere of distrust were thereby reinforced. Many solutions have been proposed by various commissions, advisory bodies, branch organisations and research institutes. The general idea of these solutions entailed that the key to alleviate these problems is to be found in R&D and innovation as well as culture change amongst stakeholders. For this purpose, PSIBouw, a scheme for process and system innovation in the construction industry, was initiated in 2004. The aim was to improve and change the industry through process and system innovation in close collaboration with the major stakeholders in the Dutch construction industry: clients, contractors, suppliers, consultants, research institutes and universities (PSIBouw, 2004).

The Dutch *PSIBouw programme* had a budget of EUR34 million and consisted of a network of construction firms, research institutes, and engineering firms. The PSIBouw initiative was launched after the Parliamentary Inquiry on collusion in the construction industry in 2003.

This section is largely based on a historical reflection by two members of the PSIBouw scientific core team (Dewulf & Noorderhaven, 2011). From its beginning, PSIBouw had struggled with the optimal balance between science and practice. The original structure was criticised by the industry for its emphasis on scientific research. A more *demand-driven* and *top-down* structured programme was later proposed and finally developed. The idea was that the industry would decide on the agenda and researchers would then execute it. However, the focus was on *quick wins* and did not result in innovative sustainable solutions. Moreover, the approach led to an even wider gap between academia and practitioners and it soon became apparent that a new approach was needed.

The final programme was the result of an iterative process. As a result of many discussions within and outside PSIBouw, scenarios of the future of the construction industry were developed. Based on these scenarios, a set of themes were developed by the board, the project office and the scientific core team. For each specific theme a community of practice was organised consisting of academic and industry leaders. These communities of practice were further responsible for developing a research agenda and for stimulating proposals for innovative research projects and practical experiments. The research proposals were evaluated by the scientific core team based on scientific quality and innovativeness, and by the project team based on value for practice and financial soundness.

The programme entailed a large variety of projects aimed at developing and implementing novel process approaches in construction. At the end of the programme, the scientific core team carried out a self-evaluation depicting various mechanisms that hindered or drove the emergence of this large R&D programme (Dewulf & Noorderhaven, 2011). The following section offers a short summary of main lessons learned.

Intrinsic motivation of academics and practitioners

A first important lesson is that all actors in the construction industry need to see clear benefits in collaboration. Collaboration between clients, construction firms and academics is needed to develop a learning atmosphere that enables new ideas and implementation of innovations. Collaboration within the PSIBouw network was often seen as a necessity to raise funds for research rather than an intrinsic motivation to develop the collaboration needed to enhance the implementation of research findings. Developing a relationship-based network of scientists and practitioners in PSIBouw also proved to be difficult because of the scepticism of construction practitioners and their lack of a learning attitude.

Short-term goals of the industry

The experience within PSibouw revealed that practitioners were primarily interested in short-term applied research. Often, a profound debate took place between researchers and scientists about what an adventurous and innovative project it was. This lesson confirmed a more general observation made about the Dutch construction industry: the reluctance to invest in long-term innovative research. The focus of the industry on *quick wins* hampered the development of innovative solutions that might only have an impact in the long term.

A focus is needed

PSIBouw wanted to be a programme for the whole industry and the interests of all parties concerned needed to be served. There was a strong pressure not to discourage participation by sectors or categories of participants from different sectors. Hence, whenever possible, initiatives were accommodated within the programme. Moreover, the way PSIBouw emerged is a clear example of the Dutch *polder* model. As in politics, the innovation programmes in the Netherlands can be characterised as consensus programmes satisfying the requirements of a multitude of actors. Recent initiatives by the Government have slowly moved away from this consensus idea, primarily forced by decreasing public budgets and the need to create competitive advantage.

Accountability rules hinder innovation

The Programme Office, acting as the operational office, although consisting of seasoned practitioners, in relative terms lacked referent power. The authority of the Programme Office was based mostly on the power to decide over resources or reward power, subject to approval by the board. Over time, the Programme Office also accumulated considerable expertise in the field of change management, thus increasing their basis of power. However, the expertise of the Programme Office, as well as that of the board, did not extend to the scientific domain, making it difficult for either of the two bodies to exercise legitimate authority over the complete PSIBouw network, including the research part. The scientific core team, by way of contrast, lacked legitimacy in the domain of practice. Moreover, PSIBouw had to work within an administrative system dictated by the conditions of the funding scheme, and this was based on institutionalised distrust. The emergence of the programme had to take place within a context of rather rigid rules. We may then conclude that a truly innovative programme, as PSIBouw aspired to be, is not well served with a regime of *upfront* definition of end results and strict control of activities, as dictated by the Government subsidy regulations.

Case study: innovation in the steel structure conservation sector (SSCS)

This case describes the analyses of the innovation system, R&D and innovative efforts in the production chain of steel structure conservation (SSC). The data were presented during the 2008 CIB conference in Dubai (Egmond, *et al.*, 2008). Drawing on innovation theories, the concepts of innovation, innovation system and actor network were used to explore:

- the building blocks of the innovation system of the steel structure conservation sector, a subsector of the construction industry
- the R&D and innovative efforts of the system
- the mechanism of innovation, namely the processes of creation, diffusion and implementation of new knowledge embodied in products and production processes, by the production chain actors of the SSCS.

The actor network is seen as the organisational structure in which R&D and innovation takes place by the network actors who combine different sets of knowledge and technological skills, and is ruled by the technological regime at the meso level and the institutional infrastructure at the macro level. The innovation system boundaries are defined by including the agents in the actor network which are:

- directly involved in a production chain, in this case the SSC production chain
- branch organisations
- knowledge institutions and organisations
- policy makers and regulating actors.

The approach led to multi-con R&D, measurement of innovation performance of firms and organisations, and to identify factors that impact the innovation system processes. This includes analyses of:

- the composition and structure of the actor network: functional relations between actors
- the institutional infrastructure: macro-level contextual factors codified in laws and regulations
- the current knowledge and technology sets in the innovation system
- the R&D and innovation performance of the SSC production chain actors
- the technological regime of the firms and organisations directly involved in the production chain of SSC.

The results of the first three analyses provided insight to the macro- and meso-level setting in which R&D and innovation at firm level takes place. The analyses at macro and meso level formed the basis for further exploration of the R&D and innovation performance of the firms and organisations involved in the SSC production chain and the features of the technological regime that have an impact on the innovation trajectory. Data were collected by means of literature review and expert interviews. A problem and disadvantage of analysing the innovation system at these levels is the broad range of knowledge and technologies that exist in the system, which make it impossible to analyse them in depth (Carlsson, *et al.*, 2002).

Thus, the decision was made to focus on the current knowledge and technology domains in the SSCS. These analyses took place at firm and organisational levels. R&D and innovation performance was defined in this research as the extent to which a firm or organisation has been able to bring competitiveness and sustainability to their professional practices by means of R&D and innovation.

A questionnaire was used in this part of the research. The key questions addressed referred to features of the technological regime, particularly professional practices, during three phases of the innovation process: creation, diffusion and implementation of new knowledge and technology. The questions related to the firms' and organisations' expectations, awareness, strategies and commitment to R&D and innovation through budget and staffing. The mechanisms used by the firms and organisations in the process of creation, diffusion and implementation of new knowledge and technologies in the production chain were also addressed. Examples of these mechanisms were: dedicated R&D; collaboration with other actors; project evaluation; literature and document reviews; conferences; meetings; publications; training; and the resulting proactive innovation-based relations between organisations and firms in the innovation system. The questionnaire was sent to a representative sample of firms and organisations operating in the production chain of SSC in the Netherlands, classified as suppliers, contractors and public and private clients. The questionnaire was validated by comparing it with similar surveys used for innovation studies at firm level such as the SMILE Innovative Audit of the Eindhoven Centre of Innovation Studies (ECIS), developed by Prof. J. van Aken, TU Eindhoven in 2001 and later used by Gras and Wijffels (2006). The questionnaire was tested and adapted to the comments from representatives of the branch organisations of firms involved in the SSCS. A total number of 56 respondents equally divided among the three classes of actors participated in the research.

The Dutch SSC actor network

The set of distinct institutions which jointly and individually contribute to R&D and innovation in the Dutch SSC innovation system are explored in this section (Figure 12.2). The three major players in the production chains of SSC that are involved in the primary processes and that collaborate, often in an ad hoc fashion, in projects based on contractual agreements that define the quality and sustainability of an envisaged steel structure are:

- suppliers: paint manufacturers and suppliers of basic chemicals and additives, responsible for new developments regarding the conservation product quality;
- contractors: steel structure builders and metal conservation firms, responsible for the conservation process and steel structure quality,

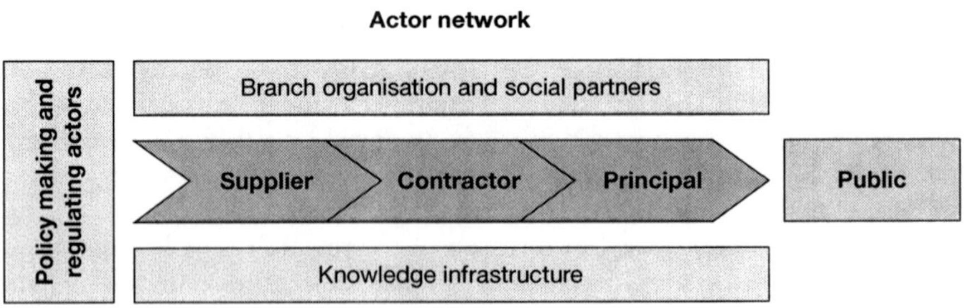

Figure 12.2 SSC actor network, Netherlands

and that have to rely on steel and conservation material producing companies;

- public and private clients, having an important role in defining the quality requirements in the contractual agreements.

The production chains are composed by integrated tendering for engineering and construction, and design and build contracts, on the basis of functional specifications by clients and selected on a best price–quality ratio. Specialist knowledge institutions such as consultancy engineering firms, R&D institutes and universities may support the production chain actors in certain cases.

Branch organisations promote the economic interests of the individual firms and organisations in the SSC production chain at national and EU levels via lobby networks, thereby acting as a connecting link between the individual firms and policy-making institutions (PSIBouw, 2005). Policy-making and regulating institutions such as various ministries, Governmental agencies, and EU institutions, formulate and implement policies to intervene in the innovation processes.

The Dutch SSC institutional infrastructure

The innovation system of the Dutch SSCS is subject to policy ambitions, institutional and social pressure, particularly reflected in the laws and regulations regarding tender procedures; occupational health and safety, and environmental protection; transport; sustainable entrepreneurship; and international norms. The construction industry's response on these matters is based on three cornerstones: innovation, transparency and price–quality ratio (Regieraad Bouw, 2005).

Dutch SSC technology and knowledge sets

The following knowledge and technology domains are mentioned to be important and relevant to achieve sustainable SSC:

- project specification and commissioning;
- steel structure design and engineering;
- conservation products;
- conservation processes;
- environmental conditions in which the production is taking place (PSIBouw, 2007).

Bottlenecks in the SSC innovation system relate to:

- knowledge of corrosion-conscious design;
- knowledge and practices regarding maintenance-monitoring systems;

- knowledge supplying institutions and organisations that do not meet the knowledge needs in the production chain;
- late involvement of consultants in new projects;
- poor formulation of inspection assignments during maintenance phase of the structure;
- required knowledge for sustainable and competitive SSC is not exchangeable, fragmented and still underdeveloped (COT, 1998; NCC, 1996).

Important developments in the SSCS, which require additional and different knowledge sets, are:

- increased project execution and delivery based on functional specifications;
- improved knowledge base among line executives;
- increased outsourcing;
- extension of the range of duties of inspectors and supervisors.

R&D and innovative performance of the Dutch SSC production chain actors

The innovative performance during the three phases of the innovation process of the SSC production chain actors, namely R&D/creation, diffusion and implementation, was measured on a scale of one to four, based on the responses to the questionnaire. The respondents represent all firms and organisations operating in the heavy steel structure conservation sector.

The innovative performance in the total production chain of SSC is rather moderate (Table 12.1).

The best scores across the production chain are observed in creation of new knowledge while the diffusion of knowledge in the production chain presents the lowest scores. The innovative performance differs between the various actors in the production chain. These findings point to suppliers as a major source of innovations. The innovative performance of the public clients scores relatively high. The level of diffusion among contractors and

Table 12.1 R&D and innovative performance in the SSC production chain

	Supplier	*Contractor Conservation*	*Contractor Steel structure*	*Private client*	*Public client*	*Total*
Creation	3.2	2.4	2.2	2.5	3.1	**2.8**
Diffusion	3.2	2.0	2.0	2.0	2.6	**2.5**
Implementation	3.0	2.4	2.1	2.4	2.6	**2.6**
Innovative performance	3.2	2.3	2.1	2.4	2.8	**2.6**

private clients is low. Also implementation lags behind among contractors and private clients.

The Dutch SSC technological regime

Table 12.2 provides an overview of major regime features that impact the R&D and innovative performance of suppliers, contractors and clients in the SSC production.

The following additional remarks can be made: a promoting regime feature for the innovative performance in the production chain is the importance attached to novelties by the actors in relation to their competitive position. Suppliers (100 per cent of respondents) and contractors (83 per cent of respondents) that have indicated to attach importance to the relevance of novelties for their competitive position, appear to have a higher innovative performance score. The same higher innovative performance score is found under suppliers and public clients who have indicated that the sustainability of the SSC, for example, through improved safety and

Table 12.2 Major influencing technological regime features on innovative performance

	Creation	*Diffusion*	*Implementation*
Promoting features	• Awareness of technological and social developments • Orientation towards innovation collaboration • Insight on own competences • Budget for innovation	• Management commitment to facilitate diffusion • Participation in meetings for innovation • Evaluation of innovation failures	• Awareness of a positive impact of innovation SSC sustainability • Orientation towards other application opportunities of novelties • Evaluation of impact of novelties • Positive expectations regarding new type of contracts; functional specification • Awareness of the pressure of current laws and regulations
Constraining features	• Limited consideration for sustainability requirements • Limited relations with knowledge institutions	• Limited combination of knowledge sets, such as marketing and engineering • Limited knowledge exchange among production chain actors on innovation failures • Lack of journal publications	• Limited opportunities to test novelties in real practice • Rigid specifications in traditional contracts • Limited involvement of clients in the implementation of novelties

health conditions and less negative environmental impact, has increased due to their innovation efforts. In contrast, the overall innovative performance level of steel structure contractors is low while they also mentioned a low sustainability increase. This is not surprising given that they generally are the risk-bearing main contractors working with low profit margins in traditional project contractual agreements.

The experience level of the leading roles of the firms and organisations in the SSC production chain is rather high: 69 per cent had over 15 years of experience. On the one hand, this may favour R&D and innovativeness in the SSC production chain. On the other hand, however, it may imply that the contractors in particular are reluctant to abandon familiar practices.

The percentage of the annual turnover dedicated to innovation is, for 18 per cent of the respondents, more than 3 per cent of the annual turnover and between 0 and 2 per cent of the annual turnover for the remaining 82 per cent. Ten per cent of the steel structure contractors and 16 per cent of the private clients indicate that the percentage to be spent on innovation will decrease; 33 per cent expect it to remain the same; and 40 per cent expect an increase. Furthermore, 70 per cent of the suppliers, 50 per cent of the conservation contractors and 36 per cent of the public clients expect an increase.

Conclusions

Increased social pressure imposed by institutional infrastructure developments, particularly in the area of environmental protection and occupational health and safety and intensified competition, have led SSC innovation system actors to carry out R&D and innovate to achieve a higher quality and sustainability of output by the production chain against lower costs and a higher added value. Developments in the institutional infrastructure highlight the stronger requirements formulated in laws and regulations regarding product design and engineering and production processes.

The social pursuit to achieve higher quality and sustainable output is reflected in the translation of the laws, regulations and norms in a requirement for an integral lifecycle approach and quality specifications in construction project contracts. The functional relations in the actor network are still ad hoc to a certain extent and not strong. Production chains are project based and composed by means of integrated tendering on the basis of functional demand specifications and a selection on best price–quality ratio.

There is a lack of central direction and promotion of the common interests of the actors in the SSC innovation system. This has counterproductive consequences for the stimulation of innovation. Quality of production output is determined by taking sustainability of steel structures as criteria, however, in this respect there is no uniformity. There is a need for uniform criteria for sustainability of products, processes and services in the SSC innovation system.

Knowledge and technology sets needed for sustainable and competitive production in the SSC innovation system are available but fragmented among different actors, not exchangeable and still require partial development.

Suppliers in the SSC production chain are the most innovative, followed by the public clients. An important point of improvement of the innovative performance of contractors and private clients relates to the diffusion of knowledge. A number of major suppliers in the production chain of steel structure conservation belong to the top 20 leading R&D firms in the Netherlands, such as AKZO-Nobel, DSM and Corus Netherlands. In 2009 Philips invested EUR707 million and had 4,269 R&D employees, and Unilever EUR149 million and employed 1,190 R&D workers (Boeters, 2010).

The improvement of competitiveness and sustainability appeared to be an important driving force for innovation.

The major bottlenecks in the technological regime of the SSC production chains are detrimental to the innovative performance and in particular to the diffusion of knowledge in the actor network. These include:

- limited focus on sustainability requirements
- limited communication, knowledge exchange and combination of different knowledge sets such as marketing and engineering in firms and organisations, through cross-industrial collaboration with knowledge institutions and clients
- lack of academic publications
- limited opportunities to test novelties in real practice
- rigid specifications in traditional contracts.

Given the case study findings and relying on evidence from the manufacturing sector, it can be stated that sustainability and continuity in the form of longer term cross-industrial collaboration based on transparent performance measurement and improved targets will benefit clients and entrepreneurs in the SSCS. New knowledge and the resulting improved operational performance will be favourable to both clients and contractors in the whole construction industry. However, the whole society will also benefit. For example, in the case of the SSC, the transport sector benefits from a higher quality physical infrastructure.

More attention should be given to sustainability of operations and economies of scale by a transition from project towards a portfolio approach. This needs the development and implementation of different forms of cross-industrial collaboration and contractual agreements between actors. Entrepreneurs will be motivated to invest in product innovation, output quality improvement and process organisational innovation if quality and innovations are highlighted in more projects simultaneously. The role of the Government in this respect should be that of the public client collaborating and stimulating R&D and innovation in construction projects as well as that of a supporting actor for R&D.

The above calls for a change in the mutually related building blocks of the innovation system: actor network; technological regime and institutional infrastructure; and technology and knowledge set. Finally, this may entail a total system innovation.

Final remarks

A variety of innovations have taken place within the construction industry in terms of process optimisation and advances in technology that can be observed in various areas, such as materials, engineering, transport and equipment, domotics, robotics and management. However, literature and results from the two case studies indicate that construction remains inefficient, labour intensive, and slow to accept, diffuse, apply, and integrate new technologies and processes. Comparing the two case studies, we find that the governing reasons hampering the successful diffusion of new technologies and processes in the construction industry are as follows:

1 As compared to other industries, the construction industry has limited knowledge workers and an ineffective innovation disbursement system. This fact is confirmed by the literature review; only 0.2 per cent of the turnover in the construction industry is directed to R&D investments, which are much lower than in other sectors such as technology and care (de Bruijn & Maas, 2005). This is also in line with employment data from technological innovation. In the construction industry, this number is well below the average of other sectors (Centraal Bureau voor de Statistiek, 2011).
2 The linkages between the various actors in the construction innovation system are generally project bound and temporary in nature (Mohammadi, 2010), where knowledge exchange among the actors is rather limited and most of the prevailing knowledge is tacit. This is detrimental to the learning and combining of knowledge for the benefit of the continuity and cumulativeness of the technical advances as well as for other innovations in related areas. Evidence taken from the two case studies indicates that more attention should be given towards the sustainability of operations and economies of scale by the transition from project-based methods to a portfolio approach. This requires the development and implementation of different forms of cross-industrial collaboration and contractual agreements between actors.
3 Moreover, there is only limited willingness for knowledge sharing, due to close collaboration with competitors and lack of trust. Protection

against imitation in construction is low, which hampers the interest and opportunity of construction stakeholders to reap profits from innovative activities and take a leadership role in innovation. Without protection, it is not possible to recoup development costs. In a sector in which competitors work closely, the lack of intellectual property protection results in less collaboration. Patents appear to be an unappreciated instrument in construction as they are perceived as a source of information for competitors. This is why only 2 per cent of the total annual number of patent applications is generated by the construction industry, even though the sector contributes to more than 10 per cent of the national GDP in the Netherlands (Centraal Bureau voor de Statistiek, 2011).This translates into dysfunctional teams; poor levels of cooperation; overlapping; inefficiency; failure costs; complex coordination; and lost opportunities for the optimum use of the available resources and innovations.

4 The culture change among stakeholders is also a key issue. In essence, communications between members of an innovation system enhance the diffusion of inventions throughout society (Rogers, 1995). Theory and case studies suggest that an innovation will only be adopted and prosper if it fits in the prevailing technological regime that characterises the professional practice of the actors in the innovation system. Cooperation between actors from various organisations involved in construction projects with high rates of interdependency, lack of alignment between actors and a low level of communication and information exchange results in a scattered process and reduction of technological innovation development, diffusion and implementation.

The Dutch Government plays a major role in the actor network as a regulator and owner of a substantial portfolio of assets, the main initiator and the financier of a significant number of innovation programmes. Evidence shows that the successful diffusion of new technologies has been accomplished due to initial protection and support initiatives. This implies the strengthening of interaction and communication in the actor network. However, while the support of the Government is of extreme importance for the stimulation of further development, it also plays an ambivalent role as a catalyst for innovation in construction. On the one hand the Government encourages innovation through subsidy programmes. On the other hand, it limits the space for a successful commercial deployment and diffusion by an array of sector-oriented regulations and standards, biased policies, and the short lifespan of political decisions and innovation programmes.

To sum up, this chapter shows that developing a positive environment towards a mature R&D-driven industry requires a fundamental shift in the market, which calls for a change in the mutually related building blocks of the innovation system: the actor network; the technological regime and institutional infrastructure; and the technology and knowledge set. A major regime shift is required.

References

Boeters, B. (2010) *R&D bij bedrijven afgenomen, bij instituten toename* (*R&D conducted by firms, increase in institutes*), The Hague: Bèta Publishers.

Bowbesluit (2007) *Bouwbesluit 1992/2003/2012* (*Decrees 1992/2003/2012*). Available at: www.onlinebouwbesluit.nl/ (accessed 10 January 2010).

Carlsson, B., Jacobsson, S., Holmén, M. & Rickne, A. (2002) 'Innovation systems: Analytical and methodological issues', *Research Policy*, 31(2): 233–245.

CBS (2011) *StatLine tables*, Centraal Bureau voor de Statistiek. Available at: http://statline.cbs.nl/StatWeb/search/?Q=construction+industry&LA=EN (accessed 1 February 2013).

COT (1998) *Rapportage IOP VERF, Inventarisatie van de kennisbehoefte in de verfkolom* (*Inventory of the need for knowledge in the paint column*), Rapportnummer LB98-255. RAP.

Dale, J. (2007) *Innovation in construction: Ideas are the currency of the future, Survey 2007*, Ascot: Chartered Institute of Building (CIOB).

de Bruijn, P.J.M. & Maas, N. (2005) *Innovatie in de bouw* (*Innovation in construction*), TNO rapport EPS 2005-13, Delft: TNO Bouw en Ondergrond.

Dewulf, G. & Noorderhaven, N. (2011) *Managing public–private innovation programs*, Gouda: Psibouw report.

Egan, J. (1998) *Rethinking construction: Report of the construction task force*, London: HMSO.

Eurostat (2008) 'Europe in figures', in *Eurostat* Yearbook 2008, Schäfer, G. (ed.). Available at: http://ec.europa.eu/eurostat (accessed 1 February 2013).

Eurostat (2009) *Construction sector statistics, European Commission*. Available at: http://epp.eurostat.ec.europa.eu/statistics_explained/index.php/Construction_sector_statistics (accessed 5 February 2013).

Franco, L.A., Cushman, M. & Rosenhead, J. (2004) 'Project review and learning in the construction industry: Embedding a problem structuring method within a partnership context', *European Journal of Operational Research*, 152: 586–601.

Gann, D.M. & Salter, A. (2000) 'Innovation in project-based, service-enhanced firms: The construction of complex products and systems', *Research Policy*, 29: 955–972.

Geels, F.W., Jacobsson, S. & Hekkert, M.P. (eds) (2010) *The dynamics of sustainable innovation journeys*, London: Routledge.

Gras, B. & Wijffels, D. (2006) *Innovatiescan voor onderzoeksinstituten, Beschrijving van de scan en resultaten van toepassing bij KIWA water research, InnoQ* (*Innovation scan for research, description of the scan results and application to KIWA water research, InnoQ*), Nieuwegein: KIWA N.V.

Human Capital in Topsectoren (2011) *Alles over human capital in de topsectoren* (*About human capital in the top sectors*). Available at: www.topsectoren.nl (accessed 11 March 2013).

Klunder, G. (2002) *Hoe milieuvriendelijk is duurzaam bouwen? De milieubelasting van woningen gekwantificeerd* (*How environmentally friendly is sustainable building? The environmental impact of housing quantified*), Delft: Delft University Press.

Lichtenberg, J.J.N. (2009) 'Nieuwe benadering van bouwen' ('New approach to building'), *Facility Management Magazine*, 22(173): 29–33.

MacMillan, I.C. & McGrath, R. (1997) 'Discovering new points of differentiation', *Harvard Business Review*, 75(4): 133–145.

Manseau A. & Seaden G. (2001) *Innovation in construction: An international review of public policies*, London and New York: Spon Press.

Mohammadi, M. (2010) *Empowering seniors through domotic homes: Integrating intelligent technology in senior citizens' homes by merging the perspectives of demand and supply*, Eindhoven: Technische Universiteit Eindhoven.

NCC (1996) *Resultaten onderzoek naar kennistekorten op het gebied van corrosie en corrosiemanagement* (*Research results into knowledge deficits in the area of corrosion and corrosion management*), Bilthoven: Nederlands Corrosie Centrum.

Nelson, R.R. & Winter, S.G. (1982) *An evolutionary theory of economic change*, Cambridge, MA: Bellknap Press.

OECD (2002) *Frascati manual. Proposed standard practice for surveys on research and experimental development*, Paris: Organisation for Economic Cooperation and Development (OECD).

OECD (2009) *Responding to the economic crisis: Fostering industrial restructuring and renewal*, Paris: OECD Secretariat. Available at: www.oecd.org/industry/ind/43387209.pdf (accessed 1 March 2013).

Pasquier, S. & Saussay, A. (2012) *Progress implementing the IEA 25 energy efficiency policy recommendations OECD/IEA*, Paris: International Energy Agency. Available at: www.iea.org/publications/insights/progress_implementing_25_ee_recommendations.pdf (accessed 14 March 2013).

Pavitt, K. (1984) 'Sectoral patterns of technical change: Towards a taxonomy and a theory', *Research Policy*, 13: 343–373.

Pries, F. & Dorée, A. (2005) 'A century of innovation in the Dutch construction industry', *Construction Management and Economics*, 23: 561–564.

PSIBouw (2004) *Inventory of international reforms in building and construction*, Gouda: PSIBouw. Available at: www.psibouw.nl/#p=results:11 (accessed 9 October 2013).

PSIBouw (2005) *Projectplan professionalisering staalconservering* (*Professionalisation steel preservation project*), Documentnummer SCON-2005-068-TCE, Gouda: PSIBouw.

PSIBouw (2007) *Resultaten dialoog professionalisering staalconservering* (*Results communication professionalisation steel preservation*), Documentnummer SCON-2007-417-TCE 17, Gouda: PSIBouw.

Regieraad Bouw (2005) *Het Jaar van de fundamenten prioriteiten 2005–2006 van de regieraad bouw Rijswijk: Quantes* (*The year of the foundation priorities 2005–2006 of Rijswijk construction council: Quantes*), The Hague: Regieraad Bouw.

Rogers, E.M. (1995) *Diffusion of innovations*, 4th edn, New York: Free Press.

Subsidies voor het MKB (2008) *Subsidies voor het MKB* (*Subsidies for SMEs*). Available at: www.subsidiemkb.nl/ (accessed 11 March 2013).

Tidd J. & Bessant, J. (2009) *Managing innovation: Integrating technological, market and organisational change*, Chichester: Wiley & Sons Ltd.

van Egmond, E.L.C. (2007) *Industrialisation innovation and prefabrication in construction*, Working paper EVE/06-2007/BCC/TUE, Paris: OECD.

van Egmond, E.L.C. (2009) 'Innovation and transfer by strategic niche management', in proceedings *Global Innovation in Construction Conference 2009*, presented at Joint Conference of University of Loughborough and CIB TG71 on Research and Innovation Transfer, Loughborough, 13–16 September.

van Egmond, E.L.C. (2010) 'Conditions for industrialisation and innovation in construction', in *New perspective on industrialisation in construction: A state-of-the-Art Report*, Girmscheid, G. & Scheublin, F.J.M. (eds), Zürich: Eigenverlag des IBB and der ETH Zürich.

van Egmond, E.L.C., Heutink, A., Bancken, S. & Barneveld, A. (2008) 'Cross-industrial collaboration for innovation in construction? On innovation in the case of the steel conservation production chain', in *Proceedings transformation through construction*, paper presented at Joint 2008 CIB W065/W055 Symposium, Dubai, 17-19 November.

van Egmond, E.L.C. & Mohammadi, M. (2011) 'Innovations in domotics: Fulfilling potential or hampered by prevailing technological regime?', *Construction Innovation*, 11(4): 470–492.

van Hal, A. (2002) *Amerikaanse toestanden: Commerciële kansen voor duurzame woningbouw (American states: Commercial opportunities for sustainable housing)*, Boxtel: Æneas.

13 New Zealand – new directions for construction R&D

Suzanne Wilkinson and Charles Ma

The New Zealand construction industry

New Zealand (NZ) is a country of 4.4 million people and is typified by its green image and dramatic landscapes. NZ has all the facilities of a developed country such as well-developed infrastructure systems, relatively good housing and reasonable transport systems. The country is viewed as having high sporting achievements internationally and has multi-faceted tourism and recreational facilities. The country also has dramatic geological features and significant natural hazards, such as volcanoes, earthquakes and floods. Auckland, the most populated city with 1.4 million people in May 2013, is built on a volcanic field. The third most populated city, Christchurch with 360,000 people, experienced major earthquakes in 2010 and 2011 that devastated large areas of the city.

Buildingvalue (2013) state that:

> [T]he building and construction sector is a major contributor to New Zealand's economy, providing more than 4 per cent of Gross Domestic Product (GDP), around the same as agriculture; the sector employs one in 12 workers in New Zealand; 21 per cent of workers are sole traders; 25 per cent of workers have no qualification, and literacy and numeracy levels in the sector are low.
>
> (Buildingvalue, 2013)

Although there are large engineering consultancy businesses, architectural practices and contractors, some of which operate on an international scale, the industry is underpinned by small contractors, specialist consultants and design practices that fall into the small to medium-sized enterprises (SMEs) category. Companies that employ fewer than 20 full-time staff make up just over 96 per cent of businesses in the NZ construction industry (Wilkinson & Scofield, 2010). Businesses that employ five people or fewer make up approximately 90 per cent of the total number of firms in the sector (Buildingvalue, 2013). SMEs are often owned and operated by the same individual, and have few specialist management staff members.

The local construction industry uses all the main techniques, technologies and processes on offer for a technically skilled and developed country. However, being a country of 4.4 million people, the industry often modifies and develops tested technologies and processes from overseas. For instance, when using new delivery mechanisms or implementing new technologies, such as building information modelling (BIM), companies will often wait to see how these technologies are progressing before implementation becomes widespread. Hence, uptake of innovation in the NZ construction industry can be slow (Farhi & Wilkinson, 2008) and the construction industry is seen as having poor innovation (Ma, 2012).

The most comprehensive investigation of the attitudes to innovation in the NZ construction sector was carried out by Fairweather and his colleagues at Lincoln University in the late 2000s (Fairweather, 2010). This focused on the path to market for *inventors*, and concluded that, though there were mechanisms for dissemination of *learnings* from established research organisations, there was minimal capture and application of small-scale innovation. It recommended the establishment of a NZ Building Innovation Centre, paralleling those found in UK and Sweden.

The mindset of the sector in respect of innovation may have been altered as a result of the 2010 and 2011 earthquakes in Christchurch. The earthquakes and ongoing after-shocks have led the Government to focus on the resilience of the built environment, bringing changes to the construction industry in terms of needing to think about new building technology, productivity, skills and management. Research and development (R&D) to improve the resilience of the built environment is expanding and is one of the key features of NZ research at present.

Some of the other emerging issues for the built environment include housing quality and affordability, renewing and improving transport infrastructure and reducing infrastructure vulnerabilities. NZ instituted a changed Building Act and performance-based building codes in the 1990s that, coupled with changes to material and designs, led to the *leaky buildings* problems and to further changes to the Building Act (Duncan, 2005). Ongoing repairs of leaky buildings now underpin a significant fraction of some building SMEs' workload in cities other than Christchurch. In Auckland, housing and new infrastructure development and the improvement of existing infrastructure occupy the majority of the current programmes of work.

The R&D investment climate

There are three main aspects to investment in the NZ R&D environment: private investment, public investment and industry–Government partnerships. Various investment strategies have been developed in response to different industry, Government and education requirements.

Private sector investment

Private sector investment in construction research is: not substantial; with a focus on specific short-term projects to improve individual businesses; and often run as consulting projects for businesses. Within the national economy, local private sector investment is about half the Organisation for Economic Cooperation and Development (OECD) average, with NZ's economy as a whole being in the lower third of the OECD (MBIE, 2013a):

> In 2010 NZD2.4 billion (USD1.9 billion) was spent on development and research in New Zealand by the public and private sectors: NZD1 billion (USD800 million) by business, NZD629 million (USD510 million) by government departments, and NZD802 million (USD650 million) by the higher education sector. The goals of this investment are many: to facilitate economic development; to address national policy issues; to educate and train our people; and to investigate the world in which we live.
>
> (Carnaby, 2011)

StatsNZ (2010) suggests that 5 per cent (NZD127 million) of gross R&D was directed to *construction and transport* in 2010 and NZD109 million to construction and transport combined from business. However, it is difficult to assess these figures accurately due to the limited information on how the funds were spent. In 2010 figures for Government R&D expenditure and higher education R&D expenditure across construction and transport were undisclosed, but are likely to be lower than most sectors if the 2008 figures are used as a guide: NZD20 million in Government and NZD15 million in higher education construction and transport combined R&D (StatsNZ, 2010).

Public sector investment

Government investment in business and the economy is managed through the Ministry of Business, Innovation and Employment (MBIE), formed in 2012 by the amalgamation of the Department of Building and Housing; the Ministry of Economic Development; the Department of Labour; and the Ministry of Science and Innovation. Prior to 2012, investment in research was predominantly managed through the Ministry of Science and Innovation (MSI) though the Department of Building Housing, which was responsible for the NZ Building Code, would research code development, and the Department of Labour did some research on construction health and safety.

The new ministry, MBIE, has a focus on improving the performance of the national economy and supporting businesses, and so will increasingly target R&D funding to meet these objectives (MBIE, 2012). MBIE's aim is to *develop and deliver policy, services, advice and regulation to support business growth and the prosperity and wellbeing of all New Zealanders* (MBIE, 2013a). As the local construction industry makes up around 4 per cent of GDP, significant investment has the potential to improve the economy, but the construction industry per se is not directly targeted with MBIE funding. Research in construction is therefore found in a range of investment portfolios addressing environment, infrastructure and manufacturing. The highest funding available to construction and building researchers was in the 2012 MBIE investment portfolio, incorporating infrastructure and hazards, with emphasis on the latter due to the Christchurch earthquakes.

A recent OECD survey described NZ as an exporter of resource-based goods and services such as dairy and tourism (OECD, 2011). The NZ National Government sees R&D as a key driver for innovation, business success and economic growth (MBIE, 2013b). The NZ National Government's focus is however on growth on *high-tech* industries, biotechnology and niche manufacturing, which has led to economic initiatives and R&D funding available in these areas. Investment portfolios in these areas are significantly larger than those available for construction, although construction R&D organisations and construction businesses are able to bid into any investment portfolio where they can meet the requirements, particularly those with a focus on improvements to business services. Small amounts of public sector investment in R&D for the construction industry can be found from different sources as agencies, such as Housing New Zealand Corporation and Energy Efficiency and Conservation Authority, need further knowledge for underpinning their activities. But the sums are small compared to those available through MBIE. With limited public sector funding, private sector funding is now being aggressively targeted by universities.

Investment in innovation

Innovation features often in the Government's research investment conversations, where, for instance, elements such as the Government's *Business Growth Agenda* (Building Innovation, 2012) direct public policy for research investment. Reports, such as the 2012 *Building Innovation Progress Report*, assist construction researchers and other research groups in formulating their own research agendas. The most recent report *targets raising the amount businesses spend on R&D from 0.54 to over one per cent of GDP, and the amount of Government R&D from 0.6 per cent to 0.8 per cent of GDP* (Building Innovation, 2012), with the explicit aim of getting businesses and research institutions to work together more effectively. Although tending to be aimed at *high-tech* manufacturing, construction researchers are able to access funding and some successes have been found in the environmental fields.

New spending of NZD115 million a year in TechNZ co-funding programmes with companies is aimed at improving business competitiveness (TechNZ, 2013). TechNZ grants have been a long running feature of Government policy, supporting partnerships between businesses and research organisations, and are aimed at improving business innovation and competitiveness. Recently released information by the Minister for Economic Development, Science and Innovation and Tertiary Education, Skills and Training (Joyce, 2012) highlights investment being made across all sectors in R&D including: the development of innovation parks; Government investment in teaching engineering and science students; undertaking a review of how university research is funded to ensure it incentivises innovation by researchers; improving the settings for intellectual property; and the development of the *national science challenges*.

Avenues for investment

The main investment processes for Government funding are carried out through the following avenues:

- *contestable funding* through an annual funding portfolio mechanism, which does not invest in every portfolio every year
- *core funding* (MBIE, 2013b) for Crown Research Institutes (CRI). There are no specific construction-related CRI's directly funded through the Government core funding mechanisms. Direct Government investment in university research is through the Performance-Based Research Fund (PBRF) mechanism. Tertiary institutions with higher PBRF scores receive more funding. In 2012 the Government announced the development and funding of a new institute aimed at improving R&D, predominantly in *high-end manufacturing*, to be called Callaghan Innovation (NZD166 million over 4 years) (Callaghan Innovation, 2013). The focus of this institute is to improve the implementation of research in the market, mainly in the high-tech and manufacturing industries. It is thus likely to have limited impact on construction industry research other than where researchers might be able to work on areas of construction product improvement and construction manufacturing processes
- *On-demand funds* are usually funds targeting business improvements. Organisations can access on-demand funding if the application can demonstrate innovations and business improvements leading to better economic performance (MBIE, 2013b). This funding is suitable for construction R&D, but has been used little to date by businesses in the sector
- *Government research funding for the tertiary sector* alongside contestable funding and private funding, direct Government investment in university research is obtained through the PBRF mechanism

(Tertiary Education Commission, 2013). Every eligible academic staff member engaged in research prepares a portfolio of research evidence, assessed every 6 years, which is graded by a panel. Grades equate to funds: the institutions with more and *higher graded* staff receive more research funding.

The *Tertiary Education Strategy 2010–2015* (Minedu, 2013) further outlines Government priorities for universities and states that:

> Research in universities needs to combine excellence with impact. In particular, we will ensure that the Performance-Based Research Fund recognises research of direct relevance to the needs of firms and its dissemination to them. We will also ensure there are further incentives for tertiary education organisations, other research organisations and firms to work together.
>
> (Minedu, 2013)

Science challenges

In 2012 the Government introduced the concept of channelling a proportion of its research investment through *science challenges*, initially NZD60 million over 4 years. The aim is to use research funding to provide answers to key scientific questions that need answering. Ten challenges were announced in early 2013 with only one (*nature's challenges*) having a minor impact on the built environment. Ultimately, the aim of the *science challenges* is to have more of a contracted public science investment focus on core scientific questions that are of importance to NZ (MBIE, 2012). By mid-2013 no funding had been distributed, and the construction industry may not be well served with science challenge funding unless researchers in this field can directly relate their work to science challenge themes.

Construction sector investment bodies

Farsighted industry leaders agreed in 1969 to the imposition of a levy on new construction projects to fund knowledge development for the sector. Under the Building Research Levy Act 1969, 0.1 per cent of the contract value of every construction project exceeding NZD20,000 put forward for building consent in NZ is payable by the builder to the Building Research Association of New Zealand (BRANZ, 2013a). This act controls the use of levy funding and the purposes to which it may be put. BRANZ applies the levy in accordance with the act for, amongst other activities:

- conducting construction industry-related research and other scientific work

- allocation of grants to people, institutions or bodies carrying out construction industry-related research
- publication of information
- establishment and maintenance of a library
- holding of lectures, seminars, exhibitions and public meetings to disseminate information about building research
- provision of an advisory service.

The amount of actual levy funds collected each year varies in response to the level of building activity. This variability of the levy poses some problems for continuity of research, as any significant new investment can be difficult in lean times. BRANZ aims to ensure that it is close to the industry that it serves and uses a range of mechanisms to obtain advice regarding the allocation of funding including: the Research Agenda Guidance Committee; the Levy Allocation Guidance Committee; and industry surveys and forums. A fraction of the funding is provided to tertiary education research institutions through PhD and Masters scholarship programmes, and to researchers in contracted industry research.

The *mood* of the industry, as embodied in the Construction Industry Council (CIC), has become more favourable to research over the past decade.

Building Research Association of New Zealand (BRANZ)

BRANZ is a key investor and delivery vehicle for new knowledge to the NZ building and construction sector. It describes itself as *an independent and impartial research, testing, consulting and information company providing services and resources for the building industry* (BRANZ, 2013a). The organisation collects and invests on behalf of the construction industry the statutorily imposed *Building Research Levy*, and provides research and knowledge delivery services for the sector. Most of the levy investment is in the wholly owned subsidiary BRANZ Ltd, which has a focus on the performance and sustainability of buildings and the construction industry. BRANZ operates as an information centre and advisory service for the local industry, including the provision of industry good practice guides (BRANZ, 2013b).

Approximately 40 per cent of BRANZ Group's total income comes from the *Building Research Levy*, 5 to 10 per cent from MBIE sources, and around 50 per cent from commercially contracted research projects for private, Government and international clients (BRANZ, 2013b). Externally viewed, in many ways BRANZ Ltd operates akin to a CRI, with approximately 100 in-house scientific staff. Research is focused on that which is relevant to improving the industry, including research into improved techniques and materials for use in the building industry (BRANZ, 2013b).

Building and Construction Sector Productivity Partnership (BCSPP)

The establishment of the BCSPP is a development for R&D investment in the NZ construction industry. This partnership between industry and Government was established in 2010 through the Department of Building and Housing, now part of MBIE, to address low productivity in the local construction industry. *The Productivity Partnership aims for a productive, safe and profitable building sector, providing a foundation for strong communities and a prosperous economy* (Buildingvalue, 2013). Funding for the partnership comes from BRANZ and MBIE.

The partnership consists of the Governance Group and four workstreams that are striving towards a deep understanding of what would make the construction industry more productive. The titles of the workstreams are: skills; evidence; procurement; and construction systems. The vision of the partnership is for an improved, productive, skilled, innovative and efficient building and construction industry that contributes to increased business profitability and quality of life for all New Zealanders (Buildingvalue, 2013), with the associated goal of increasing productivity in the industry by 20 per cent by 2020. The Evidence Workstream produced the *Research Action Plan*, which develops research questions that are aimed to have an impact on improving productivity (RAP, 2012). Case studies and best practice approaches are regularly disseminated to the construction industry via media, workshops and news bulletins. As of mid-2013, funding for the continuation of the BCSPP is uncertain.

Strategies for construction research

Recently, a group representing different sectors of the construction industry released the document *Building a Better New Zealand*, developed by BRANZ, the CIC, the Construction Strategy Group (CSG) and the MBIE. The aim of the document is to improve the local built environment, with a focus on using research and innovation to lift performance. *Building a Better New Zealand* covers broad areas of construction research such as materials performance, sustainability and productivity (Building a Better New Zealand, 2013). According to the released document the vision includes:

> A building and construction industry that delivers: well designed, built and performing homes and buildings that meet the needs of all New Zealanders and their communities now and into the future; a built environment that is affordable today but that will stand the test of time; the modernisation of our existing homes and buildings so that they meet our changing needs and aspirations; a vibrant, sustainable and productive building and construction industry that is capable of

> delivering on these aspirations and plays its part in supporting New Zealand's economic wellbeing; and a highly skilled and competent professional workforce that is passionate about and committed to a building and construction industry that delivers attractive and achievable career opportunities.
>
> (Building a Better New Zealand, 2013)

Involved in developing the document were two key industry groups: the CSG and the CIC. Both aim to influence Government policy and practice. The CSG is a self-appointed group of leaders in the national construction industry who aim *to provide leadership and strategic direction to grow a productive value-driven professional construction industry* (CSG, 2013). The CIC is a meeting of the chief executives of many groups of the NZ construction industry such as master builders, trade training organisations, trade organisations and the like and aims, among other things, to *promote the interests of the broader construction industry to central Government and create conditions in which the sector can prosper. Explicit in the CIC activities is . . . representing industry interests by lobbying Government departments and politicians* (CIC, 2013). To what extent the CSG and CIC actually influence Government policy and practice is unknown. However, such lobby groups serve to raise the profile of the industry. Working collectively with other key stakeholders, such as tertiary education providers, would strengthen the message to the Government regarding the investment needed for R&D required for the sector.

Case studies

The following case studies represent typical current R&D investment. These demonstrate current collaboration between different sectors of Government, industry and academia in the NZ construction research environment

Case study: Beacon Pathway Ltd and STIC

The NZ Government is keen to ensure that there is a good relationship of research to industry need reflected in its investment programmes. One mechanism that they have used for this has been *research consortia*, whereby programmes running for up to 6 years at a budget of a few million NZ dollars per year are funded at a one to one ratio of Government to industry investment. The joint investment is managed by specifically formed companies governed by small boards dominated by industry investor appointees but including researchers. In the built environment sphere, there have been two of these research consortia created.

The first consortium, Beacon Pathway Ltd, addressed sustainability in the residential built environment, created demonstration houses, monitored the resource use and internal environments, and showed how they could *create homes and neighbourhoods that work well into the future and do not cost the Earth*. The company's investors included major product manufacturers in NZ, BRANZ, a local authority, and a CRI. This company operated from 2004 to 2010, and when its original 6-year research contract with the Government ended, it was carried on by some of the former industry stakeholders as a small research and advisory organisation. The company was highly successful in bringing industry attitudes to sustainability from being a *fringe concept* to an issue with which the residential building industry regularly engaged (Beacon, 2013).

The second consortium, Structural Timber Innovation Company Ltd (STIC), took up a licence for a patent developed by the University of Canterbury on innovative systems to create seismically resistant laminated veneer lumber (LVL) structural frames. The industry investment was principally by LVL manufacturers and forest products sectors in NZ and Australia. STIC used research teams in the University of Canterbury, University of Auckland and University of Technology Sydney, with assistance in some aspects from BRANZ and other subcontractors, to develop the system to the point of commercialisation. STIC was established in 2008, and its Government research investment ended in June 2013. This company created a demonstration building using the technology. This survived intact the Christchurch earthquake in February 2011, and proved immensely valuable in demonstrating to designers, builders and potential clients how the system could be used. The research programme has created new means of connecting large members, and publishing design guides for large clear-span portal frames and seismic design for multi-storey buildings. At the conclusion of the Government investment the company is expected to stay in place and to continue to promote the use of the technology (STIC, 2013).

Both of these companies have been very successful, helping the industry partners to take their concepts forward while still ensuring that most of the research findings have been placed in the public domain for use by all stakeholders in the creation of better built environments. The key issues in their success might be regarded as the close partnerships created between the researchers and the industry sponsors, and especially the creation of demonstration buildings. It seems that an extremely important step in obtaining uptake of innovation by the NZ construction industry is their ability to see and feel the physical manifestation, an observation that supports the findings of Fairweather (2010).

Case study: innovation

In order to understand innovation and construction, a team of researchers at the University of Auckland undertook an industry-funded project investigating how construction companies could become more innovative. The research analysed how Fletcher Building Roof Tile Group (RTG) could become more sustainable and competitive as an organisation by improving their ability to innovate. The goal of RTG is to obtain a larger share of the market, and the group is currently undergoing major organisational changes in order to bring more innovation into the business (RTG, 2013). There has been a large change in staff, particularly in the middle to upper management level, during the past 2 years. The expectation was that new staff could potentially develop a more innovative culture in the organisation. Increases in sales and process efficiency are two areas of focus for RTG, which considers that failure to innovate has led to a lack of competitive advantage.

The research sought to understand ways in which RTG could develop and implement new ideas and monitor innovation successes. Historically, the company was using an innovation stage gate process, but were now having management level, strategic discussion through the innovation café. The innovation café allowed wider participation of company employees to engage with innovation and innovative ideas through discussion and meetings. Researchers aimed to examine ways in which RTG could innovate spontaneously as an organisation to assist with elevating their innovative practices.

The research found that three main factors seem to be hampering the innovative practices within RTG: (i) funding for innovative projects; (ii) time commitment for developing ideas; and (iii) the changing culture of the business. Through a series of interviews and questionnaires, communication and company culture were identified as the main mechanisms to improve RTG's innovation processes (Ma, 2012). On further analysis, the researchers found that an innovation team would be an appropriate and accepted mechanism to address such communication and enhance the culture within the organisation. These findings are now being examined by RTG and the group is starting to introduce systems to encourage staff to generate innovative ideas and harness them as potential projects and process improvements.

Case study: Christchurch earthquake

As the recent earthquakes in Canterbury caused significant damage to the built environment, extended research funding to improve the resilience of the built environment, improve disaster recovery and assist the construction industry with Canterbury reconstruction has been made available. The *Canterbury Resourcing Project* (RecRes, 2012) is an example of a research project aiming to provide research evidence to improve resourcing the rebuilding of Christchurch and surrounding areas. RecRes is a collaborative effort between the University of Auckland, BRANZ, the *Resilient Organisations Programme* (Resorgs, 2013) and key industry players. This project is funded by the University of Auckland and BRANZ.

The research started by researching past experiences in NZ that had demonstrated that the country had an ability to cope with small-scale natural disasters (Zuo, *et al.*, 2006). However, the larger earthquakes experienced on 4 September 2010 and 22 February 2011 in Christchurch tested NZ's ability to tackle a large-scale event. Given the amount of damage to residential property, commercial buildings and infrastructure, the repair and reconstruction in the quake affected areas is challenging in terms of resources and capacity. Resource pressures including shortages of materials, components, specialised skills and professionals are likely to play out in pricing, quality and ability to deliver projects on time (RecRes, 2013).

RecRes aims: to understand the resources that are required for the repair and rebuilding of the damaged built environment; to monitor and evaluate resource situations as the repair and reconstruction proceeds on an ongoing basis; and then to inform the construction industry and recovery planners for greater Christchurch. The purpose of RecRes is to better understand the availability of resources to meet the long-term repair and reconstruction requirements posed by the built environment during the rebuilding of Christchurch, including understanding:

- What *problematic resources* during the earthquake recovery have an impact on the reconstruction process?
- Where, in terms of sectors and project site locations, do these *problematic resources* become critical; and at what stage of the project?
- How do *problematic resources* affect disaster recovery projects, in terms of construction cost, quality, and time?
- What issues constrain availability of the varied *problematic resources*, and thus obstruct repair and rebuilding from proceeding as intended?
- What would resolve and address these resource constraints from the point of view of the construction industry and disaster recovery planners?

RecRes, which has now been running for 2 years, is an ongoing longitudinal project that will significantly add to the knowledge being developed to improve future national and international disaster recovery and reconstruction.

The future

The future of R&D investment in New Zealand is uncertain. Organisations such as BRANZ are vital for improving construction research, but their ability to invest in research is limited by building industry activity. Additionally, there is no specific MBIE portfolio dealing with the built environment, which is disappointing given the contribution construction makes to the national GDP. As previously stated, private sector investment in R&D in NZ is far below that of other comparable OECD countries. New initiatives and efforts from the BCSPP, CIC or CSG, however, coupled with increased construction industry activity for the foreseeable future, will mean that extra funds could be available to improve NZ's construction R&D.

R&D is a changing entity by nature. A coordinated effort for construction R&D in NZ is needed, fronted by a reputable organisation, such as BRANZ, with consistent and long-term Government funding provided. R&D funding needs to be placed on an equal footing with funding for other industries, such as dairy and *high-tech* industries, especially given the important financial contribution to the local economy generated by the construction industry. The recent development of an agreed research agenda for the sector, *Building a Better New Zealand*, should provide a framework in which this can be taken ahead. If the NZ construction industry is to become internationally competitive and improve its productivity then significant R&D investment is required. Changes to R&D are ahead and researchers are hopeful that future R&D will be well coordinated and adequately funded.

Acknowledgements

The authors thank John Duncan (Board Secretary, BRANZ) for his comments on the chapter. However, the content is the responsibility of the authors.

References

Beacon (2013) *Beacon*. Available at: www.beaconpathway.co.nz (accessed 6 July 2013).

BRANZ (2013a) *Building research association of New Zealand, About BRANZ*. Available at: www.branz.co.nz/ (accessed 9 March 2013).

BRANZ (2013b) *Building research association of New Zealand*. Available at: www.branz.co.nz (accessed 9 March 2013).

Building a Better New Zealand (2013) *Research strategy for the building and construction sector*. Available at: www.buildingabetternewzealand.co.nz/ (accessed 8 July 2013).

Building Innovation (2012) *Building innovation progress report 2012*. Available at: www.mbie.govt.nz/pdf-library/what-we-do/business-growth-agenda/BGA-progress-report-building-innovation-august-2012.pdf (accessed 8 July 2013).

Buildingvalue (2013) *Building and construction productivity partnership*. Available at: www.buildingvalue.co.nz/ (accessed 9 March 2013).

Callaghan Innovation (2013) *Media release. Callaghan innovation*. Available at: www.callaghaninnovation.govt.nz/ (accessed 15 March 2013).

Carnaby, G.A. (2011) *The outlook for New Zealand*, President, Royal Society of NZ. Available at: www.royalsociety.org.nz/expert-advice/information-papers/yr2011/the-outlook-for-new-zealand/ (accessed 8 July 2013).

CIC (2013) *New Zealand construction industry council*. Available at: www.nzcic.co.nz/intro.cfm (accessed 15 March 2013).

CSG (2013) *The construction strategy group: Building industry value*. Available at: www.constructionstrategygroup.org.nz/ (accessed 8 July 2013).

Duncan J.R. (2005) 'Performance-based building: Lessons from New Zealand', *Building Research and Information*, 33(2): 120–127.

Fairweather, J. (2010) *Can building and construction sector innovation be improved? A review of innovation centres and their implications for New Zealand*. Available at: www.lincoln.ac.nz/PageFiles/7859/Canbuildingandconstructioninnovationbeimproved.pdf (accessed 8 July 2013).

Farhi, C. & Wilkinson, S. (2008) 'Could the NEC be widely used in New Zealand?', in *Proceedings of the ICE – Management, Procurement and Law*, 161(3): 107–113.

Joyce, S. (2012) *Building innovation report launched*. Available at: www.beehive.govt.nz/release/building-innovation-report-launched (accessed 9 March 2013).

Ma, C. (2012) 'Innovation in construction: Roof tiles case study', Unpublished student project, University of Auckland.

MBIE (2012) *National science challenges*, Ministry of Business, Innovation and Employment. Available at: www.mbie.govt.nz/what-we-do/national-science-challenges (accessed 15 March 2013).

MBIE (2013a) *Role of MBIE*, Ministry of Business, Innovation and Employment. Available at: www.mbie.govt.nz/ (accessed 9 March 2013).

MBIE (2013b) *Business growth agenda*, Ministry of Business, Innovation and Employment. Available at: www.mbie.govt.nz/ (accessed 11 February 2013).

Minedu (2013) *The tertiary education strategy 2010–15*. Available at: www.minedu.govt.nz/NZEducation/EducationPolicies/TertiaryEducation/PolicyAndStrategy/TertiaryEducationStrategy.aspx (accessed 11 February 2013).

OECD (2011) *OECD economic surveys overview: New Zealand*, Organisation for Economic Cooperation and Development. Available at: www.oecd.org/newzealand (accessed 9 March 2013).

OECD (2013a) *Unemployment comparisons for OECD countries*, Organisation for Economic Cooperation and Development. Available at: www.oecd.org/newzealand (accessed 9 March 2013).

OECD (2013b) *Statistical profile of New Zealand*, Organisation for Economic Cooperation and Development. Available at: www.oecd.org/newzealand (accessed 9 March 2013).

Parliamentary Counsel Office (2011) *Building research levy act (1969)*, Public Act 1969 No 23, Date of assent 11 September 1969, New Zealand Legislation. Available at: www.legislation.govt.nz/act/public/1969/0023/13.0/DLM391231.html (accessed 8 July 2013).

RAP (2012) *Research action plan*, Building and Construction Productivity Partnership. Available at: www.buildingvalue.co.nz/ (accessed 9 March 2013).

RecRes (2012) *Resourcing reconstruction research project*. Available at: www.recres.org.nz/about-us (accessed 12 March 2013).

Resorgs (2013) *Research programme*, Resilient Organisations. Available at: www.resorgs.org.nz/ (accessed 12 March 2013).

RTG (2013) Fletcher building roof tiles group. Available at: www.rooftilegroup.com/ (accessed 8 July 2013).

StatsNZ (2010) *Research and development in New Zealand 2010*. Available at: www.stats.govt.nz/browse_for_stats/businesses/research_and_development.aspx (accessed 12 March 2013).

STIC (2013) Structural timber innovation company. Available at: www.expan.co.nz (accessed 8 July 2013).

TechNZ (2013) *TechNZ terms and conditions*. Available at: www.msi.govt.nz/update-me/archive/frst-funding-archive/technz-terms-and-conditions/ (accessed 1 October 2013).

Tertiary Education Commission (TEC) (2013) *Universities*. Available at: www.tec.govt.nz/Funding/Budget/Budget-2012/Universities/ (accessed 15 March 2013).

Wilkinson, S. & Scofield, S. (2010) *Management for the New Zealand construction industry*, 2nd edn, Auckland: Pearson.

Zuo, K., Wilkinson, S., Le Masurier, J. & van der Zon, J. (2006) 'Reconstruction procurement systems: The 2005 matata flood reconstruction experience', paper presented at International Conference on Post-Disaster Reconstruction: Meeting Stakeholder Interests, i-Rec, Florence, 17–19 May.

14 Norway – improved governance and innovation in construction projects

Marit Støre Valen, Knut Samset, Ole Jonny Klakegg, Torill Meistad and Anita Moum

The Norwegian context

Geography and demographics

Because of its northern location at the same latitude as Alaska, Norway is often thought to have a cold and rough climate. However, due to the warm Gulf Stream of the Atlantic Ocean, Norway has a mostly pleasant and habitable climate, and ice-free harbours, even during winter. The long coastline and steep topography make it particularly prone to coastal storms. According to scenarios for climate change, more rain and incidents of strong winds can be expected during winter (Almås, *et al.*, 2011; Lisø, 2006). Warmer climate will represent a significant challenge to the Norwegian built environment given that most of Norway's population and almost all of its cities are found along the coast and fjords.[1]

Due to the large number of children being born in the years after the Second World War up until the 1960s, there is a growing population of senior citizens, with 13 per cent of the population being over 66 years of age in 2013 (Statistisk sentralbyrå, 2013a). This section of the population is generally in good health, has high expectations about services and living standards and will require a large increase in public services and investments in the years to come.

Economy, private and public sectors

The Norwegian economy is strong, although the difficult international financial situation seen in 2011–2012 has also exerted an impact on the Norwegian economy. According to OECD (2012), Norway had a gross domestic product (GDP) per capita of 36 per cent above the national average worldwide between 2000 and 2004. In 2011 Norway still had the second highest GDP per capita calculated using purchasing power parity (USD61,900), following Luxembourg (USD89,800). As a way of comparison, the USA had a GDP per capita of USD46,600 and Japan USD33,800, while the Norwegian

GDP is increasing at a faster pace than most European countries. Norway strengthened its economic position after the international *credit crunch* in 2008, though as a whole the Norwegian economy is deeply embedded in the European economy (Regjeringen, 2012).

In 2010 Norway's exports had a total value of NOK1,046 billion (USD180 billion), with oil and gas exports accounting for 46 per cent (NOK481 billion/USD83 billion). As a whole, the country has a large surplus in trade balance compared to other countries (NOK310 billion/ USD53 billion in 2010) (Statistisk sentralbyrå, 2012a). However, the mainland economy shows a moderate deficit in trade balance, and the interest rate is also historically low.

As of 2012, Norway had a total potential workforce of 2.7 million people, with an international and historically low unemployment rate of approximately 3.6 per cent (Statistisk sentralbyrå, 2013b). The local workforce is highly educated and the rate of unskilled workers is low. Consequently, Norway relies on *guest* workers and immigration to fulfil unskilled labour roles.

The public sector forms a large section of the economy, approximately 30 per cent of those employed, and public investments in infrastructure constitute half of all infrastructure investments (BNL, 2012).

In 2011 Norway used 1.6 per cent of its GDP on research and development (R&D), which placed Norway behind many countries with weaker economies in international statistics reflecting investment in research (OECD, 2013; Statistisk Sentralbyrå, 2013b) (see Figures 14.1 and 14.2).

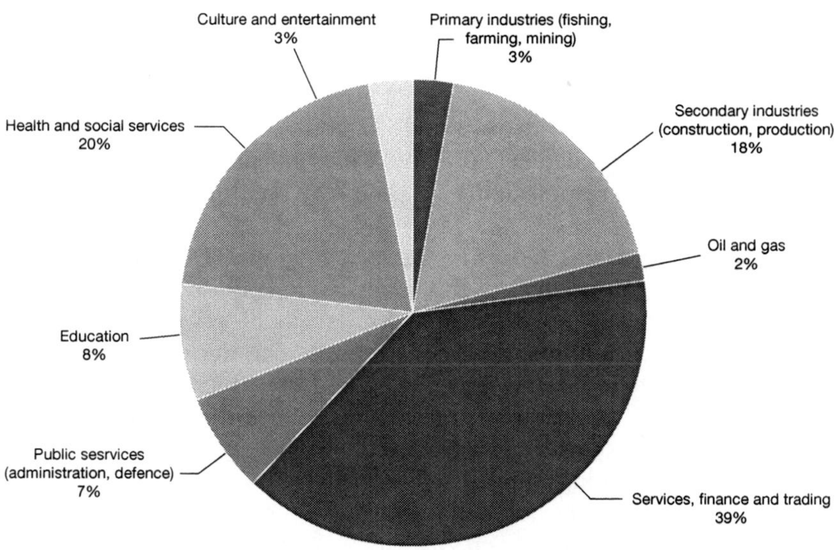

Figure 14.1 Distribution of the Norwegian workforce into different industry sectors

Source: Statistisk Sentralbyrå (2012b); BNL (2012), analysed by Klakegg, NTNU

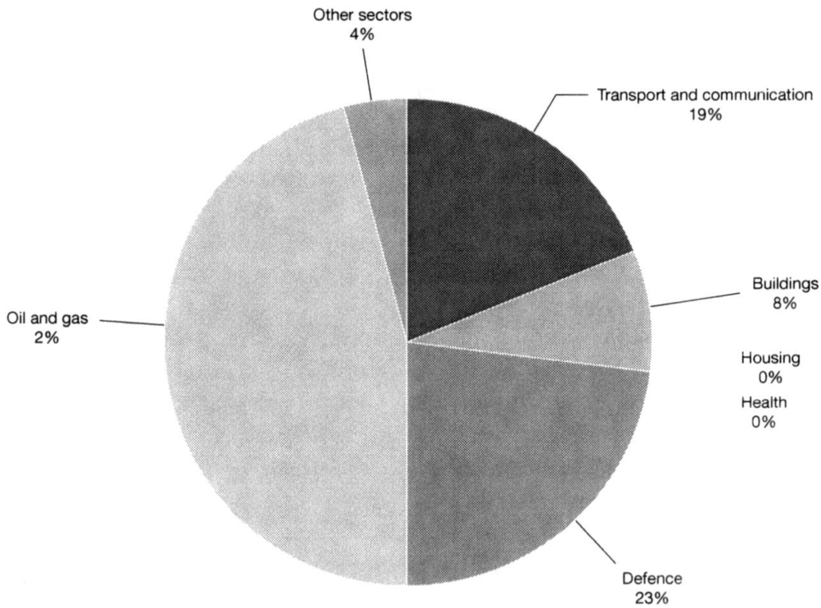

Figure 14.2 Investments in different sectors in Norway distributed among transport and communication, buildings, housing, health, defence, oil, gas and others

Source: Statistisk Sentralbyrå (2012c, 2012d), analysed by Klakegg, NTNU

Note: The construction industry is the second largest sector, employing more than 300,000 workers

Government and regulations

Norway is a constitutional democratic monarchy. The ultimate decision-making body is the Parliament and the Government is divided into 17 ministries with special responsibility for their own sector and headed by the Prime Minister.

The Ministry of Local Government and Regional Development has special relevance in the context of this book, as this ministry is responsible for Norwegian housing policy, district and regional development, local Governments and the administration of elections. This ministry is currently responsible for matters relating to housing and construction policy, rural and regional policy, local administration, local Government finances and the conducting of elections (Regjeringen, 2013a).

The Norwegian Building Authority (Regjeringen, 2013b) is a subordinate agency that is particularly influential due to its key role in construction policy implementation. This agency is the professional unit, construction competence centre and central building authority that oversees all permits required by the *Planning and Building Act* (Lovdata, 2008) for building construction, extension, large-scale renovation and purposeful change.

Permits are managed through municipal authorities to ensure compliance with the building code.

The Norwegian construction industry

Trends and challenges

Over the next 10 years, growing globalisation will promote an already increasing trend of competition among international construction companies (BNL, 2010). Additionally, Norway has to meet the following challenges:

- a growing population, expected to surpass 7 million by 2063 (Statistisk sentralbyrå, 2013b)
- an increasing trend towards centralisation will place more pressure on urban areas
- a growing elderly population with needs for health care and housing
- long cold winters and harsh climate worsened by climate change, which will lead to more floods, landslides and frequent winter storms
- more pressure on transport infrastructure
- an ever increasing immigrant workforce.

These challenges will lead to:

- high demand for new dwellings
- need for higher investment in low-energy buildings
- need for more robust and sustainable buildings and infrastructure
- need for more investment in transport infrastructure
- need for a larger workforce and recruitment in all sectors
- need for good integration programmes, development of expertise and training in relevant areas for new migrants and unskilled labour.

These challenges are similar to the challenges that the sector is facing in the rest of Europe. The following sections will expand on what has been done to meet these challenges thus far.

Research and construction policy

Several research programmes have been initiated in the building and construction industry over the last few decades. For example, *Klima 2000*, focuses on how constructions can handle a warmer climate. The *Concept Research Programme* (CRP) was initiated by the Ministry of Finance in 2002 due to several large investments projects that presented cost overruns. *Byggekostnadsprogrammet* (*The Construction Cost Programme*) was initiated by the Ministry of Local Government and Regional Development for the period 2005–2010. Lastly, there are centres conducting research within renewable and *environment-friendly energy* (FME) that are long-term research programmes (8 years) financed by the research body itself:

universities and research institutions, industry partners and the Norwegian Research Council. One such centre is the Zero Emission Buildings (ZEB) centre, which will be used to illustrate drivers for the implementation of new technology and the development of new solutions for *low-* or *zero-energy loss* housing.

Klima 2000 (2000–2007)

Concerns for implications of climate change have led to one of the largest R&D programmes within the building sector within the past decade. The purpose was to optimise the design by developing certain climate exposed construction components and features so the buildings can cope with the Norwegian climate strain. *Klima 2000* and current programme *ROBUST* have resulted in increased knowledge about the climatic impact on building components and its diffusion into the development and production of new and robust housing products such as windows and doors, and designing skills (Sintef Byggforsk, 2007).

The Concept Research Programme (CRP)

The *Concept Research Programme* was initiated in 2002 and is funded by the Ministry of Finance. The purpose is to conduct trailing research on *front-end management* of major investment projects, more specifically, on how to ensure quality at entry before the final decision to fund a project is made. Its aim is to develop *knowhow* on how to make more efficient use of resources while improving the effect of major public investments (NTNU, 2013).

This research programme was designed to explore issues within the realm of front-end management of investment projects, and to develop expertise, literature and academic training courses. The CRP is unique in that it only focuses on the *front-end phase* of a project, for which the total budget for the initial 10 years exceeded USD10 million. Approximately 40 separate studies have been completed under this programme in areas such as public management; economic analysis; project management; planning, decision making; analytical methods; risk analysis; portfolio management; contract management; the use of incentives; applied logic; and probabilistic assessment. Six textbooks and several anthologies have been produced, as well as a number of working papers and refereed scientific papers. In terms of education, the by-product has been considerable, resulting in about 20 courses with a large amount of students, about 50 master of science degrees and 10 doctoral degrees.

Byggekostnadsprogrammet (2005–2010)

This programme was initiated by the Ministry of Local Government and Regional Development and conducted in close cooperation between the construction industry, public authorities and research and education organisations. The goal was to create competence enhancement in critical areas for the users, workers and other stakeholders of the Norwegian construction industry, resulting in 39 projects and significant engagement in knowledge development and diffusion within the sector. It was suggested that this work be continued in a research and competence centre. The construction industry, represented by the associated construction organisations, research environments and large contractors and consultant companies, supported the idea. However, there was a lack of support from the Ministry of Local Government and Regional Development, which instead recommended a white paper, the first in the history of the Norwegian construction industry, as an instrument to develop a construction policy.

Project governance and major public investment projects

Project governance

In the field of project management, the focus has been on the complexity itself and the improvement of the processes and procedures involved, rather than on the governance framework that could or should give direction and help in improving the outcome of these processes. In general, the challenges are abundant: ensuring projects' viability and relevance upfront; avoiding hidden agendas during planning; the underestimation of costs and overestimation of utility; unrealistic and inconsistent assumptions; securing essential planning data; and guaranteeing adequate contract regimes among others (Flyvbjerg, *et al.*, 2003; Miller and Hobbs, 2005; Miller and Lessard, 2000; Samset, 1998).

Norway has also experienced several projects with cost overruns and time delays. The Norwegian National University Hospital, which was completed in 2000 with considerable cost overruns, is but one example. However, this was equivalent to only some months of operational costs. In 2000, the Norwegian Ministry of Finance introduced a mandatory *quality-at-entry* regime to meet the challenges described above (Berg, *et al.*, 1999).

Trialling research to improve governance of major public investment projects

Governance regimes for major investment projects comprise the processes and systems that need to be in place on behalf of the financing party to help ensure successful investments.

The *quality-at-entry* regime introduced by the Norwegian Ministry of Finance in year 2000 is shown in Figure 14.3. The system was designed to improve *analysis* and *decision making* in the front-end phase, particularly with regard to the interaction between the two. Issues addressed in the new

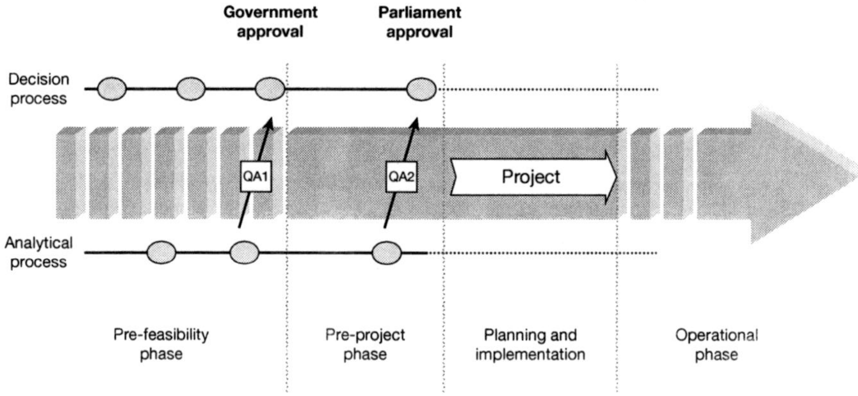

Figure 14.3 Norwegian quality-at-entry regime for major public investment projects

Source: Samset (2010)

quality assurance systems (QA), called QA1 and QA2, included that: an analysis may be biased or inadequate; decisions may be affected more by political priorities than by rational analysis; political priorities, alliances and pressure from individuals or groups of stakeholders may change over time; the amount of information is overwhelming and may be interpreted and used differently by different parties; and that the possibility for disinformation is considerable.

A response to these challenges would not be a strict and comprehensive regulatory regime. Instead, it would more likely be: (i) to establish a distinct set of milestones and decision gates that would apply to investment projects in all sectors regardless of existing practices and procedures in the various ministries or agencies involved; (ii) to ensure political control with fundamental *to go ahead or not* decisions; (iii) to ensure an adequate basis for decisions; and (iv) to focus on essential matters, and not on specific details.

Additional recommendations would be: (i) to anchor the most essential decisions in the cabinet itself; (ii) to introduce a system for quality assurance on the basis of decisions that were independent of Government and sufficiently competent to overrule the analysts; and (iii) to guarantee that the governance regime was compatible with the procedures and practices of the affected ministries and agencies.

Under the Norwegian *quality-at-entry* regime, pre-qualified external consultants are assigned to perform independent quality reviews of the decision basis in all public investment projects, with a total budget exceeding EUR100 million.

The scheme should help ensure that the choice of concept is subject to a political process of fair and rational choice. The decision is founded in the

Prime Minister's office, and if the outcome is positive, it will initiate a pre-project phase of project feasibility analysis.

As input to the QA1 review, the responsible ministries are required to explore at least two alternative concepts in addition to the *zero alternative*; *doing nothing*. They should prepare a document that includes a needs analysis; an overall strategy; an analysis of overall requirements that need to be fulfilled; and an alternatives analysis.

QA2 is performed at the end of the pre-project phase and aims to provide the responsible ministry with an independent review of decision documents before the parliamentary appropriation of funds. This is partially a final control to ensure that the budget is realistic and reasonable, and partially a forward-looking exercise to identify the managerial challenges ahead. To some degree, the results of the QA scheme can be attributed to the research being done at the Norwegian University of Science and Technology (NTNU).

A study conducted by researchers from NTNU and Stiftelsen for industriell og teknisk forskning (SINTEF, Foundation for Scientific and Industrial Research) by the *Concept Programme* in 2008, and based on a sample of 23 projects, revealed that 22 projects were completed within budget and that only one project had a cost overrun. In fact, the average cost for all these projects was close to the estimated cost (P(50)) and therefore well below the budget (P(85)). P(50) and P(85) mean that the probability that the final cost will be within the limit of the estimated cost is 50 and 85 per cent, respectively (Samset, 2010).

In a second study, 24 projects were used to evaluate the effect of the QA1 review. It was established that two of the proposed projects were entirely rejected by the Government, while the larger part, 22 projects, was approved. Nevertheless, many of the suggested alternatives that were not viable, unduly expensive or ineffective were eliminated in the process. In 17 out of 24 cases, the Government endorsed the QA reviewer's recommendation, thereby suggesting that the QA1 review also had a significant positive effect (Samset, 2010). As such, it is clearly too early to evaluate the effects of these projects, but that will be within the scope of the research programme for years to come, which will be the ultimate test of the success or failure of the scheme.

A more significant feature of the QA scheme and the research programme is the *spin-off effects* that seem to have occurred in both the Government and private sectors. There is now a growing awareness of the need for improved practices in the field of cost estimation and budgeting, risk assessment, and strategic planning. This has proliferated into the consulting and construction industry: front-end management has now become an issue within the community of project management professionals; training courses are now offered by a number of institutions and consultants; and improved practices have also been adopted and institutionalised by different Governmental agencies. This has been achieved in part because of the close interaction between public practitioners and academic researchers also reaching the international arena, which has sparked a wave of research in front-end management, placing project governance on the international research agenda (Samset, 2003, 2010).

Industry involvement and collaboration in research and education

NTNU has many research areas, while the Faculty of Engineering and Science and the Faculty of Architecture and Fine Art have two main specific areas of focus with a high degree of involvement and collaboration with industry partners from construction: (i) energy efficient and usable buildings; and (ii) design, planning and management of complex projects.

The faculties carry out extensive research in these areas and have a long tradition of broad collaboration with partners from the building and construction industry, producing master degree candidates from civil and mechanical engineering, architectural design and urban planning. At the moment, NTNU works closely with the industry to establish a common knowledge arena for the architecture, engineering and construction sector (AEC) organised as a new centre located at NTNU. This new arena can serve as a national knowledge hub for dissemination, as developing new knowledge through research, innovation and the pilot testing of interaction models in *real-life* projects demonstrates new technology and teaches the workers to use new methods.

This centre will be a national competence enhancement within project management, with a focus on the construction and design process. It has the ambition to increase the R&D effort within the built environment sector, involving industry partners to find new technical solutions and improved collaboration models. Learning from other sectors, such as the oil and gas sector and information and communication technology (ICT) sector, is important for the construction industry. The initiative leading to the new centre comes from the construction industry. Industry representatives have asked NTNU to establish a knowledge hub with a focus on the construction process, from the planning, designing and constructing to the user, as well as through the entire value chain.

The European research agenda

Two Europe-wide initiatives are worthy of note:

- The upcoming European Union (EU) framework programme for research and innovation, *Horizon 2020*, whose aim is to drive new economic growth and job creation in Europe and to secure global competitiveness through research and innovation from 2014 to 2020, has a budget of EUR80 billion. *Horizon 2020* follows from the seven previous framework programmes in Europe. Key objectives of *Horizon 2020* are to provide a boost to *top-level research* in Europe; to strengthen industrial leadership in innovation through support for small and medium-sized enterprises (SMEs) among others; to help address grand societal challenges such as climate change; to develop sustainable transport and mobility, making renewable energy more affordable, ensuring food safety and security; and coping with the challenge of an ageing population (European Commission, 2011). In Norway, a company with fewer than 100 employees is considered an SME, with 99 per cent of the companies being SMEs (Ministry of Trade and Industry, 2013).

- The *Joint Programming Initiative (JPI)* aims to gather national research programmes and pool the European countries' public R&D resources in order to address common European societal challenges that cannot be solved by national research programmes alone (European Commission, 2010). Ten JPIs have been established since 2008, and through the Norwegian Research Council and the Ministry of Education and Research, Norway is the only country involved in all 10 programmes. Relevant JPIs to the Norwegian construction sector are *Urban Europe, Cultural Heritage and Global Change* and *More Years, Better Lives*.

The European research agenda impacts on the Norwegian R&D sector in various ways: (i) Norwegian universities and research institutes are actively participating in proposals and projects with leading R&D partners and SMEs in Europe and third-party countries such as China and the USA; (ii) an increasing part of the R&D projects and activities in Norway are based on EU funding; in the last framework programme the 10 largest R&D actors received almost EUR200 million in funding; and (iii) these actors actively influence and provide input to the focus and topics for the various initiatives of the JPIs and framework programmes through, for example, participating in joint initiatives and networks such as the European Construction Technology Platform (ECTP), which aims to improve the performance and competitiveness of the construction industry by 2030 (ECTP, 2005).

To ensure a better position in the competition for EU funding and participation in international projects, this is on the strategic agenda of most Norwegian universities, research institutes and the Norwegian Research Council. In addition to being an active partner in the development of *Horizon 2020* and JPIs, the Norwegian Research Council has indicated that an increasing fraction of their resources and funding will be oriented towards the focus of the European agenda. Therefore, the European research agenda forms part of the Norwegian research agenda and vice versa. Several of the major societal challenges to be addressed by the European funding instruments match those challenges observed in the Norwegian construction industry. However, there are unique national challenges that can only be addressed by national policies and funding instruments.

Case studies

This section presents three cases of innovative construction projects, their historical background and contributions from research and political instruments. The Vennesla Library illustrates the development of wood as a building material, the Brøset Neighbourhood showcases the development of urban green living, and the Powerhouse case presents recent initiatives on energy producing buildings.

The case studies also illustrate processes through which research and development ideas are translated into industry outcomes.

Case study: Vennesla Library

Wood has a long tradition as a construction material in Norway; in the Viking age they built long houses from timber logs and long ships from oak. The tradition of building houses with timber logs lasted until the beginning of the nineteenth century when steel and concrete took over as construction materials. In the decade from 1990–2000, innovative technology allowed for the development of new wood-based products such as glulam, that were longer lasting, stronger and more versatile, thus re-starting the trend of using wood as a building material. Wood is considered a genuine and natural building material among Norwegians.[2] From an architectural and engineering perspective, many of Norway's most interesting new buildings are made of wood, reflecting the strong appeal that this material continues to hold for Norwegian designers and builders, as well as demonstrating the meaning of an integrated design and construction process (Norway the Official Site, 2012).

Historic development of wood as a building material in Norway

The Stave churches from the twelfth and thirteenth centuries are an important part of our cultural heritage and are internationally recognised, as well as old wharf houses (Figure 14.4) which have been included in the UNESCO World Heritage list.[3]

Fire has threatened Nordic wooden towns throughout history, leading to the prohibition of new wooden buildings in Norwegian cities from 1904 (Larsen, 1987). The ban was lifted following the development of fire-protecting chemicals in the 1990s.

A new era of wooden construction started with the exploration of the glulam technique. Vikingskipet ('The Viking Ship') built in 1992, officially known as Hamar Olympic Amphi (Figure 14.5), and the Oslo Airport Gardermoen (1994-98) promote the potential benefits of large wooden

Figure 14.4 The old wharf of Bergen from the sixteenth-century Hanseatic trade period

Source: Robin Strand, Norphoto AS

Figure 14.5 Hamar Olympic Amphi, also known as 'The Viking Ship', built for the Winter Olympics, Lillehammer, 1994, demonstrating potential of the new glulam technique

Source: Sonja Gussiås Viken

structures (Gravdal, 1992). As a *gateway to Norway*, the airport terminal welcomes visitors with materials representing the country's appeal. The use of laminated veneer lumber allowed for the creation of a vast, undulating roof, with 136-metre-long curved trusses that free span up to 54 metres (Sierra Business Council, 2007). The new cultural centre of Vennesla (2011) explores glulam further. The wooden ribs not only carry the load of the building, but are also integrated with the interior architecture to create an aesthetically appealing and spectacular interior.

Completion in 2005 of the five-storey student and affordable housing at Svartlamoen in Trondheim persuaded Norway's architectural community to consider wood as an exciting contemporary building material. Modern wooden bridges demonstrate wooden constructions' ability to carry the load of long free spans. Kjøllsæter Bridge,[4] built in 2006, fits 109 metric tonnes into a span of 30 metres, and is in accordance with NATO standards (Teknisk Ukeblad, 2010).

Relevant policy, research and development initiatives

These projects are the result of combined efforts from industry, Government, research and education over the last couple of decades. A national aspiration for an increased use of wood has been developing since the 1990s. For example, the *Programme for Wood Innovation* has involved the entire production chain improving knowledge and exploring construction and production methods.[5] In this programme, traditional wood material was combined with new technology to create new designs and products. Additionally, an increased use of wood as a renewable resource will contribute to reducing CO_2 emissions.

Additional efforts have been made to develop education and research programmes focusing on wood as a construction material, which has led to a renewed attention in the education of architects and engineers on wood. NTNU is also participating in shared efforts with industry partners and SINTEF[6] to: increase research on the quality of wood; improve its strength and longevity; and document effects on the natural environment and indoor climate, through the *Treprogrammet*. Research efforts have proved valuable in solving the challenges of timber multi-storey buildings, exploring load-carrying capacities in various shapes of massive wood and coping with fire hazards (NTNU, 2010).

The initiative *Norwegian Wooden Cities* combines traditional materials and modern design, stating that wood is the preferred construction material for schools and other public buildings. Trondheim and Stavanger are among Norwegian cities participating in the international wood commitment through the project *Nordic Wooden Towns – Urban Development with Wood* and the *EU Culture 2000* initiative *Wooden Towns in Europe*. In 2008 the *Norwegian Wood Programme* in Stavanger became an international showcase for the entire construction industry, which included

Figure 14.6 Pulpit Rock Mountain Lodge, situated at the trailhead leading up to cliff overhanging Lyse Fjord on Norway's west coast

Source: Kjell Helle-Olsen

residential quarters, a theatre, road and railway bridges, a visitor centre and the Pulpit Rock Mountain Lodge (Figure 14.6).

Case study: the Powerhouse

Powerhouse #1 exemplifies how constructed facilities can be turned from energy consumers into energy producers. In 2009, the non-Governmental environmental protection organisation ZERO challenged the Norwegian construction industry to construct a building that produces more energy than it consumes over its life span. As a response, the Powerhouse alliance was formed by the developers Entra Eiendom, a development and property company for office buildings with a high environmental profile; Skanska, a construction company; Snøhetta, a famous international prize winning architecture firm with a profile on a sustainable trans-disciplinary approach; aluminium company Hydro, producer of energy and sun shading building façades; and ZERO. Powerhouse #1 draws on the experience from all its alliance partners, in addition to contributions from international research partners.

Design follows environment is the slogan for the concept project. A multi-disciplinary team has designed a building where the roof and most of the wall area are covered with photovoltaic panels, with an optimal angle for sunlight conditions at a northerly latitudinal degree (Hegli, 2012a). This has proven to challenge the public opinion regarding *good* and *bad* design, and is also a test of the degree to which local regulation plans are able to include innovative solutions for energy-ambitious buildings.

Another challenge is dealing with the energy embodied in building materials, maintenance during operations and the final disposal of materials. Additionally, environmental product declarations (EPDs) are rare and difficult to compare for different alternatives (Hegli, 2012b).

Relevant policy, research and development initiatives

The National Norwegian Energy Fund (ENOVA) was established in 2001 with the purpose of providing educational and economic support for innovative and ambitious solutions for energy efficient buildings, including the aforementioned pilot projects.

The *National Programme for Low-Energy Buildings* was established in 2007[7] as a collaborative effort between Governmental institutions and the construction industry.[8] The ambition of this programme was that the industry should become one of the best in Europe regarding energy-efficient construction practices. The means used are to improve the level of knowledge to be able to deliver according to the coming 2017 passive house standard (Lavenergiprogrammet, 2012).

The National Housing Bank (Husbanken) has a long history as an operative instrument to put current housing policy into practice. In recent years, energy efficiency and renewable power have been among the criteria for qualifying for favourable banking loans from Husbanken. Affordable housing projects and student villages are among the favoured purposes.

In 2010 the energy labelling system implemented by the EU was made compulsory for services and industrial buildings in Norway. That same year, the Norwegian Green Building Council was established to promote a volunteer- and market-driven classification system to improve environmental standards in Norwegian constructions (Norwegian Green Building Council, 2012).

The *Passive House* standard is already the quality requirement for all construction contracts with public owners, such as local municipalities, the Government, hospitals, or the defence force, among others. This is due to the role of public institutions as demanding customers for their built facilities. Also, a Norwegian standard version for passive houses has been established (Standards Norway, 2013).

In 2012 leading companies in the Nordic building industry signed the *Nordic Built Charter*, which commits the Nordic building sector to join forces to deliver sustainable solutions for the Nordic region (Nordic Innovation, 2012).

Case study: the Brøset Neighbourhood

Urbanisation and a growing population are challenging Norwegian community planning.[9] Political ambitions to improve climate and environmental performance in our cities are incentives to explore new solutions and rethink existing practice. Urban green living is a new concept in Norway, and there are many questions to be answered before the first green urban neighbourhood becomes a reality. Hence, the following section presents a summary of the plans for Brøset in Trondheim and the experiences and research contributions it draws on.

Brøset is an area of 1.4 sq km located between existing settlements and the green landscape surrounding Trondheim. In combination with various health services, the area has a long farming history. However, in the near future, the area will be available for new purposes, as a preliminary estimation shows that it is possible to build approximately 1,200 new dwellings at Brøset. In 2008, local authorities in Trondheim decided to start a planning process for a carbon-neutral neighbourhood at Brøset. The vision is to combine low energy demand and *healthy* materials with a socially sustainable living environment. The latter implies that the settlement should be accessible for all types of resident, including low-income and other vulnerable groups.

Green living involves much more than buildings and transport systems. The aim is to create neighbourhoods in which residents can live, work, shop, study and find meaningful leisure activities, while simultaneously enabling them to lower their carbon footprint.

The purpose of this experimental neighbourhood is to learn how to build cities that stop global warming (Trondheim Municipality, 2012), and Trondheim will cooperate with the other cities in *Cities of the Future* and share experiences related to planning for climate change adaptation.

Public involvement is vital for the planning process, with the first public workshop on carbon neutral living arrangements held in 2007. Then, in 2009, the local Government agreed upon a programme plan for Brøset and an international urban development competition was announced, in which four teams worked in a parallel architectural process (Gansmo, *et al.*, 2011). There were three common workshops during the process open to the public, and in 2012, the municipality council presented a zoning plan for public hearing.

The plan draws on experience from previous green city projects, particularly the local Rosenborg neighbourhood and the German city of Freiburg. Rosenborg Park, completed in 2003, is one of six nationwide pioneer projects within the field of urban environmental development (Husbanken, 2003). Even though the project succeeded to a high degree, it received negative attention in public discussions due to the high density of dwellings, and the same issue is now being discussed regarding the Brøset plan. The neighbourhood in Freiburg has more than 20 years of experience with housing and community solutions similar to those planned for Brøset, and the two municipalities have agreed to cooperate and to exchange experiences during the ongoing process (Freiburg im Breisgau, 2011).

Relevant policy, development and research initiatives

When the United Nation's Intergovernmental Panel on Climate Change presented its reports in 2007, it called forth initiatives at the national level. In response to this, the Norwegian Government Climate Agreement (2008) stated that Norway would be a carbon-neutral nation by 2030. To reach this ambitious goal, Norwegian municipalities were challenged to identify what this goal meant for the development of buildings and physical infrastructure on a local scale. Several programmes have emerged as a response to this challenge, including the two discussed here:

- *Cities for the Future*: in 2008 the Norwegian Government invited all cities to apply for partnership in the new initiative, *Cities for the Future* (Norwegian Ministry of Environment, 2008). This programme was set to run from 2008 to 2014 and is a collaborative effort between the Norwegian Government and the 13 largest cities in Norway with the aim of reducing greenhouse gas emissions to make the cities better places to live in (Regjeringen, 2008). The vision is one of more densely built neighbourhoods, with more jobs, shops and leisure activities within reach by foot, bicycle and collective transportation. The *Cities of the Future* programme will help city municipalities to share their climate-friendly city development ideas with each other, as well as with the business sector, the regions and the Government.
- *FutureBuilt* is a similar programme, a cooperative partnership between four municipalities and stakeholder organisations. During a 10-year period (2010–2020), the aim is to complete a number of pilot projects, urban areas and individual buildings with the lowest possible greenhouse gas emissions. These prototypes will also contribute to a good city environment with regard to ecological cycles, health and the general impression of the city (FutureBuilt, 2010). *FutureBuilt* aims to be an arena for innovation, competence building and exchange of experiences, in addition to being a learning platform for clients, architects, advisers, entrepreneurs, municipal administrations and users (FutureBuilt, 2010).

Future directions

White paper Good Buildings for a Better Society *(18 June 2012)*

The Ministry of Local Government and Regional Development developed a white paper for the Norwegian construction industry, based on over

29 contributions from universities and research institutes, organisations, SMEs and actors representing the entire value chain of building (Ministry of Local Government and Regional Development, 2012).

The paper suggests several focus areas and measures for the future construction industry to address the current challenges of the sector and to ensure well designed, secure, energy-efficient and healthy buildings, as well as better and more effective building processes.

Some objectives stated in the white paper are:

- good architecture, handcraft and design, which contribute to attractive environments
- secure buildings that address the challenges of climate change
- buildings with a healthy indoor environment
- to increase the amount of universally designed buildings, which do not exclude users with specific needs, by 2025
- to reduce the monthly energy use of buildings until 2020
- to simplify and make legislation and related processes more user friendly
- the smart use of ICT in order to achieve better productivity and cost efficiency
- control and supervision as a driver to help reduce building failure.

In the white paper's press release, Minister Navarsete stated:

> A lot of money is involved in the building industry. We therefore need a competent industry. Too many mistakes are made in the design and practical construction of buildings. At the same time, productivity growth is low. The government will therefore invite the building industry to collaborate on a campaign to boost expertise. The agenda is called Bygg21.
>
> (Ministry of Local Government and Regional Development, 2012)

Bygg21 agenda

Bygg21 will be one of the instruments for transforming goals and objectives into action, establishing strategies, action plans and measures related to three focus areas: *R&D and innovation*; *education and development of skills competence*; *and the dissemination of knowledge and experience*. The strategies and measures will be developed with key actors in the construction industry and related R&D environments. Bygg21 is intended to be a long-term interactive and collaborative arena between industry and public authorities (DIBK, 2012).

Other instruments for driving necessary change and improvements include simplifying the legislative and procedural framework for building projects, thereby further enhancing the role of the public client as a change agent and role model in relation to requiring new knowledge and acting as a driver in the construction industry.

The Ministry of Local Government and Regional Development is responsible for initiating Bygg21 and coordinating initiatives with other ministries. This process was initiated in 2012 and is planned to be finished in 2014 with recommendations and priorities. The process will be governed by the Agency for Construction Quality (DIBK, 2012) and followed up by strategies and actions that will be carried out by a joint effort between the construction industry and the Government. It is expected that Bygg21 will create measures that may lead to a need for restructuring the construction industry towards 2030.

A national strategy and common platform for future investment is expected to have an impact on all actors in the construction industry, but will only be binding after the strategies are rooted in the competent bodies of both Government and industry.

The establishment of a national centre for project enterprises and construction processes

This is NTNU's strategic focus on building a national competence enhancement within project management, with a focus on the construction and design process. It has an ambition to increase the R&D effort within the built environment sector involving industry partners to find new technical solutions and improved collaboration models. Learning from other sectors, such as the oil and gas sector and ICT sector, is important for the construction industry. Industry representatives have asked NTNU to establish a knowledge hub with a focus on the construction process from the planning, designing and constructing to the user, as well as through the entire value chain. This structure will be established during 2013 and serve as a knowledge hub and a bridge between the industry and academia.

Summary – how R&D investments translate into industry outcomes

The rethinking process in the Norwegian construction industry was described on a general level and exemplified in the case studies.

The Norwegian Government is now required to produce realistic preliminary cost estimates at an early stage, in which alternative concepts are being considered. The implementation of the QA1 regime in all public investments has no doubt had a disciplinary effect on analysts, thus far reducing large cost overruns in public mega projects through a close interaction between public sector practitioners and academic researchers. This has proliferated beyond the national arena and has sparked a wave of research in front-end management, putting project governance on the research agenda internationally, all of which has served to illustrate how mutual goals and cooperation are crucial for translating R&D outcomes to industry practices. Researchers and industry actors mutually challenging and inspiring each other has led to innovation and development in the Norwegian construction industry.

In summary:

1 Public hearings and debates among industrial enterprises and organisations, as well as other stakeholders regarding new regulations and standards for the construction industry, have helped to develop a mutual trust and exchange of knowledge, thereby preparing the ground for the implementation of changes in practice.
2 As it has been observed through the *Concept Programme* and implementation of the QA regime in the upfront phase, Government policy plays an important role.
3 Programmes that combine financial and expertise support, the national housing bank and ENOVA, have roles as financial instruments to implement national policy regarding environmental and energy targets in Norway.
4 Long-term research programmes with both industry and international research partners are important due to the need for development and dissemination of new knowledge and innovation in the construction industry.

Notes

1 Norway is well known for the spectacular fjords and mountains. For more information visit: www.visitnorway.com/.
2 Ninety-four per cent of Norwegians surveyed consider wood a genuine and natural building material (Riis-Johansen, 2006).
3 One example is the wooden church of Urnes (the *stavkirke*), which stands in the natural setting of Sogn og Fjordane (UNESCO, 2012).
4 Kjøllsæter Bridge (Norwegian Kjøllsæterbrua) is 158 metres (518 feet) long and crosses the Renaelva River in Åmot, Hedmark in Norway
5 Speech made by the Minister for Agriculture presenting the renewed strategy for wood, Bergen 2006. Available at www.regjeringen.no/nb/dep/lmd.
6 More information on SINTEF is available at: www.sintef.no/home/About-us/History/.
7 Following the Intergovernmental Panel on Climate Change (IPCC) Fourth Assessment (Intergovernmental Panel on Climate Change, 2007).
8 Partners are Byggenæringens Landsforening, Arkitektbedriftene, Husbanken, ENOVA, Direktoratet for byggkvalitet, Norges vassdrags- og energidirektorat og Statsbygg (Lavenergiprogrammet, 2012).
9 Today, cities are home to half the world's population. Cities are already the biggest consumer of energy and are responsible for 80 per cent of the world's greenhouse gas emissions (Intergovernmental Panel on Climate Change, 2007).

References

Almås, A.-J., Lisø, K.R., Hyfen, H.P., Øien, C.F. & Thue, J.V. (2011) 'An approach to impact assessments of buildings in a changing climate', *Building Research & Information*, 39(3): 227–238.

Berg, P., *et al.* (1999) *Styring av statlige investeringer* (*Governance of public investments*), Final report from Steering Committee of *Projects for Governance of Public Investments*, Oslo: Ministry of Finance.

BNL (2010) *Visjon, forretningside og mål for byggenæringens landsforening 2010–2020* (*Vision, mission and goals for BNL 2010–2020*). Available at: www.bnl.no/category.php/category/BNLs%20strategi%202010-2020/?categoryID=288 (accessed 17 December 2012).

BNL (2012) *Byggenæringen I tall 2011* (*The building sector in 2011 figures*). Available at: www.nrl.no/getfile.php/PDF/Brosjyrer/byggenaringen_tall011_korr2.pdf (accessed 27 March 2013).

DIBK (2012) *The bygg21 program*. Available at: www.dibk.no/bygg21 (accessed 1 February 2013).

ECTP (2005) *European construction technology platform, challenging and changing Europe's built environment – A vision for a sustainable and competitive construction sector by 2030*. Available at: www.ectp.org/documentation/ECTP-Vision2030-25Feb2005.pdf (accessed 22 March 2013).

European Commission (2010) *Joint programming*. Available at: http://ec.europa.eu/research/era/areas/programming/joint_programming_en.htm (accessed 20 December 2012).

European Commission (2011) *Horizon 2020*, official version 7 December 2011. Available at: http://ec.europa.eu/research/horizon2020/index_en.cfm?pg=h2020 (accessed 23 March 2013).

Freiburg im Breisgau (2011) *Green city Freiburg. Approaches to sustainability*. Available at: www.fwtm.freiburg.de/servlet/PB/show/1199617_l2/GreenCity.pdf (accessed 17 December 2012).

Flyvbjerg, B., Bruzelius, N. & Rothengatter, W. (2003) *Megaprojects and risk: An anatomy of ambition*, Cambridge: Cambridge University Press.

FutureBuilt (2010) *What is FutureBuilt?* Available at: www.futurebuilt.no/?nid=206235&lcid=1044 (accessed 17 December 2012).

Gansmo, H.J., Larssæther, S. & Thomsen, J. (2011) *På vei til Brøset – Evaluering av det åpne parallelle oppdraget* (*On the way to Brøset – Evaluation of the open parallel assignment)*, Trondheim: Sintef/NTNU. Available at: http://brozed.wordpress.com/ (accessed 17 December 2012).

Gravdal, G. (1992) 'Ny ol-kontrakt til moelven' ('New olympic contract to moelven'), *Aftenposten*, 19 August.

Hegli, T. (2012a) 'PowerHouse #1 at Brattørkaia in Trondheim', paper presented at ZEB Conference, Oslo, 9 September. Available at: www.zeb.no/index.php/news-and-events/39-zeb-konferanse-artikler (accessed 14 December 2012).

Hegli, T. (2012b) 'Byggebransjens månelanding?' ('Moon landing for the construction industry?'), presented at BM-dagen, Trondheim, 6 November.

Husbanken (2003) *Husbanken halverer energibehovet – 6 prosjekter, 600 boliger* (*The Norwegian housing bank bisects energy requirements – 6 projects, 600 dwellings*), Oslo: Husbanken.

Intergovernmental Panel on Climate Change (2007) *Climate change 2007 – Mitigation of climate change*, Working Group III contribution to the Fourth Assessment Report of the Intergovernmental Panel on Climate Change (IPCC). Available at: www.ipcc.ch/report/ar5/wg1/ (accessed 4 October 2013).

Larsen, K.E. (1987) 'Some aspects of the development of the wooden towns in the Nordic countries until the 20th-century', in *Old Cultures in New Worlds*, presented at 8th ICOMOS General Assembly and International Symposium, Washington: Programme report – Compte rendu, US/ICOMOS.

Lavenergiprogrammet (2012) *Om lavenergiprogrammet* (*About the programme*). Available at: www.lavenergiprogrammet.no/om-lavenergiprogrammet/ (accessed 24 March 2013).

Lisø, K.R. (2006) 'Integrated approach to risk management of future climate change impacts', *Building Research & Information*, 34(1): 1–10.

Lovdata (2008) *Lov om planlegging og byggesaksbehandling (plan- og bygningsloven)* (*Law on planning and building regulations, planning and building*). Available at: www.lovdata.no/all/nl-20080627-071.html (accessed 12 March 2013).

Miller, R. & Hobbs, B. (2005) 'Governance regimes for large complex projects', *Project Management Journal*, 36(3): 42–50.

Miller, R. & Lessard, D. R. (2000) *The strategic management of large engineering projects: Shaping institutions, risk and governance*, Cambridge, MA: MIT Press.

Ministry of Local Government and Regional Development (2012) *Gode bygg for eit betre samfunn. Ein framtidsretta bygningspolitikk* (*White paper: Good buildings for a better society*), Oslo: Ministry of Local Government and Regional Development. Available at: www.regjeringen.no/pages/37918068/PDFS/STM 201120120028000DDDPDFS.pdf (accessed 20 August 2012).

Ministry of Trade and Industry (2013) *Små og mellomstore bedrifter* (*Small and medium business*). Available at: www.regjeringen.no/en/dep/nhd/selected-topics/simplification-for-business/sma-og-mellomstore-bedrifter.html?id=614069 (accessed 30 April 2013).

Nordic Innovation (Norden) (2012) *Leading companies in the Nordic building industry signed Nordic built charter*. Available at: http://nordicinnovation.org/news/leading-companies-in-the-nordic-building-industry-signed-nordic-built-charter-/ (accessed 17 December 2012).

Norway the Official Site (2012) *Stavanger: European capital of culture 2008*. Available at: www.norway.org/ARCHIVE/Miscellaneous/stavanger08/ (accessed 14 December 2012).

Norwegian Green Building Council (2012) *Norwegian green building council*. Available at: www.ngbc.no/ (accessed 17 December 2012).

Norwegian Ministry of Environment (2008) *Invitation for participation in cities of the future*, Oslo: Government of Norway.

NTNU (2010) *TRE – I bygninger, konstruksjoner, produkter* (*Wood – In buildings, constructions, products*), Bringing the initiative further at NTNU, Trondheim: NTNU.

NTNU (2013) *Concept research programme*. Available at: www.concept.ntnu.no/english (accessed 28 March 2013).

OECD (2012) *OECD data lab*. Available at: www.oecd.org/dataoecd/21/57/3957 4824.xls (accessed 8 September 2012).

OECD (2013) *Gross domestic expenditure on R&D as a percentage of GDP*, OECD library, Science and technology, Key tables from OECD, 1. Available at:

www.oecd-ilibrary.org/science-and-technology/gross-domestic-expenditure-on-r-d_2075843x-table1 (accessed 30 September 2013).

Regjeringen (2008) *Cities of the future*. Available at: www.regjeringen.no/en/sub/framtidensbyer/cities-of-the-future.html?id=548028 (accessed 17 December 2012).

Regjeringen (2012) *Økonomisk integrasjon og avhengighet* (*Economic integration and dependency*). Available at: www.regjeringen.no/nb/dep/ud/dok/nou-er/2012/nou-2012-2/15/4.html?id=669582 (accessed 8 September 2012).

Regjeringen (2013a) *Ministry of local government and modernisation*. Available at: www.regjeringen.no/en/dep/krd.html?id=504 (accessed 12 March 2013).

Regjeringen (2013b) *The Norwegian building authority*. Available at: www.regjeringen.no/en/dep/krd/min/Subordinate-agencies-and-institutions/national-office-of-building-technology-a.html?id=85812 (accessed 12 March 2013).

Riis-Johansen, T. (2006) *Regjerningens tresatsing*, speech given at the Conference Nordic Wooden Cities, Bergen, 30 May. Available at: www.regjeringen.no/en/dep/lmd/aktuelt/taler_artikler/ministeren/tidligere_landbruks_og_matminister_riis_/2006/regjeringens-tresatsing.html?id=113604 (accessed 4 October 2013).

Samset, K. (1998) 'Project management in a high-uncertainty situation. Uncertainty, risk and project management in international development projects', PhD dissertation, Norwegian University of Science and Technology.

Samset, K. (2003) *Project evaluation. Making investments succeed*, Trondheim: Tapir Academic Press.

Samset, K. (2010) *Early project appraisal. Making the initial choices*, Basingstoke: Palgrave Macmillan.

Sierra Business Council (2007) *Gardermoen airport*. Available at: http://sierrabusiness.org/what-we-do/publications/125-Oslo--Gardermoen-Aiport (accessed 14 December 2012).

Sintef Byggforsk (2007) *The R&D programme klima 2000 – Climatic adaption of the building construction*. Available at: www.sintef.no/home/Environment/Klimatilpasning/ (accessed 30 August 2013).

Standards Norway (2013) *NS 3700 Kriterier for passivhus og lavenergibygninger* (*NS 3700 Criteria for passive and low-energy buildings*). Available at: www.standard.no/passivhus (accessed 30 August 2013).

Statistisk sentralbyrå (2012a) *Nøkkeltall for utenriksøkonomi* (*Key figures for external balance*). Available at: www.ssb.no/ur_okonomi/ (accessed 8 September 2012).

Statistisk sentralbyrå (2012b) *Arbeidskraftundersøkelsen*, 4 kvartal 2012 (*Labour force survey, 4th quarter 2012*). Available at: www.ssb.no/aku/tab-2012-08-02-07.html (accessed 8 September 2012).

Statistisk sentralbyrå (2012c) *Omsetning i bygge- og anleggsvirksomhet, 6. termin 2011* (*Trading in building and construction, 6 forward 2011*). Available at: www.ssb.no/bygg-bolig-og-eiendom/statistikker/bygganloms/termin/2012-05-03 (accessed 12 March 2013).

Statistisk sentralbyrå (2012d) *Stabil høy anleggsvirksomhet* (*Stable high construction*). Available at: www.ssb.no/bygganlprod/ (accessed 8 September 2012).

Statistisk sentralbyrå (2013a) *Population by age, sex, marital status and citizenship, 01 January 2013*. Available at: www.ssb.no/en/befolkning/statistikker/folkemengde (accessed 30 October 2013).

Statistisk sentralbyrå (2013b) *Nøkkeltall konjunkturer* (*Key figures market trends*). Available at: www.ssb.no/nasjonalregnskap-og-konjunkturer/nokkeltall/konjunkturer-statistikk-analyser-og-prognoser/ (accessed 12 March 2013).

Teknisk Ukeblad (2010) *Tror på flere trebruer* (*Think of several wooden bridges*). Available at: www.tu.no/bygg/2010/09/27/tror-pa-flere-trebruer (accessed 17 December 2012).

Trondheim Municipality (2012) *Grønn by på Brøset* (*Green city at Brøset*). Available at: www.trondheim.kommune.no/gronnbybroset/ (accessed 17 December 2012).

UNESCO (2012) *Urnes Stave Church*. Available at: http://whc.unesco.org/en/list/58 (accessed 12 September 2012).

15 Sweden – collaboration for competitiveness

Anna Kadefors and Jan Bröchner

The national context

Background, institutions and culture

Construction in Sweden is affected fundamentally by the dictates of a cold climate, which it shares with Canada, Finland, Norway and Russia but with hardly any other country. The climate explains the low population density, which leads to high per capita expenditure on transportation infrastructure, and also the need for solid structures that offer protection from nature. Thus, Sweden has traditionally been leading in areas such as energy efficiency and indoor climate, both in terms of research and practice. This position is less obvious today, as sustainability concerns have also prompted other countries to tighten their building regulations. However, there is also a historic reduced Government engagement in building sector policies and subsidies in general, including research for the sector. A major shift in this respect took place in the early 1990s, when a deep recession was accompanied by the Swedish entry to the European Community.

Although Sweden remained neutral during the Second World War, the country emerged with a heavily regulated economy in 1945. Over time, successive deregulation has both caused construction crises and created new opportunities. More recently, financial deregulation during the 1980s and the lowering of international barriers contributed to inflated real estate values. When the crisis came in the early 1990s, construction suffered and many firms were bankrupted, along with a number of financial institutions. It was only around 1998 that real estate prices approached the pre-crisis level and construction volumes began to increase again.

Between 1974 and 1991, there was a Ministry of Housing, but today construction issues are handled by numerous ministries, and Government clients have seldom been required to take leadership in industry development. However, this does not mean that cost and quality in the construction industry have been seen as free of problems. Reports from the Government Construction Cost Delegation (Byggkostnadsdelegationen, 2000), the

Government Construction Commission (Byggkommissionen, 2002) and the Swedish Agency for Public Management (Statskontoret, 2009) stated that building costs were high, emphasising difficulties for young people and low income households to pay for accommodation in large cities.

In 2012, a third Government commission dealt with issues of low productivity in infrastructure construction and the role of the largest central Government client: the Swedish Transport Administration (Produktivitetskommittén, 2012). Other issues currently debated relate to moisture in external walls in housing as well as snow-induced roof collapses of large hall structures. All these concerns have not led to any major reform programmes; collaborative contracting or *partnering*, for example, has not been promoted by Government despite the Swedish collaborative culture (Bröchner, *et al.*, 2002).

In parallel with the financial crisis in the early 1990s there was a change in the style and logic of building regulations. The former Government subsidies had been tied to specific designs, often quite detailed, and contributed to standardisation, but were also thought to raise building costs unnecessarily. In 1994, new building codes based on performance requirements were issued. The action space for the construction industry was thus expanded. This technical deregulation also made large Swedish contractors interested in raising their technology capabilities and developing their collaborative relations with universities (Miozzo & Dewick, 2002).

National construction industry overview

The Swedish industry has many multinational firms and in construction a small number of leading contractors hold strong positions. However, professional service firms are weaker than in most countries. Additionally, large Swedish consultancy firms are active abroad, and a significant proportion of them have foreign owners.

From the 1990s, important changes have taken place within public client functions. Before 1992, there was one large Government property owner for buildings: the National Building Agency. This agency had a large development division with nation-wide impact in areas such as specification of requirements and project management. Partly to fulfil requirements for access to the European Community, the agency was split into a number of specialised organisations, some of which were privatised and some were restructured as Government corporations. The total research and development (R&D) resources were significantly reduced, as well as most of the in-house project management staff. Since the 1990s, construction and design branches of the public infrastructure authorities have also been separated successively from their client functions and privatised. Client functions in municipal social housing companies were also reduced, given that it has become more difficult to finance residential construction for rent within the Swedish system for rent control.

As the financial situation slowly improved throughout the 1990s, culminating in the information technology (IT) boom around the year 2000, new actors emerged. Perhaps most importantly, residential construction became dominated by new developer units within the large construction contractors. This meant that contractors acquired deeper knowledge about user preferences and quality. They produced blocks of flats for new housing cooperatives, rather than for the decreasing rental market and single family dwellings. Today, construction activities have benefitted from more than a decade of historically low interest rates, although growing concern with the rising volume of household mortgage debt has led to Government pressure on financial institutions to reduce lending.

Trends in R&D investment from 1990 to 2010

No official statistics for Swedish construction research and innovation exist. Nevertheless, Forskningsrådet för miljö, areella näringar och samhällsbyggande (Formas, the Research Council for Environment, Agricultural Sciences and Spatial Planning) and IQ Samhällsbyggnad (IQS, Centre for Innovation and Quality in the Built Environment) carried out analyses for the year 2011.

Trends in the private sector

Little is known about the volume of private sector R&D in the construction industry. Data can be found primarily for the joint schemes intended to produce publicly available results. Svenska Byggbranschens Utvecklingsfond (SBUF, the Development Fund of the Swedish Construction Industry), is the largest source of private sector grants. SBUF, founded in 1983 (Bröchner and Grandinson, 1992), is financed by approximately 5,000 contractors, based on agreements between construction labour market organisations, and about half of the funds go back to member firms for development projects.

Since 1993, SBUF has also contributed to the funding of many research projects in universities, primarily directed towards projects that lead to research degrees in interaction with member firms. According to Formas (2012), SBUF funded university research with SEK37 million (Swedish Krona (SEK), USD5.4 million) in 2011. According to IQS (2011), consultancy companies and their associated foundations spent about SEK50 million (USD7.4 million). Other foundations were estimated at SEK20 million (USD3 million). Furthermore, development and innovation projects led by firms currently require up to 50 per cent of cash or in-kind contribution from the private sector in order to obtain public financial support. IQS (2011) estimates that these contributions amount to SEK60 million (USD9 million) for such projects co-funded by Formas, VINNOVA (the Swedish Government Agency for Innovation Systems) and the Energy Agency.

As mentioned above, private companies and primarily the larger contractors (Bröchner, 2010) have increased their engagement in R&D in parallel with Government retrenchment. There is no official overview of the direction of these activities. A hint can be found in a SBUF 2006 survey of construction contractors, which indicated that large firms were engaged in materials and technology innovations, whereas smaller firms would report only immaterial innovations (Bröchner, 2010). Despite the lack of reliable statistics for construction R&D in private industry, it is obvious that private construction R&D is significantly lower as a percentage of turnover than for the major export firms in the Swedish manufacturing industry.

R&D trends in the public sector

The four oldest technical universities (Chalmers University of Technology, Luleå University of Technology, Lund University and the Royal Institute of Technology) are still dominant in built environment R&D. During the 1990s, new regional universities were established, and there is applied research in several of these. Basic and applied research relating to the built environment is also carried out by the universities in Göteborg, Stockholm, Lund, Linköping and Uppsala as well as by the Swedish Agricultural University.

The Swedish research institute sector is weak by international comparison. During recent years, however, more Government funding has been allocated to the institutes and they have been coordinated under one umbrella organisation: Research Institutes of Sweden (RISE). For the built environment, important research institutes are SP Technical Research Institute of Sweden which has research units for wood, fire, materials and energy as well as CBI Betonginstitutet (Swedish Cement and Concrete Research Institute) and Glafo (the Glass Research Institute). In the infrastructure sector, the Swedish National Road and Transport Research Institute (VTI) is also important.

The stated objective of current Swedish research and innovation policy is for Sweden to be a prominent research nation in which research and innovation are conducted at a high quality, contributing to the development of society and the competitiveness of industry. Today, the focus is either on research excellence proved by highly cited publications or on the commercialisation of research-based products that can be sold globally. There has been a strong policy trend away from using R&D for solving national problems in various sectors. Before 2001, there was a system of sectoral support, where the Swedish Council for Building Research was responsible for support of R&D in its sector (Bröchner and Sjöström, 2003). The Council supported R&D in both firms and universities. It was during the sectoral policy period, dating back to the 1940s, that the private sector SBUF contractor scheme for collaborative R&D was set up in 1983 (Bröchner and Grandinson, 1992).

Formas is the only research council that has an explicit responsibility for research related to the built environment. The mission of Formas is to promote and support basic research and need-driven research of the highest scientific quality and relevance. Funded research should have further high sustainability relevance. Development and innovation projects are funded through VINNOVA, and the Energy Agency funds both basic and applied research related to energy production and consumption. Since 2009, VINNOVA has run the *Bygginnovationen* (*Construction Innovation*) programme. The overall purpose of this programme is to develop a strong and lasting innovation environment for Swedish construction.

A Formas survey shows that public spending on built environment research amounted to about SEK500 million, excluding direct university funding from the central Government (Formas, 2012). Formas is the largest funding body for built environment research in Sweden with about SEK150 million. The other important sources of funding are the Transport Administration with SEK122 million, the Energy Agency (SEK59 million) and VINNOVA (SEK52 million). The contribution from *European Framework Programmes* was estimated at SEK140 million for built environment research during 2011. In addition, a number of semiprivate foundations came into existence in 1994 (Benner & Sörlin, 2007); two of these, Stiftelsen för Miljöstrategisk Forskning (Mistra, Swedish Foundation for Strategic Environmental Research) and Stiftelsen för Strategisk Forskning (SSF, Foundation for Strategic Research), would support research of interest for the construction industry, including the Competitive Building Graduate Research School of the building sector, operating 1998–2003. Mistra Urban Futures, a centre for sustainable urban development, has received support since 2010.

According to VINNOVA (Elg & Håkansson, 2011) the funding landscape in applied research has become more complex in the past 20 years, where VINNOVA is a smaller organisation when compared to its precursors who were major technology support actors: Närings- och teknikutvecklingsverket (NUTEK, National Board for Industrial and Technical Development) and Styrelsen för Teknisk Utveckling (STU, National Board for Technical Development). New regional actors, often with European Union (EU) support, have also entered the funding scene.

As approved by the Riksdag, Sweden's national legislative assembly, in February 2013, the *Government Research and Innovation Bill* for the 2013–2016 period includes a special appropriation for research on sustainable urban development, as an area of strategic research for business and society (Government Offices of Sweden, 2012). In steps up to 2016, Formas will receive an extra SEK150 million specifically dedicated to this area, which includes technical building-related research (Government Bill, 2012a). The Swedish Energy Agency will receive a total of SEK120 million during 2014–2016 to support demonstration projects within zero-energy building (Government Bill, 2012b).

Case studies

Three case studies are presented here, illustrating how three of the major funding bodies, Formas, the Energy Agency and VINNOVA, operate.

Case study: the Formas-BIC/IQS Programme

Since 2003, Formas and the Byggsektorns Innovationscentrum (BIC, Swedish Construction Sector Innovation Centre) had organised a number of joint calls for research grant applications. This initiative followed the Government instruction to Formas to develop a joint research strategy for sustainable development in the built environment. In 2010, IQS was created by merging BIC and Rådet för Byggkvalitet (BQR, Council for Constructing Excellence). Formas' collaboration with the Swedish construction industry continues within the framework of IQS.

Up to 2012, 14 joint requests for grant applications had been completed, half of them European within *Erabuild* and *Eracobuild*. The Formas-BIC/IQS collaboration has a dual goal to produce scientific results and to promote implementation and dissemination of research results. Applicants have been required to present an implementation plan, and each project must also have an implementation manager. The requirement is that the projects should have 50 per cent co-funding from the industry, where 30 per cent can be in-kind through, for example, person hours or equipment, and a minimum of 20 per cent should be in cash. In total, 120 projects have been funded, most of them within the national calls for proposals. The total Formas funding for the programme has been SEK181 million, and thus the total volume including industry co-funding is about SEK360 million (Formas, 2011).

The following description is based on a Formas-led evaluation of the five calls for proposals issued between 2003 and 2006, comprising a total of 71 projects (Formas, 2011). The calls herein evaluated have been directed towards specified areas, each with a combination of subareas such as energy, materials, indoor climate, environment and lifecycle analysis (LCA), information and communication technology (ICT), the role of the building owner, building processes, and industrialisation. In the evaluation, the 71 projects have been grouped into thematic areas: building technology, health and indoor climate, building processes, ICT in the building sector, environmental quality of buildings and building energy. The evaluation was performed by an international panel formed by five active researchers from the engineering research community and is based on the application documents, project documentation and a self-evaluation questionnaire for project leaders.

The evaluation committee found the results acceptable in terms of both scientific quality and benefits gained by the companies involved. There were differences in quality and benefits between themes, but the overall impression was similar. However, the report also points to important weaknesses, in both scientific dissemination by means of international publication, where scores are significantly below what was considered acceptable, and implementation, which in general did not reach the goals set in the implementation plan. The evaluation committee concluded that one problem might be that projects were below the critical size. It was further suggested that the project partners, project leader and implementation manager should be evaluated in terms of their previous experience and qualifications regarding implementation and that the academic qualifications of the implementation manager should be considered. Another recommendation was that the implementation plan should be evaluated upfront together with the research plan.

Case study: the CERBOF Programme and building-related energy research in general

Centrum för Energi- och Resurseffektivitet i Byggande och Förvaltning (CERBOF, the Centre for Energy and Resource Efficiency in the Built Environment) was a research and innovation programme initiated by the Swedish Energy Agency in 2007, managed by IQS and terminated in 2012. CERBOF's vision was *efficient and sustainable exploitation of energy and resources in the construction sector, as well as buildings with high quality indoor environment* (Centre for Energy and Resource Efficiency in the Built Environment, 2013). CERBOF's activities were to be conducive to the commercialisation of research findings and results, and to contribute towards the attainment of national energy and environmental objectives, as well as to the strengthening of the competitiveness of Swedish trade and industry. The board had seven members: one from academia and six from industry, funding bodies and Government agencies.

CERBOF funded projects within two main areas: (i) behaviour, processes and control mechanisms (30 per cent); and (ii) the building as an energy system (70 per cent). Both research teams and industry could apply for funding. Project proposals were evaluated by evaluation committees with external specialists. Projects should be co-funded by industry by a minimum of 60 per cent with 20 per cent in cash. Between 2007 and 2011, there have been six calls with different themes. In total 47 projects have been funded, with a total project volume of SEK130 million, including SEK52 million from CERBOF.

The CERBOF programme has had three goals, of which the most important is to develop collaboration between industry, universities and public authorities. The two other goals were to preserve and develop the competence of the established university-based research groups and to develop more efficient communication and dissemination of results to relevant stakeholders.

The CERBOF programme was evaluated in 2011 by a consultancy firm (Profu, 2011). A questionnaire was sent to project managers, 23 interviews were performed with board members, members of application evaluation committees, administrative staff and researchers, and 10 final reports from funded projects were peer reviewed by a group of independent experts. The evaluation results were mainly positive: goals were considered relevant, the collaboration was perceived as good although with some variation between projects, and industry found university participation to add value. The administration was also quite highly rated, especially for dissemination activities. Respondents were more critical of how projects were followed up. Final reports were lacking for many projects and not regularly checked by CERBOF. Several reports reviewed during the evaluation were found unsatisfactory and none met the criteria for scientific publication.

Another problem was the low quality of applications, especially those from industry. The requirement for co-funding was criticised: co-funding was seen as desirable to ensure industry commitment, but 60 per cent seemed to be too high for many companies. In general, it was easier to find co-funding for technical projects, especially from contractors. The behaviour and processes area was characterised by a lower quality of applications, a more fragmented research community and higher difficulty in finding co-funding. Some suggestions that regard improved administration activities to support more and better applications were: workshops, involvement of evaluation committees, reduction of co-funding to 50–50, and a better long-term planning of calls.

The CERBOF programme has also been discussed in a recent overview of all energy-related building research funded by Swedish bodies (Jansson, *et al.*, 2012). This analysis, carried out by consultants for the Swedish Energy Agency, is intended to serve as a basis for the Agency's internal planning. The mapping and synthesis is based on project lists, interviews with 40 researchers, representatives of funding agencies, end users and purchasers of research, and two hearings/seminars.

This report states that the basis for long-term energy policies is a well-established vision about the transition to a sustainable energy system. In a well-functioning innovation system, research, development, demonstration and commercialisation are related to each other and placed in a broader context of measures, actions and instruments. The conclusion is that the current instruments are insufficient to support implementation of new research results, much due to the high fragmentation in the funding of energy-related building research and the lack of a system perspective that enables identification and elimination of important gaps in the innovation/implementation process.

The mapping identified a coordination issue within and between funding agencies and research councils. The Swedish Energy Agency (SEA) is the largest funding body in this area, but Formas and VINNOVA are also important. Within the SEA there is a multitude of programmes, often administered by a variety of external parties. The CERBOF was seen in this context as a role model since it was the only coordinated programme in the area. However, the report points at a silo mentality and absence of an organisation that has an overall responsibility to ensure that no essential links in the innovation chain are under-financed. Important concerns were raised regarding funding allocation not being sufficient to maintain research skills at the major university-based research environments. Furthermore, implementation of new technologies may be delayed or obstructed because the interaction between researchers and industry is insufficiently developed or there is a lack of relevant knowledge about user behaviour or the impact of policy instruments, management and operation. Lack of funds for large experiments and demonstrators may also make many valuable contributions less effective than they would otherwise have been.

Case study: Bygginnovationen – a VINNOVA programme

Bygginnovationen (Bygginnovationen, 2012) is an innovation programme that runs between 2011 and 2014. It is based on an agreement between VINNOVA, the Governmental Agency for Innovation Systems, and a consortium of 19 firms in the built environment industry. The total budget is SEK42 million; half from VINNOVA and half from Swedish industry. The co-financing from the consortium firms is primarily in kind through work by employees. Three types of grant are offered to firms, which need not be members of the consortium: innovation vouchers, planning grants and development grants.

The overall purpose of the programme is to develop a strong and lasting innovation environment for Swedish construction: bridging the academia/industry gap and promoting the commercialisation of knowledge, solutions and research results. Prioritised areas are information and communication technologies, efficient processes as well as sustainable growth.

One of the strengths of Bygginnovationen is the business advisory committee, a group of 40-odd industry specialists with considerable practical experience of developing products, technologies and processes in the industry. Each application is assessed by at least three members of this committee before the programme board recommends a grant decision to VINNOVA. In many cases, the assessment includes an element of advice that is of value to the applicant, even if the application happens to be rejected. Another strong point and novelty is that all participants involved in the processing of applications are bound by confidentiality agreements regarding the information provided by applicants.

Small and medium-sized enterprises (SMEs) have a prominent position among projects supported by Bygginnovationen. The fact that applications can be submitted as late as 15 workdays before board meetings explains why applicants consider the process to be rapid, efficient and non-bureaucratic.

Programme activities are expected to produce: (i) short-term results such as new collaborative relations, more demand-driven research, new solutions/prototypes/software, new commercial contacts, and increased knowledge and competence in firms; and (ii) medium-term results including new products, processes and services developed by the firms, reducing costs, raising productivity and contributing to increased sustainability, increased demand for research graduates within firms, and attractive research environments within universities and industry. Ultimately, the overall goals are green growth that generates increased employment and turnover, stronger competitiveness of Swedish construction and closing the gap between industry and the university and research institute sector. In general, there is more emphasis on creating stronger mechanisms for the Swedish construction innovation system than on financial returns from projects within the programme.

The programme has links to the Swedish University of the Built Environment, an association where four universities of technology cooperate. Joint efforts include the annual *Research to Construction Business* PhD course and a 2012 anthology on construction innovation.

Of the 33 grants awarded 2011–2012, 24 grants engaged universities and 30 grants SMEs. The main orientation of eight grants was *process*, six grants *ICT*, while the majority, 19 grants, can be classified under *sustainability*. The awarding of over SEK1 million to four development projects indicates the wide range of issues covered by the programme:

- training simulator for virtual concrete spraying
- road-marking paint robot
- software for synchronised construction site activities
- freezing of wastewater sludge.

There is a routine for an initial expert assessment identifying industries that probably will increase their productivity because of these projects and the categories of outputs and inputs that will be directly affected (Bröchner & Olofsson, 2012). This assessment identifies effects also outside the construction industry itself, notably the potential for raising the productivity of construction clients. Predicted increases in total productivity concerned the construction industry (18.5 grants), real estate (14.5), manufacturing (11.5), architectural and consultancy services (3), information and communications (1.5) and energy supply (0.5).

These figures sum up to more than 33 since some projects are expected to raise productivity in more than one industry. Total productivity is expected to rise because of output and input effects. Among the secondary output effects, 11 grants are expected to reduce client energy use, reduce use of other resources for operations and maintenance (8.5), raise user comfort (6), reduce interruption of client activities (5), reduce client risks (5) and raise architectural quality (2). The input effects are to reduce use of labour (13.5 grants), reduce use of materials (9), improve occupational health and safety (2), reduce negative external effects (1.5), reduce energy use in construction (1) and reduce use of services (0.5).

An unpublished early programme evaluation, carried out independently by IQS, identified five strengths of the programme:

- an important, previously lacking, link in the innovation chain to the sector
- the stepwise arrangement with three types of grant works well for the intended target group
- the continuous call for applications and the rapid decision process support an active environment for innovation
- the business advisory committee contributes to a secure, rapid and competent assessment
- the industry consortium gives credibility to the joint private–public effort.

IQS also found areas for potential improvement:

- stronger communication of programme activities
- simplify and clarify the VINNOVA grant application process
- add supporting and coaching activities to what the business advisory committee offers
- add further support in early and late stages to the three types of grant
- develop the tools used for monitoring and measurement.

The programme board of Bygginnovationen has formulated a long-term innovation strategy for the Swedish built environment industry (Bygginnovationen, 2012). The strategy reflects challenges that have been identified for the firms in the built environment industry and also includes background analyses of a customer-driven innovation climate, interaction for innovation, and encouragement of collaborative innovation.

R&D impact and measurement

More formal evaluation of Swedish construction research projects was a practice that emerged in the 1980s. The Building Research Council selected research topics and engaged Swedish and foreign experts who analysed publications and visited research groups. Emphasis was on ongoing projects, the quality of the research and publication practices. Starting in 1989 the private sector SBUF also began conducting reviews of groups of funded projects, although the focus in these reviews has been on barriers to diffusion and implementation (Bröchner, *et al.*, 1989; Widén & Hansson, 2007).

The reviews of the Formas-BIC/IQS programme and the CERBOF programme, summarised in the above sections, are both strongly based on project documentation and self-evaluations by project managers. From some projects, very little documentation is available, and some project managers have not answered the questions. Furthermore, the Formas-BIC/

IQS survey questions were formulated so that project managers should rate the performance of their projects in relation to what they would expect in this kind of collaboration. Thus, subjectivity enters both in the grading of one's own performance and in defining expected performance. In the CERBOF programme, the grading scale runs from poor to very good, but it is likely that the frame of reference in this case is also based on previous experience from other collaborations in the area. Much of the information is obtained by interviews.

The overview for the Energy Agency produced by Jansson, *et al.* (2012) is not a formal evaluation and is focused on programmes rather than projects. However, it uses a methodology similar to those of the other evaluations: interviews, seminars and documentation, but no self-assessment questionnaires. The results point at important shortcomings and gaps in the current strategy for building-related energy research and innovation, but also suggest some difficulties with evaluating the performance of single programmes in general: it is not only that success may be hard to assess, but it may also depend on weaknesses in other parts of the system. Thus, an important implication is that a meaningful evaluation needs to comprise a wider knowledge base than only that of the specific programme in focus.

A recent meta-analysis of a VINNOVA series of impact studies (Elg & Håkansson, 2011) has highlighted that public investment in research and innovation is generally efficient from a social cost-benefit viewpoint. However, a timescale of decades is needed to capture many effects and Government-supported projects cannot be isolated from a broader industrial context. It was concluded that the most important effects have been in the form of competence development and organisational learning, in other words improvement of organisational innovation capabilities and development of innovation systems. What is supported by VINNOVA and other funding agencies as a project or even a whole programme is typically only a part of longer term innovation processes within companies as well as universities. Obviously, this makes it difficult to discern effects of single funding initiatives.

Strategies for the future

There is no overarching attempt currently to provide a full roadmap for the future of Swedish construction R&D. However, the programme board of Bygginnovationen has formulated a long-term innovation strategy for the Swedish built environment industry (Bygginnovationen, 2012). This strategy is seen as a basis for discussion and detailing of action plans for continued efforts for innovation by the stakeholders within the built environment sector.

The innovation strategy is based primarily on interviews and questionnaires with representatives of the firms within Bygginnovationen and a selection of organisations active within the built environment sector, as well as with representatives of public and private sources of research funding. The strategy reflects challenges that have been identified for the firms in the built environment sector and also includes background analyses of a customer-driven innovation climate, interaction for innovation and encouragement of collaborative innovation.

VINNOVA has recently begun supporting *strategic innovation areas* (SIOs) where various research and innovation activities, with funding from different sources, related to the same industry or problem area are to be integrated into a national innovation programme. It is expected that a couple of such SIOs will be related to the built environment sector, since VINNOVA supports the preparation of strategic innovation agenda documents for building information modelling (BIM) and industrialised building, written jointly by representatives of firms and universities. Also Formas is preparing a new strategy for how to handle additional funding to urban development research.

Concluding remarks

The general trend in Sweden is towards an increasing focus on university–industry collaboration in research and innovation. Industry co-financing will be required not only by VINNOVA and the Swedish Energy Agency but also for the new funds allocated to Formas according to the *Government Research and Innovation Bill*. Second, there is a focus on *grand challenges* for society. A third trend is collaboration and resource integration: VINNOVA aims for national collaborative programmes to strengthen international competitiveness and Formas will initially fund larger trans-disciplinary research groups taking an integrative, triple bottom line perspective on issues of urban development. Altogether, this new, complex and partly contradictory funding landscape is not easy to navigate through for potential partners from the industry and universities. The national cooperative organisation, Swedish Universities of the Built Environment, has highlighted the need for a holistic perspective on the funding of research and innovation activities, including the entire innovation chain in various subareas and education aspects at bachelor and master level. Their ideas are in line with the SEA review of energy-related research (Jansson, *et al.*, 2012), but also stress that a share of free and flexible funding is important in order to support long-term research and innovation strategies in a volatile environment.

References

Benner, M. & Sörlin, S. (2007) 'Shaping strategic research: Power, resources, and interests in Swedish research policy', *Minerva*, 45: 31–48.

Bröchner, J. (2010) 'Construction contractors as service innovators', *Building Research and Information*, 38: 235–246.

Bröchner, J. & Grandinson, B. (1992) 'R&D cooperation by Swedish contractors', *Journal of Construction Engineering and Management*, 118: 3–16.

Bröchner, J., Gustafsson, G. & Rahm, H.G. (1989) 'Utvecklingsprojekt med bidrag från SBUF' ('Development projects with support from SBUF') in *Royal Institute of Technology Bulletin No. 24*, Stockholm: Department of Building Economics and Organisation.

Bröchner, J., Josephson, P.-E. & Kadefors, A. (2002) 'Swedish construction culture, management and collaborative quality practice', *Building Research and Information*, 30: 392–400.

Bröchner, J. & Olofsson, T. (2012) 'Construction productivity measures for innovation projects', *Journal of Construction Engineering and Management*, 138: 670–677.

Bröchner, J. & Sjöström, C. (2003) 'Quality and coordination: Internationalizing Swedish building research', *Building Research and Information*, 31: 479–484.

Bygginnovationen (2012) *Innovation strategy for the Swedish built environment industry*. Available at: www.bygginnovationen.se/documents/Bygginnovationen/Miscellaneous/Innovation_strategy_20121126.pdf (accessed 12 December 2012).

Byggkommissionen (2002) *Skärpning gubbar! Om konkurrensen, kvaliteten, kostnaderna och kompetensen i byggsektorn* (*Attention guys! On competition, quality, cost and capability in the construction sector*), Statens offentliga utredningar (SOU) 2002: 115, Stockholm: Byggkommissionen, Socialdepartementet.

Byggkostnadsdelegationen (2000) *Från byggsekt till byggsektor* (*From construction sect to construction sector*), SOU 2000: 44. Stockholm: Byggkommissionen, Socialdepartementet.

Centre for Energy and Resource Efficiency in the Built Environment (2013) *Centre for energy and resource efficiency in the built environment (CERBOF)*. Available at: www.cerbof.se/sa/node.asp?node=50 (accessed 24 September 2013).

Elg, L. & Håkansson, S. (2011) *När staten spelat roll – Lärdomar av VINNOVAs effektstudier* (*When the state has played a role: Lessons from VINNOVA effect studies*), VINNOVA Analys VA 2011: 10. Available at: www.vinnova.se/upload/EPiStorePDF/va-11-10.pdf (accessed 18 December 2012).

Formas (2011) *Research projects initiated by Formas-BIC 2003–2006*. Evaluation report 6, Stockholm: Formas. Available at: wwwold.formas.se/upload/EPiStorePDF/Research_projects_initiated_by_Formas_BIC_2003_2006_Evaluation_Report_R6_2011/Formas_R6_211.pdf (accessed 21 March 2013).

Formas (2012) *Kartläggning av forskningen inom samhällsbyggnadsområdet* (*Mapping of research in the built environment area*). Available at: www.formas.se/PageFiles/4859/FormasKartl%C3%A4ggning_samh%C3%A4llsbyggande120430.pdf (accessed 21 March 2013).

Government Bill (2012a) *Forskning och innovation* (*Research and innovation*), Prop. 2012/13:30, Stockholm: Utbildningsdepartementet.

Government Bill (2012b) *Forskning och innovation för ett långsiktigt hållbart energisystem* (*Research and innovation for a long-term sustainable energy system*), Prop. 2012/13: 21, Stockholm: Utbildningsdepartementet.

Government Offices of Sweden (2012) *Research and innovation. A summary of government bill* 2012/13: 30. Available at: www.regeringen.se/content/1/c6/20/70/30/775db39c.pdf (accessed 21 March 2013).

IQS (2011) *Kompletterande PM angående finansiering av FoU inom samhällsbyggnadsområdet* (*Additional memorandum on R&D funding within the built environment field*), unpublished memorandum, IQ Samhällsbyggnad.

Jansson, T., Faugert, S., Håkansson, A., Stern, P. & Terrell, M. (2012) *Energirelaterad byggforskning i Sverige* (*Energy related building research in Sweden*). Available at: www.energimyndigheten.se/Global/Forskning/Bygg/FaugertoCo_Byggforskning%20i%20Sverige_slutrapport%20120329.pdf (accessed 18 December 2012).

Miozzo, M. & Dewick, P. (2002) 'Building competitive advantage: Innovation and corporate governance in European construction', *Research Policy*, 31: 989–1008.

Produktivitetskommittén (2012) *Vägar till förbättrad produktivitet och innovationsgrad i anläggningsbranschen* (*Roads to improved productivity and degree of innovation in the civil engineering industry*), Statens Offentliga Utredningar (SOU) 2012: 39, Stockholm: Produktivitetskommittén, Näringsdepartementet.

Profu (2011) *Utvärdering av CERBOF-programmet (Evaluation of the CERBOF programme)*. Available at: www.energimyndigheten.se/Global/Forskning/Bygg/CERBOF/Profu_Utv%C3%A4rdering_Cerbof_110315rev0406.pdf (accessed 18 December 2012).

Statskontoret (2009) *Sega gubbar? En uppföljning av Byggkommissionens betänkande 'Skärpning gubbar!'* (*Tough guys? A sequel to the 'Attention guys!'*), Byggkommissionen report, Statskontoret 2009: 6, Stockholm: Statskontoret.

Widén, K. & Hansson, B. (2007) 'Diffusion characteristics of private sector-financed innovation in Sweden', *Construction Management and Economics*, 25: 467–475.

16 USA – characteristics, impacts and future directions

Sarah Slaughter, Douglas Thomas and Robert Chapman

Snapshot of the US construction industry

Investment in infrastructure, plant and facilities in the form of construction activity, provides the basis for civil society, that is for the health, safety, and wellbeing of citizens and the viability of communities, as well as commercial activity: from the production of products to the delivery of goods and services, and the movement of individuals. Investment in the built environment provides benefits across all sectors of the US economy. For example, expenditures in housing accommodate new households and allow existing households to improve or expand their living conditions, while creating a demand pull throughout the manufacturing sector for materials, products and systems. It is clear that construction activities affect nearly every aspect of the US quality of life and the economy, and that this industry is vital to the continued development of the national economy. In 2011, the latest year for which construction data are available, the construction industry's contribution to gross domestic product (GDP) was USD520 billion, or 3.4 per cent of GDP (Table 16.1).

In 2010 construction volume, that is the value of construction put in place plus expenditures for maintenance and repair, was over USD900 billion (Table 16.2). Approximately 38 per cent of the value of construction volume, equivalent to USD307 billion, was due to the demand for manufactured products, materials, components and systems. Constructed facilities increase the demand for manufactured goods once they go into use. Consequently, the construction industry generated over USD400 billion in demand for manufactured products in 2010. For example, USD97 billion was spent on contents, namely carpets, curtains, appliances, lighting fixtures, audio/video equipment, and furnishings in 2010. Once constructed facilities go into use, they generate demands for energy, water and services. In 2010 these demands amounted to USD552 billion, of which USD430 billion was for energy.

Construction also has a major impact on US employment. In 2011 over 9 million people were employed in the construction industry, and construction employment has equalled between 6.5 and 8 per cent of total US employment in recent years (Table 16.3). The composition of the

Table 16.1 US Construction industry as a proportion of GDP, value added to GDP and value put in place

	2004	*2005*	*2006*	*2007*	*2008*	*2009*	*2010*	*2011*
Construction as % GDP	4.7	4.9	4.9	4.7	4.3	3.9	3.5	3.4
Construction value added to GDP (billion USD)	554.2	612.5	651.0	653.8	614.2	541.9	511.6	520.3
Construction annual value put in place (billion USD)	991.4	1,104.1	1,167.2	1,152.4	1,067.6	903.2	804.6	778.2

Source: U.S. Bureau of Economic Analysis (2012); U.S. Census Bureau (2012)

Table 16.2 Construction expenditures (billion USD), 2010, USA

	New construction	*Renovation*	*Maintenance and repair*	*Contents and furnishing*	*Energy, water, services*	*Total*
Construction volume	547	257	133			937
Manufactured goods	----------------	--------307--------	----------------	97		404*
Operations					552	552
					Total	**1,586******

Source: Thomas (2010)

Note: Sources and methods used to calculate these values are described in D. S. Thomas, 'Methodology for Calculating Construction Industry Supply Chain Statistics', NIST Special Publication 1116, September 2010

* In manufactured goods, USD307 billion is included in construction volume

** To prevent double counting, the total equals the sum of 937, 97 and 552.

Table 16.3 Construction industry employment as proportion of total employment, construction industry employment and total employment, 2004–2011, USA

	2004	*2005*	*2006*	*2007*	*2008*	*2009*	*2010*	*2011*
Construction as % of total employment	7.7	7.9	8.1	8.1	7.5	6.9	6.5	6.5
Construction employment ('000)	10,768	11,197	11,749	11,856	10,974	9,702	9,077	9,039
Total employment ('000)	**139,252**	**141,730**	**144,427**	**146,047**	**145,362**	**139,877**	**139,064**	**139,869**

Source: Bureau of Labor Statistics (2012)

construction workforce differs from much of the US workforce due to the large number of self-employed workers, sole proprietorships and partnerships. Within the construction industry, there are 1.7 million self-employed workers. In contrast, manufacturing, which employs 14.1 million workers, has only 304,000 self-employed workers.

US investment in construction R&D

US annual expenditures for construction-related research and development (R&D) are approximately USD2 billion, with major expenditures from the US Federal Government. Recently published data from the National Science Foundation (NSF), covering both Federal-funded construction-related R&D and private and other than Federal-funded construction-related R&D, provide a view of shifts experienced over the past two decades. Federal-funded construction-related R&D (in millions of 2010 dollars) vary widely by department and by year, reflecting some of the changing budget priorities between 1992 and 2009, the latest year for which published data from NSF are available (Figure 16.1).

Private companies and other organisations also support construction-related R&D (Table 16.4). The data from NSF on these R&D expenditures are more limited than for Federal funding sources. As a result, only data for 2001 through 2006 are reported. Note that funding levels between 2001 and 2003 were considerably less than the funding levels for 2004 through 2006.

The US Federal Government has a number of Federal research laboratories contributing to construction R&D. The National Institute of Standards and Technology (NIST) has been a major participant, with the Building and Fire Research Laboratory, now part of the Engineering Laboratory, focused specifically on this area. Other laboratories include the National

Table 16.4 Private and other than Federal-funded construction-related R&D levels (millions of 2010 USD), 2001–2006, USA

Year	*Software development*	*Materials synthesis and processing*	*Other*	*Total R&D*
2001	–	–	394[e]	**394[e]**
2002	15	27	158	**199**
2003	14	17	364	**395**
2004	59	–	1,599	**1,710***
2005	–	74	1,345	**1,419[e]**
2006	9	148	1,336	**1,492**

Source: National Science Foundation (2012a)

Note: [e] Estimated

* Some data values were suppressed to avoid disclosure of confidential information.

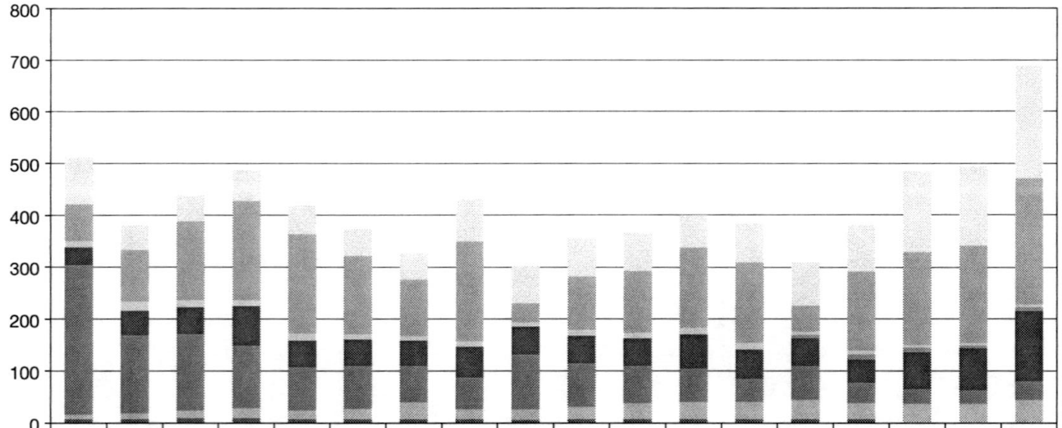

	1992	1993	1994	1995	1996	1997	1998	1999	2000	2001	2002	2003	2004	2005	2006	2007	2008	2009
Other	90	45	48	99	52	52	51	80	73	72	74	64	71	86	88	155	153	218
Transportation	75	105	157	193	194	153	110	196	38	106	122	159	159	52	157	183	189	248
Interior	8	15	11	11	12	10	6	7	8	9	8	8	9	6	5	2	2	2
Homeland Security	0	0	0	0	0	0	0	0	0	0	0	0	2	6	9	8	5	10
Energy	40	49	34	75	53	50	51	60	53	56	57	69	58	56	47	73	85	133
Defense	286	150	147	120	83	83	69	61	106	83	70	63	44	63	36	29	25	38
Commerce	7	9	13	19	16	19	31	20	19	23	29	32	32	35	32	34	33	41
Agriculture	8	8	9	9	7	7	8	6	6	6	7	7	7	7	7	2	1	1

Figure 16.1 Federally funded construction-related R&D levels (millions of 2010 USD), 1992–2009, USA

Source: National Science Foundation (2012b)

Renewable Energy Laboratory (NREL), Lawrence Berkeley Laboratory (LBL) and the Pacific Northwest National Laboratory (PNNL).

Universities contribute to construction R&D through sponsored research from Federal agencies, including the NSF, and various Federal agencies, including the Department of Energy (DOE), Environmental Protection Agency (EPA), Department of Transportation (DOT) and the Department of Commerce, as well as research contracts from State and local Governments, companies, and other organisations. In addition, research organisations within companies and industry associations, such as FIATECH and Construction Industry Institute (CII), also perform R&D related to the construction industry.

The American Society of Civil Engineers (ASCE) Civil Engineering Research Foundation (CERF) conducted a nationwide survey, co-sponsored by NSF, to examine how research efforts vary by funding source, programme area and type of research (Civil Engineering Research Foundation, 1993). The CERF study included information from Federal agencies, industry, academia, State and local Government, and non-profit organisations. Research programme areas included structures, such as buildings, bridges, industrial facilities, among others; materials; energy; environmental; and natural hazards. The types of research covered included basic, applied, development, demonstration, and others. The CERF study is particularly illuminating because it allows breakdowns of expenditures by sector: Federal agencies, industry, academia, and other; and type of research: basic, applied, development, demonstration and other. In this study, nearly two-thirds (63 per cent) of research expenditures were by Federal agencies, while industry accounted for 16 per cent, academia for 12 per cent and other organisations 9 per cent (Table 16.5).

Industry is strongly focused on moving research into the marketplace; it invests three-quarters of its research funding in development and demonstration, while it invests less than 10 per cent of its research funding in the basic and applied stages of research (Figure 16.2). Academia invests more than three-quarters of its research funding in the basic and applied stages of research.

Table 16.5 Research expenditures by funding source and type of research, USA (millions of 1992 USD)

Type of research	*Federal*	*Industry*	*Academia*	*Other*	*Total*	*Proportion of total*
Basic + applied	687	31	204	115	1,037	49%
Development + demonstration	634	312	54	65	1,065	51%
Total	**1,322**	**343**	**258**	**180**	**2,103**	**100%**

Source: Civil Engineering Research Foundation (1993)

Note: The sum of the totals may not equal the total shown due to rounding.

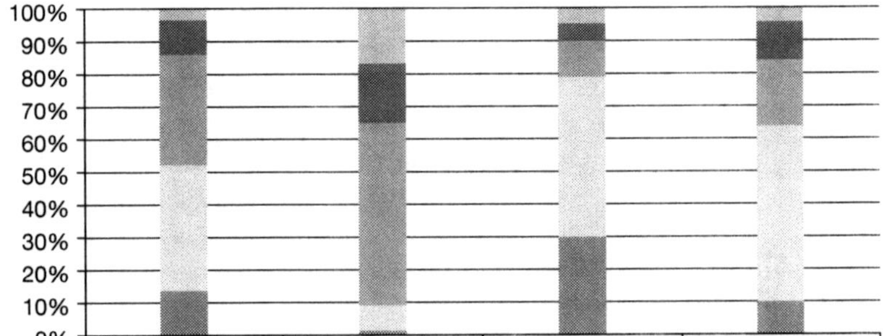

	Federal	Industry	Academia	Other
Other	3%	16%	5%	4%
Demonstration	11%	19%	5%	12%
Development	34%	56%	12%	20%
Applied	38%	8%	48%	54%
Basic	14%	1%	31%	10%

Figure 16.2 Construction-related research by sector and type of research, 1992, USA

Source: Civil Engineering Research Foundation (1993)

Note: The sum of the values may not equal the total shown due to rounding.

Federal agencies allocated roughly half of the Federal research funds to the basic and applied stages of research and about 45 per cent to the development and demonstration stages. Federal agencies are the dominant funder of basic and applied research, accounting for two-thirds of the total for all basic and applied research. Federal agencies are thus ideally situated to complement shortfalls in industry research in the basic and applied stages and academia in the development and demonstration stages.

The focus of construction-related R&D funding over the last 20 years has included:

- strength and durability of materials under normal and extreme conditions, particularly that of steel, concrete, carbon-fibre based materials and wood
- strength and durability of systems, including structural, communications, transportation, and energy; under normal and extreme conditions
- computer-aided design, manufacturing, and operations of facilities, including facility information models, interoperability of components and controls and facility management systems
- construction means and methods, including equipment improvements, robotics and site materials management
- condition assessment, including non-destructive testing, remote sensing and real-time condition monitoring

- energy efficiency and renewable energy
- environmental impacts, mitigation and recovery.

Challenges in the US construction industry and opportunities for R&D

Given the demonstrated large impact of construction on the nation's macroeconomic objectives, effective construction research becomes critical to the economy. As noted earlier, the combined annual expenditure for construction-related R&D is approximately USD2 billion, which is approximately 0.25 per cent of the value of construction put in place. The CERF study estimated approximately USD4 billion in annual R&D expenditures, which is approximately 0.5 per cent of the value of construction put in place. These proportions are low for an industry that is one of the largest contributors to US GDP. For instance, private sector R&D investments in manufacturing totalled nearly USD155 billion in 2006 (USD168 billion in 2010 dollars). Total R&D investments in construction in 2010 were even surpassed by segments of the manufacturing industry: USD9.7 billion (USD10.5 billion in 2010 dollars) for machinery, a mature segment of the industry, and USD48.3 billion (USD52.2 billion in 2010 dollars) for computers and electronic products, a high-technology segment of the industry (National Science Foundation, 2011).

Underinvesting reduces the potential for innovations that contribute to substantial national benefits; namely, constructed facilities that enhance occupant health, safety and wellbeing, contribute to a community's viability, are more user friendly, affordable, productive, and that are easier, faster, and more lifecycle cost effective to build, operate and maintain. Given the impact of construction spending on the economy's health, and that construction research helps make construction globally competitive and profitable, construction research becomes a critical variable in generating economic development. Current key drivers for change in construction research are:

- sustainability and environmental security
- competition due to globalisation and offshoring
- homeland security and disaster resilience
- infrastructure renewal
- demand for better, faster and less costly construction
- management and control of complex, interdependent, large-scale systems.

Finding a solution to the underinvestment problem is difficult because the construction market fails to provide sufficient incentives to encourage industry research. Multiple reasons contribute to market failure. First, the contracting systems used for many projects, particularly design/bid/

build contracts, often create conflicting incentives for developing and implementing advancements in the industry. For instance, one party may bear most of the costs and risks for introducing an innovation while another party may appropriate the majority of the benefits. This misalignment of incentives often stifles innovation. In addition, the design/bid/build system of contracting separates the design and construction phases of a project, which means that construction knowledge cannot be fully considered during design, which can increase the time, cost and effort associated with introducing an innovation.

Second, profit margins in the construction industry are very tight, effectively excluding many firms from any kind of R&D activity. For example, net income as a percentage of expenditures in the construction industry averaged only 11.2 per cent in 2007 compared to 19.7 per cent in manufacturing (U.S. Census Bureau, 2010). These factors, namely the design/bid/build system of contracting and modest profit margins, are exacerbated by the fragmented nature of the construction industry. Both the residential and non-residential construction markets have relatively few large establishments that can afford research and new technology risks: small establishments, firms with nine or fewer employees, comprise 80 per cent of the total. However, the larger establishments, firms with 10 or more employees, account for 82 per cent of total business with 77 per cent of total employees (U.S. Census Bureau, 2010).

The large number of small construction establishments and the concentration of self-employed workers complicate the adoption of new technologies and practices. Since construction employment is affected by both the weather and the business cycle, year-to-year changes in employment can be substantial, resulting in major layoffs and hiring surges. The cyclical nature of construction employment produces shortages in many highly skilled trades. These shortages adversely impact productivity in the construction industry and reduce the availability of skilled tradesman to adapt and adopt the innovation as needed to enhance diffusion across the industry.

Teicholz (2004) highlights the magnitude of the construction industry's perceived decline in productivity (Figure 16.3). As measured by constant dollars of new construction work per field work hour, labour productivity in the construction industry has trended downward over the past 40 years at an average compound rate of –0.6 per cent per year. That is, construction projects have required significantly more field work hours per constant dollar of contract. This is particularly alarming when compared to the increasing labour productivity in all non-farm industries, which has trended upward at an average compound rate of 1.8 per cent per year. This trend indicates that the construction industry lags behind other industries in developing labour saving ideas and in finding ways to substitute equipment for labour. Teicholz focuses on three topics that have an adverse impact on construction labour productivity: the lack of R&D spending, fragmentation

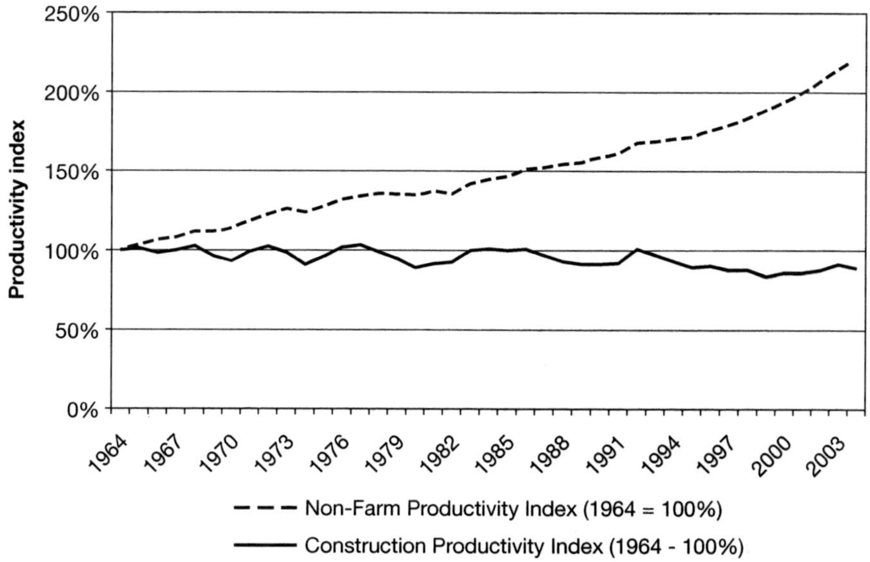

Figure 16.3 Labour Productivity Index for the US construction industry and all non-farm industries, 1964–2003

Source: Teicholz (2004)

within the industry and declining real wage rates. He also notes that despite the fact that there has been a significant adoption of new information technology by the construction industry over the past 35 years, these applications tend to run in a standalone mode that does not permit improved collaboration among the project team members.

Significant opportunities exist to rapidly improve the construction industry, which may require additional research expenditures as well as new approaches to develop and deploy advancements.

A recent study by the US National Research Council, sponsored by NIST, identified five interrelated activities that could lead to breakthrough improvements in construction efficiency and productivity in 2 to 10 years, noting *if implemented throughout the capital facilities sector, these activities could significantly advance construction efficiency and improve the quality, timeliness, cost-effectiveness, and sustainability of construction projects* (National Research Council, 2009a). The five activities are:

1 widespread deployment and use of interoperable technology applications, also called building information modelling (BIM)
2 improved job-site efficiency through more effective interfacing of people, processes, materials, equipment and information

3 greater use of prefabrication, preassembly, modularisation and off-site fabrication techniques and processes
4 innovative, widespread use of demonstration installations
5 effective performance measurement to drive efficiency and support innovation.

A follow-up study, co-sponsored by NIST and McGraw-Hill Construction, focused on the opportunities associated with prefabrication and modularisation and the challenges that must be overcome for these techniques to be applied at the level needed to improve efficiency and productivity in the industry (McGraw-Hill Construction, 2011). The objectives of the study were to explain why the use of prefabrication and modularisation has not become as widespread in the commercial construction marketplace as it has in the industrial sector and to offer recommendations for how the industry can move forward to maximise these practices in the future. Commonly used prefabricated and modular building elements include: mechanical, electrical, and plumbing systems; exterior walls; building superstructures; roofing; floors; and interior room modules.

The report demonstrated that prefabrication and modularisation are already yielding real business benefits to users. Out of a sample of over 800 architecture, engineering and construction (AEC) professionals, 66 per cent reported improved project schedules, 65 per cent reported decreased project costs and 77 per cent reported reduced construction waste. Increased adoption of BIM is also fuelling the re-emergence of prefabrication and modularisation as a critical new trend and BIM and prefabrication/modular construction in combination improve both worksite productivity and overall project return on investment. Additional sections of the report explored potential new cutting-edge research.

Additional opportunities exist to expand construction-related research to address critical issues. For instance, the US National Institute of Occupational Safety and Health (NIOSH), with the centres for disease control (CDC), have formulated a *National Occupational Research Agenda* to improve construction safety and health (Centers for Disease Control and Prevention, 2008), and have been pursuing research in this area. The National Science and Technology Council, Subcommittee on Buildings Technology Research and Development, developed a Federal R&D agenda for *Net-Zero Energy, High Performance Green Buildings* (National Science and Technology Council, 2008), which included recommendations for coordinated research on:

- integrated, performance-based design, construction and operation
- net-zero energy building technologies and strategies
- water use and rainwater retention
- material utilisation, waste and lifecycle environmental impacts
- occupant health and performance
- overcoming barriers to implementation.

Work is also proceeding on these topics, with additional opportunities for significant improvements.

Additional high social-return, growth-supporting research opportunities abound. If some of the Federal funding commitment can be tied into construction industry roadmaps and can further foster public–private partnerships, then the combined efforts can result in enabling technologies that promote both more cost-effective constructed facilities and a more productive construction workforce. A number of leaders in construction-related research, including NIST, Stanford University's Center for Integrated Facilities Engineering (CIFE), Massachusetts Institute of Technology (MIT), CII, and FIATECH, have participated in public–private partnerships aimed at improving the competitiveness and efficiency of the US construction industry.

Although the challenges of changing such a large industry are formidable, there are several noteworthy successes that have resulted in breakthrough improvements. The following case studies explore the impact and potential for construction-related R&D to address critical challenges and opportunities in the industry.

Case studies

A recent series of NIST impact studies determined that specific Government research projects in construction technology resulted in rates of return ranging from 13 to 57 per cent, and averaging 27 per cent, indicating extraordinary returns to be had from construction-related research. NIST and its predecessor organisation, the National Bureau of Standards, have conducted periodical studies of the economic impacts of its research investments on the built environment. These studies covered both retrospective analyses in which data from actual industry applications were used to estimate the impacts of NIST's research efforts and prospective studies in which industry stakeholders were contacted about their perceptions of future benefits associated with NIST's research efforts. Four studies are briefly summarised below; the first two are retrospective and the last two are prospective. In all four studies, NIST's contribution helped bring new technologies into the marketplace on an accelerated basis. The economic impact studies measure the value of NIST's contribution to the accelerated adoption of the new technologies.

Case study: ASHRAE 90-75

Even before the 1973 oil embargo, energy consumption in buildings was a major concern and the subject of NIST research because an increasing number of brownouts in various regions of the USA led many States to consider energy conservation standards. In early 1973 the National Conference of States on Building Codes and Standards (NCSBCS) asked NIST to develop recommendations for building code provisions for energy conservation. NCSBCS's goal was to produce building code provisions that would be acceptable to industry in general and lead to consistent standards for all States. The results of NIST's research, first published in February 1974, were adopted as industry consensus standards by the American Society for Heating, Refrigerating, and Air-Conditioning Engineers (ASHRAE) in August 1975. Based on published data, the ASHRAE 90-75 standard produced significant energy cost savings in single-family residences nationwide. The return on NIST's investment in performing the research that underpins the ASHRAE 90-75 standard was estimated to be 57 per cent (Chapman & Fuller, 1996).

Case study: fire safety in health care facilities

Concern over fire safety in hospitals and nursing homes caused Congress to make compliance with the *Life Safety Code* a requirement for participation in the Medicare and Medicaid programmes. Because the Life Safety Code was a prescriptive code, many health care facilities were faced with costly renovations. In 1975, the Department of Health and Human Services sought technical support from NIST. This support led to the development of the Fire Safety Evaluation System (FSES), a framework for determining how combinations of several widely accepted fire safety measures could be used to provide a level of safety equivalent to that prescribed by the Life Safety Code. Optimisation software developed by NIST built on the FSES equivalency approach, providing facility managers a menu of solutions, many of which resulted in significant cost savings. NIST's active role in the development of the FSES led to its acceptance into the Life Safety Code in 1981. Without NIST's participation in the development of the FSES, it is likely that the acceptance into the Life Safety Code would have been delayed by at least four years. The return on NIST's FSES-related research investments was estimated to be 35 per cent (Chapman & Weber, 1996).

Case study: cybernetic building systems (CBS)

A CBS is defined as a multi-system configuration able to communicate information and control functions simultaneously and seamlessly at multiple levels. The multiple levels of communication and control are based on the building automation and control networks (BACnet) layered protocol architecture. NIST's focused research on BACnet, under the auspices of ASHRAE, led to BACnet's acceptance as a national standard in 1995. Complementary efforts in fire technology, mechanical and electrical systems, fault detection and diagnostic systems, and economic analysis led NIST to form an integrated CBS project team in 1997. NIST was uniquely positioned to collaborate with industry on the development of CBS products and services and to provide a forum for conducting interoperability testing. This collaborative effort led to significant savings in energy costs, while maintaining or improving system performance. This prospective study of NIST's CBS-related research had an estimated rate of return of 16 per cent (Chapman, 1999).

Case study: construction systems integration and automation technologies (CONSIAT)

NIST's focused research on Plant STEP, construction metrology, economic analysis, and its collaboration with the CII, led to the creation of an integrated CONSIAT project team in 1997. The team worked closely with building owners, contractors, equipment and systems manufacturers and service providers, software developers, professional societies, university researchers and other Government agencies on the development of a first generation of fully integrated and automated project processes. A prospective economic impact study carried out in collaboration with CII and covering capital facilities, namely industrial facilities and commercial buildings, estimated NIST's CONSIAT-related rate of return as 17 per cent (Chapman, 2000, 2001, 2002).

Measuring the costs of inadequate interoperability

Research by NIST and other construction industry leaders stimulated interest in interoperability. One-time data entry and the seamless flow of information to all project participants throughout the project lifecycle is a goal of the interoperable construction environment. Interoperability among computer-aided design, engineering and construction software systems offers the potential for revolutionary change in the effectiveness with which construction-related activities are executed and in the value they add to construction industry stakeholders. The lack of quantitative measures of the annual cost burden, however, has hampered efforts to promote change within the industry.

Renewed interest in the costs of inadequate interoperability was sparked by the creation of FIATECH, and the need for an unbiased effort at measuring the magnitude of those costs, which led to NIST's decision to embark on a formal study. The study, published as NIST GCR 04-867, produced a USD15.8 billion estimate of the annual cost burden due to inadequate interoperability in the capital facilities segment of the US construction industry (Gallaher, *et al.*, 2004). These cost impacts are of interest to multiple stakeholders, namely owners and operators of capital facilities; design, construction, operation and maintenance, and other providers of professional services in the capital facilities industry; and public and private sector research organisations engaged in developing interoperability solutions.

Interoperability problems affect an array of stakeholders, which in this study focused on architects and engineers, general contractors, specialty fabricators and suppliers, and owners and operators; throughout the lifecycle phases of planning, engineering and design, construction, operations and maintenance (O&M), and decommissioning. Examples of efficiency losses associated with current business activities include manual re-entry of data, duplication of business functions, and the continued reliance of paper-based information management systems. These efficiency losses are indicative of three generic cost categories: avoidance costs, mitigation costs and delay costs. Avoidance costs are related to activities undertaken to prevent or minimise the impact of interoperability problems before they occur. Mitigation costs stem from activities responding to interoperability problems. Most mitigation costs result from electronic or paper files that have to be re-entered manually into multiple systems and from searching paper archives. Delay costs arise from interoperability problems that delay the completion of a project or the length of time a facility is not in normal operation. NIST GCR 04-867 (Gallaher, *et al.*, 2004) shows clearly that the majority of estimated costs are borne by owners and operators and that the O&M phase has higher costs associated with it than other lifecycle phases as information management and accessibility hurdles hamper efficient facilities operation (Table 16.6).

Owners and operators bore about two-thirds of the total estimated costs. Architects and engineers had the lowest interoperability costs, and general contractors and specialty fabricators/suppliers bore the balance of costs (Gallaher, *et al.*, 2004). Costs increase sharply from the planning, engineering and design phase, through the construction phase, to the O&M phase. The cost burdens borne by the various stakeholder groups also shift. In the planning, engineering and design phase, architects and engineers bear the majority of the cost burden while the general contractors and specialty fabricators and suppliers shoulder most of the cost burden during the construction phase. O&M phase costs are borne almost exclusively by owners and operators (Gallaher, *et al.*, 2004).

The USD15.8 billion annual cost burden demonstrates that significant opportunities exist both for increasing the efficiency of the project delivery process and for better resource allocation in the O&M phase. However, these opportunities face major challenges that must be addressed to generate change.

Table 16.6 Estimated cost of inadequate interoperability, USA

Stakeholder	*Estimated costs (billion USD)*	*Inefficiency loss*	*Lifecycle phase*
Architect/engineer	1.2	1.1%	Plan/eng/design
General contractor	1.8	0.9%	Construction
Specialty fabricators/suppliers	2.2	1.2%	Construction
Owners/occupants	10.6	2.8%	Operation
Total	**15.8**		

Source: Gallaher, *et al.* (2004)

The four stakeholder groups, namely architects and engineers (A&Es); general contractors (GCs); specialty fabricators and suppliers (SF&Ss); and owners and operators (O&Os), have differing measures of economic performance against which to compare their cost burden. For A&Es, GCs and SF&Ss, the key measure is the annual receipts for the work they perform. When these measures of economic performance are compared to the annual cost burdens, the impact of inadequate interoperability on the corporate balance sheet begins to emerge. These inefficiency losses are expressed as a percentage of the economic measure of performance (Gallaher, *et al.*, 2004). In an industry in which profit margins are thin, these inefficiency losses represent a genuine opportunity for improvement. The key for initiating change, however, is to identify specific inefficiencies and to find opportunities for collaboration both within individual firms and across firms and stakeholder groups in the capital facilities industry, which will result in mutual gains.

The NIST interoperability study considered the effect of inadequate interoperability on internal business functions. Specific questions in each stakeholder group's survey instrument were used to estimate the magnitude of inefficient business process management costs. These survey questions were also useful in identifying the top business processes that were the prime areas of concern. Three business processes, namely project management, document management and information request processing, were identified as problem areas across all four stakeholder groups.

Because business process management functions affect both internal efficiency between different units within the firm, and information exchange between project participants external to the firm, reducing their cost burden offers potential for mutual gains. Inefficient business process management costs range from almost USD400 million for A&Es to USD2.6 billion for O&Os. Overall, inefficient business process management costs are USD6 billion or nearly 40 per cent of the USD15.8 billion annual cost burden. O&Os have the lowest percentage share at slightly less than 25 per cent of their USD10.6 billion total. In the case of SF&Ss, almost 85 per cent of their USD2.2 billion total is due to inefficient business process management costs (Gallaher, *et al.*, 2004).

From the study, it is clear that inadequate interoperability imposes a significant cost burden on all stakeholders in the capital facilities industry. Having a measure of that cost burden creates opportunities for change. O&Os have come to the realisation that costs compound as one moves forward in the facility lifecycle and that their heavy costs in the O&M phase are a result of disconnects in the design and construction phases. A&Es, GCs and SF&Ss also view interoperability costs as stemming from disconnects as well as a lack of incentive to improve interoperability, both within and among firms. Thus, areas that offer the potential for mutual gains are candidates for bringing about change within the industry, and business process management functions offer a rich opportunity for mutual benefits.

A new vision for the future

The driving objectives of construction R&D are to develop new processes and means to effectively and efficiently build and renew the built environment that provides critical services to communities. These advancements can enhance the health, safety and wellbeing of individuals; improve the viability of communities under normal and extreme conditions; regenerate natural systems; and develop the basis for commerce and economic development in the construction industry and throughout the economy.

Construction R&D can be on the frontline to realise the promise of breakthrough developments in various science, engineering and social science fields. These advances may pose disruptive changes in materials, components, systems and processes for the built environment; and construction-specific R&D is needed to ensure that these new systems can be built, maintained and renewed as needed to provide the necessary quality and quantity of those services. These advancements have the promise to significantly improve performance while reducing resource use and environmental impacts. They may also reduce operations and maintenance costs over extended service lives, such as through providing real-time condition information on single components as well as performance throughout multiple systems. In addition, construction R&D will be needed to incorporate these advancements into the very structure and function of current and new systems and processes. For example, R&D programmes can expedite the effective diffusion of these innovations throughout local markets and across the value-adding chain.

For example, research in biology and chemistry is exploring new modes to purify water through biochemical processes, such as the development of microbial fuel cells that take liquid organic waste and directly produce electricity and clean water (Microbial Fuel Cells, 2012). Other research has used photosynthesis as the inspiration for storing energy by splitting water molecules (Sun Catalytix, 2012). In material science and physics, recent research has developed new materials with self-diagnosing and self-healing properties, such as a *bio-inspired coating substrate* (Toohey, *et al.*, 2007) that help the component repair itself from cracks and fractures. Research in maths and computer science is focusing on data collection and control systems for multi-modal interdependent systems, including *smart grids* for on-site energy production linked to wider energy distribution systems (Gharavi & Xu, n.d.) and improved water management systems (Kinzli, *et al.*, 2011).

Recent research in civil, mechanical, aeronautical, electrical and other engineering fields has focused on new systems and processes, often utilising novel materials and components, to achieve higher performance. For instance, the field of bio-inspired materials includes systems that respond to external stimuli, such as light, to change their properties, such as a window that changes its opacity to block direct sunlight (Biologically Inspired Materials Institute, 2012). Additional developments include self-assembling materials and continuous material casting and fusing to create on-site structures (Hoyt, 2012). Work in chemical and industrial engineering has developed new processes to more effectively use and re-use inputs to improve production cost and availability, such as the re-use or redirection of waste into construction materials. There is also an increasing focus on *sustainable building material manufacturing*, which includes reducing or eliminating the use of toxic materials during manufacturing and construction (Oak Ridge National Laboratory, n.d.).

Recent research in medicine and sociology explores the impact of the built environment on the health of humans, such as the impact of green schools on student health (National Research Council, 2006; Oak Ridge National Laboratory, n.d.). This work is influencing research in urban planning and architecture to explore new forms that enhance health and wellbeing (American Planning Association, 2012). Related research in management, law, political science and economics analyses new forms of financing, governing and managing public investments in critical infrastructure systems, such as the use of property-assessed clean energy (PACE) bonds and other modes to finance new energy generation and efficiency investments (Ameli & Kammen, 2012).

A recent study by the US National Research Council (National Research Council, 2009b) presented a new framework for sustainable critical infrastructure systems. Specifically, the study concluded that the focus of infrastructure investments, including R&D, should shift towards the provision of critical services, rather than the physical means to provide those services such as pipes and girders, which can open the field to new science-based solutions for achieving radical improvements. The study also concluded that future developments will need to explicitly address the interdependencies of critical infrastructure systems and the utilisation of collaborative, systems-based approaches at regional scales. Further, the study recommended extended efforts to develop, apply and refine performance measures to ensure the effective implementation and diffusion of advanced solutions and to address issues of risk and social equity.

To most effectively realise this vision, and to conduct the necessary construction research, new partnerships are needed across the value-adding chain to access the necessary expertise, resources, and experience. For instance, the Construction Users Roundtable (CURT) has established an initiative to bring together owners, contractors and labour unions to identify problems in the industry and develop new solutions. Additional programmes can be developed to expand these partnerships. For example, new construction materials and systems will need to specifically address construction safety and productivity, and the direct involvement of skilled labour groups could enrich and speed the development and deployment of results. In the same way, engaging the financial, insurance and regulatory organisations in the development of new management, information and compliance systems, as well as new materials, components and infrastructure systems, can provide insights that directly inform the results.

New programmes could speed the movement of research results from the laboratory to the field through structured and expanded field tests and demonstration projects. For example, the US DOD and the EPA recently signed a memorandum of understanding to establish joint demonstration projects to evaluate new advancements related to sustainability (U.S. Environmental Protection Agency, 2012) and DOD recently expanded its Environmental Security Technology Certification Program (ESTCP). The US General Services Administration (GSA) established a programme to demonstrate and evaluate promising developments in Federal buildings (U.S. General Services Administration, 2012). The US Department of Housing and Urban Development has established a demonstration programme in Native American communities (U.S. Department of Housing and Urban Development, n.d.), and several States and towns are establishing similar programmes, such as the programme in Missoula, MT (Missoula Urban Demonstration Project, 2012) and the Energy Innovation Corridor in St Paul, MN (Energy Innovation Corridor, 2012).

The nation, and the construction industry itself, are demanding new capabilities to improve effectiveness, efficiency and productivity in the creation and renewal of the built environment. Additional investments in construction-related R&D can meet this need and overcome existing challenges by directly engaging with breakthrough research in other fields and expanding its partnerships throughout the value-adding chain, which can enhance civil society and natural environments, and expedite advancements in the field.

References

Ameli, N. & Kammen, D.M. (2012) 'Clean energy deployment: Addressing financing cost', *Environmental Research Letters*, 7. Available at: http://iopscience.iop.org/1748-9326/7/3/034008 (accessed 17 December 2012).

American Planning Association (2012) *Planning for public health*. Available at: www.planning.org/research/publichealth/ (accessed 17 December 2012).

Biologically Inspired Materials Institute (2012) *The biologically inspired material institute*. Available at: www.bimat.org/ (accessed 17 December 2012).

Bureau of Labor Statistics (2012) *Current population survey*. Available at: www.bls.gov/cps/tables.htm (accessed 9 October 2012).

Centers for Disease Control and Prevention (2008) *National construction agenda*. Available at: www.cdc.gov/niosh/nora/comment/agendas/construction/ (accessed 18 December 2012).

Chapman, R.E. (1999) *Benefits and costs of research: A case study of cybernetic building systems*, NISTIR 6303, Gaithersburg, MD: National Institute of

Standards and Technology. Available at: http://fire.nist.gov/bfrlpubs/build99/art003.html (accessed 8 November 2012).

Chapman, R.E. (2000) *Benefits and costs of research: A case study of construction systems integration and automation technologies in industrial facilities*, NISTIR 6501, Gaithersburg, MD: National Institute of Standards and Technology. Available at: http://fire.nist.gov/bfrlpubs/ build00/art025.html (accessed 8 November 2012).

Chapman, R.E. (2001) *Benefits and costs of research: A case study of construction systems integration and automation technologies in commercial buildings*, NISTIR 6763, Gaithersburg, MD: National Institute of Standards and Technology. Available at: http://fire.nist.gov/bfrlpubs/build01/art070.html (accessed 8 November 2012).

Chapman, R.E. (2002) 'An economic assessment of selected integration and automation technologies', paper presented at National Institute of Standards and Technology 19th International Symposium on Automation and Robotics in Construction, NIST SP 989, Gaithersburg, 23–25 September.

Chapman, R.E. & Fuller, S.K. (1996) *Benefits and costs of research: Two case studies in building technology*, NISTIR 5840, Gaithersburg, MD: National Institute of Standards and Technology. Available at: http://fire.nist.gov/bfrlpubs/build96/art113.html (accessed 8 November 2012).

Chapman, R.E. & Weber, S.F. (1996) *Benefits and costs of research: A case study of the fire safety evaluation system*, NISTIR 5863, Gaithersburg, MD: National Institute of Standards and Technology. Available at: http://fire.nist.gov/bfrlpubs/build96/art114.html (accessed 8 November 2012).

Civil Engineering Research Foundation (1993) *A nationwide survey of civil engineering-related R&D*, CERF Report #93-5006, Washington, DC: Civil Engineering Research Foundation.

Energy Innovation Corridor (2012) *Energy innovation corridor. Partnering to develop a cleaner energy and transportation future for Minnesota*. Available at: www.energyinnovationcorridor.com/page/ (accessed 17 December 2012).

Gallaher, M.P., O'Connor, A.C., Dettbarn, J.L. & Gilday, L.T. (2004) *Cost analysis of inadequate interoperability in the U.S. capital facilities industry*, NIST GCR 04-867, Gaithersburg, MD: National Institute of Standards and Technology. Available at: www.nist.gov/manuscript-publication-search.cfm?pub_id=101287 (accessed 8 November 2012).

Gharavi, H. & Xu, C. (n.d.) *Analysis of traffic scheduling technique for smart grid mesh networks*, Gaithersburg, MD: National Institute of Standards and Technology. Available at: http://w3.antd.nist.gov/wctg/adhocvideo/html/Analysis.pdf (accessed 17 December 2012).

Hoyt, R. (2012) *Process for on-orbit construction kilometer-scale apertures*, National Aeronautics and Space Administration. Available at: www.nasa.gov/offices/oct/stp/niac/2012_phase_I_fellows_hoyt_spiderfab.html (accessed 17 December 2012).

Kinzli, K., Gensler, D. & Oad, R. (2011) 'Linking a developed decision support system with advanced methodologies for optimized agricultural water delivery', in *Efficient decision support systems – Practice and challenges in multidisciplinary domains,* Jao, C. (ed.). Available at: www.intechopen.com/books/efficient-decision-support-systems-practice-and-challenges-in-multidisciplinary-domains/linking-a-developed-decision-support-system-with-advanced-methodologies-for-optimized-agricultural-w (accessed 17 December 2012).

McGraw-Hill Construction (2011) *Prefabrication and modularization: Increasing productivity in the construction industry*, SmartMarket Report. Available at: http://construction.com/market_research/freereport/prefabsmr/ (accessed 8 November 2012).

Microbial Fuel Cells (2012) *From waste to power in one step!*, Microbial fuel cells. Available at: www.microbialfuelcell.org/www (accessed 17 December 2012).

Missoula Urban Demonstration Project (2012) *Missoula urban demonstration project (MUD)*. Available at: www.mudproject.org/ (accessed 17 December 2012).

National Research Council (2006) *Review and assessment of the health and productivity benefits of green schools: An interim report*, Committee to Review and Assess the Health and Productivity Benefits of Green Schools, Washington DC: National Academies Press. Available at: www.nap.edu/catalog.php?record_id=11574 (accessed 17 December 2012).

National Research Council (2009a) *Advancing the competitiveness and efficiency of the U.S. construction industry*, Washington, DC: National Academies Press.

National Research Council (2009b) *Sustainable critical infrastructure systems*, Washington DC: National Academies Press. Available at: www.nap.edu/catalog.php?record_id=12638 (accessed 17 December 2012).

National Science Foundation (2011) *Research and development in industry: 2006–07*. Available at: www.nsf.gov/statistics/nsf11301/pdf/nsf11301.pdf (accessed 8 November 2012).

National Science Foundation (2012a) *Business and industrial R&D*. Available at: www.nsf.gov/statistics/industry (accessed 8 November 2012).

National Science Foundation (2012b) *Federal funds for R&D*. Available at: www.nsf.gov/statistics/fedfunds/ (accessed 8 November 2012).

National Science and Technology Council (2008) *Federal research and development agenda for net-zero energy, high-performance green buildings*, Committee on Technology, Report of the Subcommittee on Buildings Technology Research and Development. Available at: www.bfrl.nist.gov/buildingtechnology/documents/FederalRDAgendaforNetZeroEnergyHighPerformanceGreenBuildings.pdf (accessed 18 December 2012).

Oak Ridge National Laboratory (n.d.) *Sustainable manufacturing*. Available at: www.ornl.gov/sci/eere/sustainable_manufacturing.shtml (accessed 17 December 2012).

Sun Catalytix (2012) *Our technology*. Available at: www.suncatalytix.com/tech.html (accessed 17 December 2012).

Teicholz, P. (2004) 'Labor productivity declines in the construction industry: Causes and remedies', *AECbytes Viewpoint*, Issue 4. Available at: www.aecbytes.com/viewpoint/2004/issue_4.html (accessed 17 December 2012).

Thomas, D.S. (2010) *Methodology for calculating construction industry supply chain statistics*, NIST Special Publication 1116. Available at: www.nist.gov/manuscript-publication-search.cfm?pub_id=906651 (accessed 8 November 2012).

Toohey, K.S., Sottos, N.R., Lewis, J.A., Moore, J.S. & White, S.R. (2007) 'Self-healing materials with microvascular networks', *Nature Materials*, 6: 581–585. Available at: www.imechanica.org/files/Self-healing%20materials%20with%20microvascular%20networks.pdf (accessed 17 December 2012).

U.S. Bureau of Economic Analysis (2012) *GDP by industry*. Available at: www.bea.gov/industry/gdpbyind_data.htm (accessed 8 November 2012).

U.S. Census Bureau (2010) *2007 economic census*. Available at: www.census.gov/econ/census07 (accessed 11 September 2012).

U.S. Census Bureau (2012) *Construction spending*. Available at: www.census.gov/const/C30/total.pdf (accessed 18 August 2011).

U.S. Department of Housing and Urban Development (n.d.) *Frequently asked questions: Sustainable construction in Indian country*. Available at: www.huduser.org/portal/publications/pdf/na_constr_FAQ.pdf (accessed 17 December 2012).

U.S. Environmental Protection Agency (2012) *Memorandum of understanding between the U.S. environmental protection agency office of research and development and the office of the deputy undersecretary of defense for installations and environment*. Available at: www.epa.gov/ORD/memo_of_understanding.pdf (accessed 17 December 2012).

U.S. General Services Administration (2012) *Demonstration research projects*. Available at: www.gsa.gov/portal/content/124159 (accessed 17 December 2012).

17 What next? Future directions for R&D investment

Judy A. Kraatz, Keith D. Hampson, Rachel L. Parker and Göran Roos

Introduction

Professor Peter Barrett at the 2013 CIB World Building Congress[1] (WBC13) presented a timely context for the future of research and development (R&D) investment in the global construction industry (Barrett, 2013). He called for a shift in the focus from *lessons learned and doing things better* to *what is the right thing to do* and developing a new paradigm for achieving this. This shift requires *empathy with industry and users; a desire to generate and transmit knowledge; an opportunity to study deeply and over the long term; and with an objective stance towards positive and negative findings*. This shift includes the creation of *standards for the holistic impact of spaces through exemplary pilot projects creating evidence for policy makers and clients* (Barrett, 2013).

In this context, the *value* of the construction industry to a nation and to the global community, and the R&D which informs it, can be considered in terms of both *value to* and *value in* the economy (Ruddock & Ruddock, 2009). The former is seen as *a driver for growth and a catalyst for other industries to develop*, and the latter is seen as a measure of its *contribution to the national economy*. Such value, as discussed in the preceding chapters, is unique to each country. For example, in Brazil (Chapter 4) and India (Chapter 11), a current priority is the development of infrastructure including housing and transportation, whereas in New Zealand (Chapter 13) and the USA (Chapter 16), there is a strong focus on the resilience of existing assets while *future-proofing* new infrastructure. Meanwhile in Europe there is a similar focus on how their historic building stock can be adapted to withstand the impacts of climate change.

Barrett (2013) further reminds us that this industry continues to have unrealised potential compared with others considered central to national wellbeing. He calls for *double-digit improvements, through evidence-based design innovations* to deliver additional value through innovation that enhances quality of life, especially for the more vulnerable. This requires a broader consideration of *value* beyond that typically measured by GDP calculations. These broader issues are widely addressed in the preceding chapters and often highlighted as specific country-based research priorities. Some are longstanding issues, such as those being addressed by Germany's housing research programme (Chapter 9), while others originate in response to more recent social pressures: sustainable and cleaner production (the Netherlands, Chapter 12); accessing investment capital for housing and a focus on integrating infrastructure systems with housing (Brazil, Chapter 4); developing sustainable communities and user-oriented spaces (Finland, Chapter 7); and better accommodating an ageing population (Norway, Chapter 14).

To address these challenges the key issues identified in the preceding country analyses that need to be addressed into the future are:

- *Building meaningful collaboration*: between and across the supply chain (including SMEs); between academia, industry and the community; and bridging the historic gap between divergent drivers for public policy and private practice, and the short-term project focus of industry and a longer term academic focus on knowledge building.
- *Building more durable R&D infrastructure*: within and between organisations, across the entire supply chain and between countries.
- *Improving the dissemination and impact of research findings*: as highlighted by Carole Le Gall (Chief Executive Director of the Centre Scientifique et Technique du Bâtiment) at the WBC13 who called for a move to a *diffusion-oriented policy* aimed at *disseminating capabilities throughout industry supply chain*. This can be achieved in part by a greater focus on improving industry-wide skills levels and training opportunities, especially for SMEs (Le Gall, 2013).

Better understanding the construction industry

To ensure benefits from R&D investment are realised, it is necessary to understand the broader structural issues that impact on productivity. For example, Miozzo and Dewick (2002) highlight the role that governance structures play in a contractor's ability to undertake R&D and to innovate. They discuss the differences in structures between Anglo-Saxon- and Germanic-based models. The former is *market oriented* with technical development occurring through radical innovation, making use of human capital and external knowledge, while the latter is *network driven* focused on incremental innovation using internal knowledge and resources. They also highlight the impact of corporate governance structures on contractors' involvement in R&D. They note national differences between countries with a *strong pattern of cross-holding with suppliers and clients* that supports collaboration and *the dissemination of innovation across a project supply chain* (Germany, Sweden and France) and countries in which there is limited or no evidence of such patterns (Denmark and the UK). Ruddock and Ruddock (2009) reinforce the benefit of company diversification across the supply chain, both upstream to extractive industries for raw materials and downstream to facilities management. This emphasises the need to understand the diversity of corporate governance when setting R&D policy, establishing industry-based programmes to stimulate and support R&D, and developing effective dissemination pathways.

Chapter 2 identifies some of the shared characteristics that define this industry globally including its project-based nature, varying skills levels and the predominance of SMEs. Roos and Pike (2011) suggest that this last characteristic increases the importance of research and technology organisations which have *insight into the firm's reality and insight into the research result space*. Other key issues, which have been highlighted in the preceding chapters, include:

- *Vulnerability to economic cycles*: a close synergy exists between global and national economies and the construction industry with the most recent economic downturn having serious consequences in most nations. Again, the impact has been varied across the globe: Dewulf, *et al.* (Chapter 12) note that this recent crisis had significant impacts across the EU including a *decline of output by 14.2 per cent between the first quarter of 2008 and the third quarter of 2009* and a reduction in the EU27 employment index for construction from *8.8 per cent between the first quarter of 2008 and the second quarter of 2009*. In India, impacts were evident in a reduction in migrant workers and foreign investment important for meeting their expanding need for infrastructure (Chapter 11).
- *Slow uptake of new technology and approaches*: in Finland, a key development has been *the Vera Technology Programme, which has allowed the continued translation of research outputs into practice through an open access model* (Chapter 7). This programme has likely contributed to this country being a global leader in the uptake of technology in this industry. The importance of new technology has been recognised in Australia by the Allen Consulting Group (2010), which projected that the accelerated widespread adoption of building information modelling (BIM) technology would enhance the productivity of different players in the buildings network and would have a significant expansionary effect on the Australian economy. It indicates that by 2025 GDP is estimated to be 5 basis points higher, when compared with a *business-as-usual* scenario.
- *Applied nature of R&D*: Bougrain (Chapter 8) highlights that most contractor-based innovations are made at the job site and are not classified as R&D in line with OECD definitions (OECD, 2002). This has several impacts including how R&D is recorded in national statistics and contributes to the tension between academics and practitioners. This tension is further heightened by the need for industry

organisations to address several dimensions (technology, design, business models, efficiency and effectiveness) simultaneously to achieve impact (Roos & Pike, 2011). Thus, partnerships need to ensure that the short-term and diverse industry needs are met while at the same time providing academics with access to funding for investigations into longer term issues of strategic importance to the industry.

- *Under-investment in R&D*: Slaughter, *et al.* (Chapter 16) indicate that the construction market in the USA fails to provide sufficient incentives to encourage industry research. They attribute this to: conflicting incentives created between design/bid/build contracts and those for developing and implementing innovations; and the separation between the design and construction phases of a project that inhibits the flow of construction knowledge. One-off or ad hoc subcontracting also means that project profit can be maximised by limiting investment in subcontractor training and R&D, since neither is likely to provide future benefit.
- *Improving industry productivity and competitiveness*: is often highlighted as an inherent structural challenge (Chapter 2). New Zealand has longstanding co-funding programmes that support partnerships between business and researchers, such as TechNZ. This is aimed specifically at improving business competitiveness with investment being made in innovation parks, engineering teaching, innovation-focused incentives; improving settings for intellectual property; and developing the New Zealand National Science Challenges (Chapter 13).

It follows that, for construction R&D to address current issues in the industry, it is necessary to establish both a strategic and an operational focus. The former R&D being that which inputs into both the industry and an organisation's longer term strategic development, while the latter focuses on project-specific innovations and knowledge diffusion both within the firm and throughout the supply chain (Miozzo & Dewick, 2002). This lens can provide some clarity when considering the diversity of activity presented in the preceding chapters.

Key drivers for R&D that have emerged from the preceding chapters to provide a focus on the broader purpose and value of the industry include:

- *Urban infrastructure development and renewal*: the nature of which is dependent on a nation's current economic priorities and the age of the existing infrastructure.
- *Contribution to the economy*: the status and role of the industry, policy development, funding levels for R&D and the capacity to undertake R&D and effectively diffuse outcomes.
- *Environmental sustainability*: including a focus on materials such as upcycling waste (Sweden, Norway, Finland and India) and energy savings (Denmark, Germany, France and the USA).

- *Social wellbeing*: dependent on the profile of the population. For example, in Brazil, (Chapter 4) the focus is on housing a growing population, while in Finland (Chapter 8), the focus is on providing services for an ageing population.
- *Standards and code development*: including the development of regulations to boost investment in eco-friendly infrastructure in the Netherlands, and the development of codes as prioritised by the Brazilian Chamber of Industry and Commerce (CBIC).

Mechanisms for future advantage

> By neglecting to conduct our own R&D, we not only reduce the chances that we will discover new ideas and develop new inventions before our competitors, we also limit our abilities to accept and use those new inventions that are developed elsewhere.
>
> (Chubb, 2013)

Two strong mechanisms have emerged for delivering better R&D outcomes into the future: (i) collaboration throughout and across the supply chain and across national boundaries; and (ii) leveraging new technology to transform practice.

Collaboration

> Increasingly, more obscure areas of basic research are being found to have applications in construction: bullet trains in Japan are designed on an understanding of the hummingbirds' aerodynamics to reduce noise; [and] the incredible water repellence of lotus leaves has been applied to airplane coatings and exterior paints to repel rain and keep surfaces clean.
>
> (Chubb, 2013)

Collaboration has been repeatedly cited throughout this international research publication as an important mechanism for enhancing the R&D capacity of the construction industry. One way in which this provides benefit is through creating a learning environment that contributes to the capacity of both the industry as a whole, and individual organisations, to absorb knowledge and thereby innovate. Dewulf, *et al.* (Chapter 12) note this correlation with *50 per cent of all innovations have been established in collaboration between firms*. This operates at many layers. It occurs within the industry between both project and R&D partners, and across the diverse supply chain. It also occurs with other industries such as new technology and the development of new materials, and between countries, most notably within the European Union (EU). Such collaboration brings together diverse views and approaches and provides an important opportunity for *innovative and comprehensive solutions* to ongoing issues confronting this industry (Chapter 5). The benefits from collaborating in clusters have been quantified as 14 percentage points higher value added growth, 7 percentage points higher profitability growth and 2 percentage points higher wages per employee, the later considered a proxy for productivity.[2]

Within the construction industry

Their remains a continuing need to build bridges between industry and academia and across the supply chain to maximise the impact of R&D.

Slaughter, *et al.* (Chapter 16) indicate that industry in the USA *invests three-quarters of its research funding in development and demonstration,* while *academia invests more than three-quarters of its research funding in the basic and applied stages of research*. They note that the former tends to be more short-term in focus, while the latter tends towards longer term targets. These differences highlight the need for strong collaboration to ensure that both practical and strategic research activities are equally valued.

Norway (Chapter 14) reports a strong tradition of academia/industry collaboration in both realms. Collaboration between industry and researchers on practical innovation occurs, for example, in the context of wood, energy efficiency and the *powerhouse* concept. More strategic collaboration occurs in the training of master's degree candidates and in the formation of a new centre at the Norwegian University of Science and Technology (NTNU). This centre will *serve as a national knowledge hub for dissemination, as developing new knowledge through research, innovation and the pilot testing of interaction models in real-life projects demonstrate new technology and teach the workers to use new methods*.

Similar initiatives are reported in other countries. In Sweden (Chapter 15) the Centrum för Energi- och Resurseffektivitet i Byggande och Förvaltning (CERBOF, Centre for Energy and Resource Efficiency in the Built Environment) is a programme established to foster collaboration and the dissemination of knowledge between industry and academia. In Brazil (Chapter 4), the Associação Nacional de Tecnologia do Ambiente Construído (ANTAC, National Association of Technology of the Built Environment) defined five priority areas focused on bringing researchers, industry and public agencies together, relating to: systems, processes and management; quality; resources; and urban infrastructure.

The growing role for R&D brokers is also enabling both industry and researchers to work together to better leverage investment and address concerns regarding issues such as confidentiality and competitiveness. These organisations can help overcome the cost and efficiency barriers often confronted by contractors (Miozzo & Dewick, 2002). Keast and Hampson (2007) examine the important role of relationship management in such collaborations. The role the Australian Cooperative Research Centre for Construction Innovation (CRC CI, 2002–2009) and the current Sustainable Built Environment National Research Centre (SBEnrc) as such a broker is explored in the context of motivating supply chain firms to improve their organisational capabilities in order to acquire, assimilate, transfer and exploit R&D outcomes to their advantage, and to create broader industry and national benefits (Kraatz & Hampson, 2013). In Canada (Chapter 5), the brokering activities of the University of New Brunswick's Construction Engineering and Management Group (UNB-CEM) focuses on finding better ways to support greater collaboration using technology to improve knowledge dissemination activities.

International collaboration

International collaboration provides a further avenue for building the innovation capacity and R&D credentials of the industry. Productive global partnerships require strong leadership and coordination.

Internationally, as mentioned in Chapter 1, the International Council for Research and Innovation in Building and Construction (CIB) was established in 1953 to stimulate and facilitate international cooperation and information exchange between governmental research institutes in the building and construction sector (CIB, 2013). This publication being one outcome of this active approach to building international partnerships through the *CIB Task Group 85: R&D Investment and Impact.*

In Australia, both the CRC CI, through the International Construction Research Alliance (ICALL), and now SBEnrc have developed and maintained extensive international research networks with representation across the globe. This includes the ongoing development of a close working relationship developed between Australia and one of New Zealand's peak bodies, BRANZ.

In Canada, where research topics have a broad impact on industry, *industry members work with the NRC* (National Research Council of Canada) *to undertake 'precompetitive'-type research*, with results facilitating the creation or broadening of a market. Key outcomes of this programme inform not only national codes and standards, but also international standards.

In Europe, the European Innovation Partnerships (EIP) work across the research supply chain, at EU, national and regional levels on projects such as *Smart Cities and Communities* and *Raw Materials*. The intent being to: (i) *step up research and development efforts;* (ii) *coordinate investments in demonstration and pilots;* (iii) *anticipate and fast track any necessary regulation and standards;* and (iv) *mobilise 'demand' in particular through better coordinated public procurement to ensure that any breakthroughs are quickly brought to market* (European Commission, 2013).

Leveraging technology

It is widely acknowledged that the adoption of new technologies has the potential to generate transformative change in the industry in terms of both productivity (Allen Consulting Group, 2010) and enabling innovative solutions to emerging challenges, such as urban systems modelling (OECD, 2011). A past focus on short-term gains as a result of technology has often proved unsuccessful due to the lack of focus on required associated changes in procedures, culture, habits and attitudes.

Several countries are providing global leadership in the uptake of technology within the construction industry. In the USA, *the widespread deployment and use of interoperable technology applications, also called building information modelling (BIM)* has been identified by the US National Research Council as one of *five interrelated activities that could lead to breakthrough improvements in construction efficiency and productivity in 2 to 10 years*. In Sweden, while infrastructure consultants are only legally bound to deliver paper drawings almost every project team delivers 3D models (Vianova, 2013):

> Only innovations that are compatible with the existing technological regime will be adapted. Regime shifts are therefore often needed to make innovations sustainable in the long run. Changes in rules and procedures but also changes in culture, habits and attitudes, are needed to improve the emergence of sustainable research and investments programmes.
>
> (Dewulf, *et al.*, Chapter 12)

The technology perspective

A clear strategy is required within organisations to maximise the benefits of new technologies. This requires the rethinking of practices and the development of a technology-based component for future strategy. The EU has identified several *key enabling technology* areas (European Commission, 2009) that enable process, goods and service innovation throughout the economy. They are multi-disciplinary and cut across many technology areas with a trend towards convergence and integration. These include: *nanotechnology* with applications relating to concrete, steel, wood, glass, coatings and fire protection/detection, as for example through CIPET in India; *micro- and nanoelectronics* with applications relating to energy savings and intelligent buildings; *photonics* used in electrical and electronic processes such as sensors and real-time infrastructure monitoring in Finland; *advanced materials* that possess different types of internal structure and exhibit special properties such as improved performance in demanding environments, such as wood in Norway; *biotechnology* such as applied in the recycling of waste and production processes of raw materials, for example, upcycling in India; and *advanced manufacturing technologies* including automation, robotics technology and additive manufacturing.[3]

In addition, ICT-enabled developments that can also make an unprecedented contribution to the industry include: *social technologies* with new uses, technical advances and social business models are continually evolving and provide potential for new value creation (Chui, *et al.*, 2012); *the internet of things*[4] capturing how the internet will expand as consumer devices and physical assets are connected to it using embedded sensors, image recognition technologies and near-field communications; and *big data*, being the masses of data collected from a wide variety of sources, resulting in giant and diverse data volumes and types that are enabling new insights through the application of advanced analytics (Russom, 2011).

Recent research (Ahlqvist, *et al.*, 2011; Kraatz, *et al.*, 2012) has identified R&D priorities that acknowledge this broad range of technological imperatives for Australia's construction industry. In Denmark, the Digital Construction Task Force on Building Research, established in 2000, has had its mandate extended until 2014. And in Hong Kong, the Innovation and Technology Fund aims to assist enterprises in that country to *promote their technological level and introduce innovation to their businesses* (Chapter 9).

The strategy perspective

Given the global pressure for productivity improvements and innovation, it is critical that both existing and emerging technologies are rapidly assimilated, successfully deployed and converted to business advantage. This is especially crucial given that recent research demonstrates that technology is the key long-term driver of productivity growth (Tassey, 2010). This requires a well-considered *tech strategy* to guide deployment through the increasingly complex and technology intensive nature of global competition that is characterised by compressing technology lifecycles and increasingly narrow windows of opportunity (Tassey, 2012). Tassey (2010) further considers that early and substantial investment in process technologies and achieving economies of scale is essential to attaining large market shares in this technology-based global economy.

One of the most challenging aspects of technology-based competition is the *transition from the current to the next technology lifecycle*. Failure to plan for and efficiently executed lifecycle transitions can bring about the demise of organisations. Failure to invest in eco-systems, efficiently integrated supply-chain structures and supporting technology infrastructures will lead to slow market penetration in the early phase of the technology's lifecycle and almost guarantee low market share in the middle portion of that cycle (Tassey, 2012). The dominant view in technology strategy has been that displacement of established firms and technologies by new ones is driven by the superior performance offered by newcomers combined with established players' difficulties in matching the newcomers' performance and capabilities (Foster, 1986). By identifying the possibility that technologies with inferior performance can displace established incumbents, the notion of *disruptive technologies* (Bower & Christensen, 1995) has had a profound effect on the way in which managers approach technology and has prompted a reassessment of the ways in which firms approach technological threats and opportunities (Adner, 2002).

The term *disruptive technologies* refers to products, solutions and systems, which in their early phases aim to serve specific niche markets, but over time manage to outperform their competitors on mainstream markets (Bower & Christensen, 1995). By examining how technology is evaluated and how this evaluation changes as performance improves, new insight into the impact of structure of demand environment on competitive dynamics is enabled. In general, incumbent firms tend not to be at the forefront of disruptive technologies. Instead, it is often entrepreneurs and start-ups that enjoy first-mover advantage based on their relatively high risk orientation and *low path* dependency (Afuah & Tucci, 2003). Disruptive technologies hence make established technologies obsolete and therefore destroy the value of the investments that incumbents have made in those technologies (Danneels, 2004).

In order to be able to benefit from disruptive technologies, the drivers and implications of such disruptions must be understood so that these dynamics can be made to work to the advantage of the firm. The challenge is to manage the dynamics of innovation that underlie both disruptive and sustaining innovations, which is partly about understanding the interaction between needs and technologies (Paap & Katz, 2004). These managerial challenges require a balancing act between constantly looking forward and backward. Leitner and Guldenberg (2010) recently reported on findings that show that such a strategy positively influenced profitability, and employment and turnover growth.

Another important concept is *convergence*, that is, the growing together of previously separate domains into one new homogenous knowledge domain that can form the basis for new industrial activities. Organisations thus need to develop a specific dynamic capability of continuously balancing between exploring and exploiting current and emerging opportunities of inter-industry spillover (Hacklin, 2008). They need the ability to understand the clear distinction between short-term and long-term strategies, or pre-convergence and post-convergence policies (Hacklin, 2008). Hacklin suggests that as a consequence of convergence, firms need to be able to adapt and extend organisational structures in such a way that the internal knowledge bases of the firm are able to dynamically anticipate, follow and match the new structures of the industry.

Research has observed six common principles that innovative companies seem to follow in their approach technology-based business strategy development (Berman & Hagan, 2006):

1 *Consider technology a core input*: equal to other variables such as customers, markets and competitors. Sweden's adoption of BIM is a key example of this.
2 *Revisit strategy and technology context regularly*: continuously manage and revise strategy to proactively take advantage of the evolving technological environment.
3 *Manage emerging business opportunities separately*: separate organisational procedures, structures and policies to manage emerging business opportunities differently than their core businesses. For example, in France, *Bpifrance has always focused its attention on the smallest and youngest enterprises that experience greater difficulty in gaining ready access to financing* (Chapter 8).
4 *Plan for disruptions*: understanding the power of technology to change business assumptions, and better anticipate market changes and disruptions. BIM is commonly regarded as a disruptive technology. Slaughter, *et al.* (Chapter 16) recognise that *construction R&D can be on the frontlines to realise the promise of breakthrough developments in various science, engineering, and social science fields,* which *may pose disruptive changes in materials, components, systems and processes for the built environment.* They suggest that R&D programmes *can expedite the effective diffusion of these innovations throughout local markets and across the value-adding chain.*
5 *Manage for the present and future context*: manage a diversified portfolio of capabilities, comprised of both sustaining technologies and emerging technologies. This challenge is evidenced in the Indian and Brazilian cases presented in this publication.
6 *Focus technology on the customers' priorities*: rather than focusing exclusively on technology-enabled internal efficiencies, also concentrate on problems affecting customers and identify technologies and business models impacting on those issues. Reflecting on the need for both the development and renewal of our urban infrastructure in this context could provide a powerful framework for technology-enabled innovation in the construction industry (OECD, 2011).

Collectively, these principles change the traditional strategy development approach. Instead of being an implementation issue, technology becomes a catalyst at the very initial stages of strategic planning, merging with market insights to produce innovative concepts and services. Furthermore, the strategy process needs also to encompass the associated changes in the business model. The rule of thumb is that introducing disruptive technology without changing the existing business model tends to lead to business failure (Roos, 2013).

Dewulf, *et al.* (Chapter 12) highlight several factors that, based on their case study, limit the uptake of new technologies including: limited knowledge among workers; ineffective diffusion; project-based relationships; limited sharing of knowledge; practice embedded in local laws; and the current culture. Slaughter, *et al.* (Chapter16) go on to highlight the associated cost to the industry of one element, being inadequate interoperability. Highlighting and quantifying such impediments provides an important evidence basis to support active change.

Robinson Fayek, *et al.* (Chapter 5) provide an important example of a proactive strategy to address such issues through the *Digital Technology Adoption Pilot Program* (DTAPP), which aims to increase adoption of technology skills among SMEs, through the development of supportive networks.

Building infrastructure to support R&D

The preceding chapters support the view that investment in R&D in the construction industry is not appropriately aligned with the contribution it makes to national economies and the key role it plays in shaping the communities and the environment. Additionally, activity in the industry is vulnerable to economic conditions, which has a sometimes direct impact on the extent of funding available for research. A multi-faceted approach is required to improve research spillover. Roos and Pike (2011) define this *as the means by which new knowledge developed by one firm becomes potentially available to others*, with the absorptive capacity of those in receipt of this knowledge determining *the extent to which the knowledge is incorporated.* This multi-faceted approach should include:

- Building industry capacity to undertake, uptake and assimilate research through:
    - internal structures that support R&D activity and engagement, and build absorptive capacity
    - supply chain structures and mechanisms that strengthen collaboration in order to ensure both short-term project needs and longer term knowledge-building needs are addressed.
- Improving knowledge diffusion and dissemination through:
    - better targeted contribution to industry skills development, especially with regards to new technology; and improved access to training, especially for micro firms and SMEs
    - demonstration projects and prototyping, both physical and virtual, to: develop skills; demonstrate new solutions, technologies,

processes and materials; and increase consumer awareness of the value-added contribution of the industry.

This view further reinforces the importance of establishing a culture of R&D as a central pillar within the construction industry, and within individual organisations, to ensure the capacity to absorb new knowledge and processes. Knowledge-rich clients have an important role to play in developing R&D infrastructure in this industry due to their position of power in the supply chain (BEIIC, 2012; Chapter 12).

Building capacity within organisations

The case studies presented in this book reveal a variety of strategies organisations have in place to maximise the benefits of R&D to their organisation. For example, firms in Brazil are developing a culture of innovation through: (i) the *Programa Andrade Gutierrez de Inovação Tecnológica* (PAGIT, *Technology Innovation Programme*) established by Andrade Gutierrez (large-sized firm) with an internal focus on fostering a culture of innovation; and (ii) Tecnisa, a medium-sized real estate company, which established a *Technological Development Programme* with the support of academic researchers. And, in India, Tata Housing has an industry-led R&D programme to reduce manual labour in response to the extreme difficulty in meeting the demand for new infrastructure through using manual construction techniques.

Miozzo and Dewick (2002) highlight additional mechanisms for building internal R&D infrastructure. In France, Group GTM has a process of diffusing knowledge through an intranet database, technical days and an innovation awards scheme. In Germany, Holzmann and Hochtief created competence centres and R&D support units to facilitate dissemination across geographically separated division. While in Sweden, Skanska created Skanska Teknik as a central R&D agency that integrates the firm's technical expertise and disseminates knowledge and experience across the firm.

Recent case study research in Australia of private sector R&D activity reveals the central role of R&D to maintaining market share (Kraatz & Hampson, 2014). Ampac Advanced Warning Systems, founded in 1974, designs, manufactures and exports fire protection systems. Eight per cent of their workforce is directly involved in R&D to maintain their position of market leadership. Hames Sharley is a medium-sized architectural practice that has embedded an *evidence-based design* approach into the delivery of its hospital design services. This involves drawing on both in-house and external sources of R&D, and building projects that include architects specialising in health, health planners, health service planners, health administrators, nurse planners and other medical people in order to provide a comprehensive response to the client in a professional manner.

Building capacity across organisations

Many inter-organisational and cross-sectorial network structures exist that facilitate the triple helix model for effective interaction between government, research institutions and industry. These structures provide *accelerated opportunities for information and knowledge sharing coupled with the capacity to synthesise and leverage these learnings into new and innovative outcomes* (Keast & Hampson, 2007).

The previous chapters discuss formal structures that exist within countries to develop R&D infrastructure. These can be classified in line with the schema offered by Keast and Hampson (2007) as hierarchical, typically government established and administered; market based, such as industry consortia; networks, such as many university-led initiatives; and hybrids thereof (Table 17.1).

Strong Government-led programmes in all countries provide leadership through the more traditional hierarchical agencies. In Brazil, the Fundação de Amparo à Pesquisa do Estado de São Paulo (FAPESP, Foundation for Research Support of the State of São Paulo) was founded to address housing and sector modernisation and several other public bodies. The Finnish Funding Agency for Technology and Innovation, Tekes, is recognised as a world leading innovation agency. In France, Bpifrance provides assistance and financial support to micro-sized firms and SMEs, and the *Programme de Recherché sur l'Energie dans le Bâtiment* (PREBAT, *Research and Experimental Programme on Energy in Building*), established in 2007, focuses on sustainable construction and demonstrates the positive impact of building strong communication channels between researchers. The *Bygg21 Agenda* (Norway) facilitates interaction and collaboration for industry and public authorities to provide research leadership in that country, and in Sweden, *Bygginnovationen*, a VINNOVA programme, serves a similar purpose. In the USA, various activities of the National Institute for Standards and Technologies (NIST) operate across the supply chain to deliver programmes.

Table 17.1 Governance structures for R&D infrastructure

Governance mode	*Hierarchy*	*Market*	*Networks*
Integration relationship orientation	Authority relations	Exchange relationships	Social/communal relationships
Key integration mechanisms	Centralised and legitimate authority, rules, regulations, procedures and legislation	Formalised, legal contractual arrangements, arms' length transactions, bargaining	Interpersonal trust, mutuality and reciprocity
Management focus	Administrative management	Contractual management	Relational management

Source: Keast & Hampson (2007)

Schiele and Krummaker (2011) define research consortia as a body of practitioners and academics that work together to define research questions and search for answers. Joint ventures and consortia provide examples of market structures that are facilitating R&D outcomes. In Hong Kong, in 2005, Sun Hung Kai Properties (SHKP) provided funding to the Hong Kong Polytechnic University, with the aim of enhancing *collaboration with the tertiary institutions and increas[ing] the competitiveness in technology innovation and construction practice, encouraging practically oriented research for the industry in general*. In New Zealand, the research consortium Beacon Pathway Ltd was established with equal Government and industry funds to develop demonstration houses.

Process and Systems Innovation in the Construction Industry (PSIBouw) in the Netherlands, was a network of industry and researchers that was established in 2004 to address issues arising from a parliamentary investigation into tendering processes. In Australia, the CRC CI and SBEnrc facilitate networks that enable innovation through providing a neutral space for engagement, assist with implementing outcomes, provide validation and assist with diffusion (Winch, 2005). Relationship management plays an important part of effective networks. Such an approach still requires effective and accountable management; Keast and Hampson (2007) provide a framework which identifies the characteristics of this approach (Table 17.2).

Improving knowledge diffusion and dissemination

The project-based and decentralised nature of much of the innovation activity in this industry is widely acknowledged as a barrier to continual learning and the effective dissemination of knowledge. So too is the high proportion of micro and small enterprises with limited time and funds to invest in research and skilling. Collaborative learning and research that engages industry partners in the process of creating new knowledge, rather than the more traditional approach of transferring knowledge on completion, is thus emerging as an important avenue for improving R&D impact. *Knowledge transfer is becoming institutionalised and seen as a key role conferred on the university rather than on individual university researchers* (Roos & Pike, 2011). This is seen as contributing to the emergence of new approaches such as *industry–university research centres, research joint ventures, and technology consultancies*.

Knowledge diffusion, skills and training

> For firms to acquire new knowledge through interactions with other firms and knowledge institutions, they need to have the capability to learn, sometimes described as *absorptive capacity* (Cohen & Levinthal,

Table 17.2 Relational management framework

Relational management roles and focus	*Task components*
Activating Forming membership and accessing resources	• Identify and select relevant network members • Access and gain agreement to devote skills, knowledge and resources to the network – the *buy-in* • Establish appropriate structural arrangement • Introduce new actors and resources to renew interest and change non-performing dynamics • Deactivate or disconnect non-contributing members
Framing Shifting orientation from single to collective	• Establish values, norms and rules – new terms of engagement • Introduce and champion new ideas • Encouraging members to view issues from another's perspective • Stress the benefit of working together
Mobilising Securing commitment to whole or collective identity	• Establish common vision, mission • Secure agreement on scale and scope of action • Forge coalitions and subgroups for specific actions • Drive action for outcomes • Identify and foster champions and sponsors
Synthesising Building and maintaining relationships	• Check level of involvement and sense of engagement • Monitor relationships and activities • Leverage resources toward collaborative advantage and collective benefit • Establish network and innovation culture • Deal constructively with conflict • Build communication processes

Source: Keast & Hampson (2007)

> 1990; Rodríguez-Castellanos, *et al.*, 2010; Zahra & George, 2002) but also more broadly connected with the concept of the *learning organisation* (Brown & Duguid, 1991; Nonaka, 1994).
>
> Parker and Hine (2013)

Parker and Hine discuss this in the context of the role of knowledge intermediaries in developing the networks necessary to facilitate knowledge flows from knowledge production organisations, such as research institutions, universities and training organisations, to knowledge users. This may also require the firms themselves to develop and master the art of *co-opetition*: cooperation with competitors (Peng, *et al.*, 2012).

The uptake of new knowledge and skills is particularly problematic for SMEs. *We can expect that the number and qualification of the employees of many of these firms fall below a critical mass necessary to engage in open innovation through absorptive capacity, let alone set up an independent R&D unit* (Spithoven, *et al.*, 2010). Bougrain and Haudeville (2002) point out that internal R&D capacity could take the form of a *design office*, and as such could *enhance the firm's ability to cooperate and to carry its project to success*. Building capacity in this group is receiving much international attention. For example, Chapter 8 notes that SMEs in France are an important focus of recent innovation policy and programmes. In Canada (Chapter 5), DTAPP was designed to speed up the rate at which local SMEs adopt digital technology and build digital skills, through providing access to expertise in the digital technology adoption field. In Sweden (Chapter 15), the *Bygg programme* seeks to strengthen the innovation environment for SMEs through providing *rapid, efficient and non-bureaucratic application processes*.

The need to revitalise the industry's approach to training and skilling is particularly acute in the uptake and use of new technologies and materials. Slaughter, *et al.* (Chapter 16) address this need in terms of the associated impacts on safety and productivity of new construction materials and systems, suggesting that the direct involvement of the skilled labour groups could enrich and speed the development and deployment of results.

This publication reveals several different approaches to both industry- and academia-led skilling and training. Nenonen, *et al.* (Chapter 7) highlight the role of funding from Tekes technology programmes, where *approximately half of total funding is granted to companies, universities and research institutes through programmes that support business operations, such as shared visions, seminars, training programmes and international visits* (Hyytiäinen, *et al.*, 2012). Robinson Fayek, *et al.* (Chapter 5) note several approaches to addressing skilled labour shortages in Canada. This includes the *Natural Sciences and Engineering Research Council (NSERC) of Canada's Industrial Research Chairs programme*, which facilitates *access to the specialised knowledge and resources available at Canadian universities and train*[s] *students in skills needed by industry*. Similarly the University of New Brunswick's Construction Engineering and Management Group aims to develop strategies to overcome the regional shortage of skilled workers at trade, management and supervisory skills levels. Kashikar (Chapter 11) reports on the critical importance of skills development in India's industry with several institutions taking responsibility for specific aspects of skills development. This is often coupled with demonstration projects that are used for the dual purpose of both providing infrastructure and developing a local and regional workforce. New Zealand (Chapter 13) has established industry–government partnerships to achieve an *improved, productive, skilled, innovative and efficient building and construction industry that contributes to increased business profitability and quality of life for all New Zealanders* (Buildingvalue, 2013), with a goal of increasing productivity in the industry by 20 per cent by 2020.

While in Europe, the European Network of Construction Companies for Research and Development[5] (ENCORD) is a network of 19 construction industry participants whose aim is to increase awareness of the potential of industry-led R&D, and facilitate workshops for the exchange of information on *state of the art in construction research and to set the agenda for future activities*. One key priority is *to incorporate the results and ideas of the research into everyday work, with a focus on measuring the value of research results, their use in enterprise policy decision making and maximising exploitation strategies* (ENCORD, 2009).

Demonstration projects, pilots and virtual prototypes

There is widespread recognition in the industry of the value of demonstration projects as a bridge between research and practice:

- Australia: the Queensland Government Department of Public Works undertook a series of such projects in which 3D and 4D digital building models were shared with contractors and subcontractors on a *no-risk* basis to promote industry understanding of these practices (Hampson & Kraatz, 2013).
- Denmark: the Government took an industry leadership role through demonstration projects (1994–1997) that promoted vertical collaboration. *NCC Denmark's project (comfort house) focused on developing a light build house (using steel and gypsum) suitable for industrial production, implementing new forms of integrated cooperation between the contractor and the building consultants* (Miozzo & Dewick, 2002).
- France: *experimental* building projects are used to *test new technologies and new systems in order to anticipate the forthcoming change of thermal regulation* with those choosing to do so receiving support and incentives (Chapter 8).
- Finland: the Vera Technology programme[6] explores the use of open-access digital models to allow *the continued translation of research outputs into practice* (Chapter 7).
- Hong Kong: the successful industry–university collaboration in which SHKP supplemented university laboratory and computing facilities by making available to the researchers the facilities at actual buildings for testing research theories (Chapter 10).
- India: demonstration units that are self-funded by the beneficiaries, collaborating organisations and agencies are used for training purposes to demonstrate how innovative technology can address housing problems. One example is the *Innovative Building Materials and Housing* programme, carried out through the Central Building Research Institute.[7] This included *train the trainers programmes* and the construction of demonstration units. During the course of this 5-year

programme 34,000 housing units were constructed and 30,000 people were trained (Chapter 11).

- USA: industry spends three-quarters of their R&D investment on *development and demonstration* projects while Federal agencies spend *45 per cent on the development and demonstration stages* (Chapter 16).

Such demonstration projects often provide a tangible example of the impact of R&D activity. Another of the five key activities identified in the USA NRC report was *effective performance measurement to drive efficiency and support innovation*. Similarly in 2009 in Europe, ENCORD stated *an industry goal as being the need to achieve a clear measurable link between research and results achieved* (ENCORD, 2009).

Towards breakthrough improvement – increasing research impact

Roos and Pike (2011) examine the impact of university-based R&D in the context of the return on investment to external partners including industry. They cite various reports that highlight several pertinent facts including that: (i) *studies over 30 years of work find a rate of return to public R&D of between 20% and 50%*; (ii) *10% of innovations could not have happened, at least not without significant delay, without the support of academic research*; and (iii) *that 20% of private sector innovations are based to some extent on public sector research.*

Bridging the divide between researchers and industry

Most of the country cases published here highlight an increased awareness from Government of the role that knowledge and research play in economic competitiveness. More specifically, research policy is premised on the view that universities and research institutes play a central role in economic competitiveness. The triple helix model has emphasised the blurring of the role of universities in contemporary economies, as they take on functions traditionally associated with Government and industry, in stimulating economic activity. These include engaging in commercial ventures and entrepreneurial activities and through the formation of spinoffs and incubators (Dzisah & Etzkowitz, 2011). As such, universities and research institutes face increased pressure to ensure that their research has a positive economic impact and contributes to economic development.

Collaboration and impact

Positive impact is found to occur through: collaborative learning; policy-stimulated innovation; interactive knowledge transfer; the co-production of knowledge; and technology transfer.

The Swedish case discusses the role of *collaborative learning and systems approaches* to knowledge generation and research impact. The Swedish *Bygginnovationen* programme involving the Swedish government agency for innovation, VINNOVA and a consortium of 19 firms in the built environment industry has the intention of building a positive innovation system for this sector. The authors highlight that an important mechanism for ensuring the relevance and applicability of research to industry is the *role of the business advisory committee* (Chapter 15). This committee comprises industry representatives with experience in product and technology development, which evaluates and assesses research applications in the programme. Policy programmes that have sought to stimulate the innovation system, such as the Bygginnovationen programme, have focused on how to facilitate the interactions and linkages within an economy between knowledge users such as industry organisations, and knowledge producers such as universities and research arms, for the purpose of stimulating learning and innovation. These types of programme place an emphasis on the role of the university in interactive learning between a firm and its environment, suggesting that it is impossible to separate knowledge generation from research impact. As such, linkages between universities and industry, involving feedback mechanisms, loops or incremental change, error and modification result in the creation of new knowledge through its application in practice (Edquist, 1997; Etzkowitz & Leydesdorff, 2000).

The three Canadian cases collectively demonstrate the critical interactions between the construction industry, universities/research institutes and Government in achieving research impact. The Industrial Research Chairs (IRC) and the *Collaborative Research and Development (CRD) Programme* are part of the NSERC's effort to stimulate partnerships between universities and industry in Canada. These authors explain that these programmes have been taken up at the Hole School of Construction Engineering at the University of Alberta, which disseminates its research and commercialises its research products and technology development through annual construction innovation forums and workshops for industry partners focused on products developed by the IRC and CRD programmes. While dissemination of research occurs through newsletters and websites, the engagement is much deeper than just *knowledge transfer*, which typically occurs at the end of the research process. It involves industry partners early in the process so that *researchers participate in industry events and committees to help shape industrial, as opposed to academic, research*, ensuring that the research problems being addressed by the academics are relevant to industry.

In order for university/research institute and industry collaborations to be effective, they need to involve *co-production of knowledge* by industry and universities, not just knowledge dissemination by universities at the end of the research process. Co-production involves joint identification of research problems by academics and industry, which helps to ensure the relevance to industry of problems being addressed through research efforts. Effective co-production will also typically involve multi-disciplinary research teams, as many *real-world* problems are highly complex and cannot be resolved from the input of researchers in any one field. Research impact also depends on researchers valuing the contribution of industry throughout the research process rather than treating industry as *funders of research* or as *data collection sites* (Bruneel, *et al.*, 2010; van de Ven, 2007).

Many of the research impact cases presented in this publication demonstrate the role in research funding policy. These cases focus on facilitating interactions between industry organisations and knowledge institutions particularly in terms of technology and the provision of transaction services, identifying commercialisation opportunities, and assisting with technology exploitation (Wright, *et al.*, 2008). The Hong Kong case of the *Enhanced Energy Efficiency for Office Building A/C System* was premised on the view that close collaboration between industry and academia would result in *practically oriented research* that *moves more quickly and is more to the point.* As Shen and Hong (Chapter 10) explain, the research associated with this project did not occur exclusively in university laboratories at Hong Kong Polytechnic University, but was instead introduced in buildings provided by SHKP as test beds. In addition, the case involved the development and transfer of new software that was interfaced with existing building automation systems and monitoring instruments, ensuring that the new product was tailored to the specific SHKP user context.

Measuring research impact

As mentioned in Chapter 2, many countries are now grappling with the policy challenge of finding ways in which to objectively measure the impact of research. In the UK, the Research Councils UK (2012) has adopted a very broad definition of research impact to include benefits across many diverse economic, social and environmental areas of concern including quality of life. In terms of impacts on the economy, measures relating to economic growth, productivity and reduced costs have introduced potential metrics into the process. The Australian Research Council (ARC) has begun to discuss *pathways to impact* as including industry funding of research and the training of higher degree research students (Australian Research Council, 2012). Meanwhile, Garnett, *et al.* (2008) report on a case study from Charles Darwin University in Australia using *conjoint value hierarchy* technique in determining the impact of university research in terms of stakeholder value. The intent being to provide a reliable, repeatable and *comprehensive, transparent and agreeable assessment of a university's research value as seen by the stakeholders who sponsor research* and including measurement of value against 33 attributes.

Many of the cases in this publication indicate the diverse way in which research impact in the context of the built environment is being measured in different countries. In Brazil, specific reference is made to impacts across a range of areas including an increase in the number of patents, improvements in building products, and the conservation of potable water supplies. Longer term impacts associated with investment in younger scientists contributing to this industry are yet to be realised.

In Canada, the impact of the NSERC-funded programmes include findings that support proposed changes in the national model codes for the built environment or other national or international standards, which may, in turn, create markets for new innovative products.

In France, PREBAT has contributed to developing a stronger and better coordinated research community; and more tangibly, the development of a significant number of low-energy buildings that improved knowledge levels and led to the development of new practices.

In Hong Kong, in the case of the *Enhanced Energy Efficiency for Office Building A/C System*, research impacts were measured in terms of energy savings and the production and transfer of information technology from university to industry.

In India, the CBRI has registered 93 Indian and two international patents for technical developments in the construction industry; and the impact of the Science & Technology Extension Programme extends well beyond tangible outcomes such as building materials and housing to putting in place processes that bring about a *change of worldview of the poor.*

In the Bygginnovationen programme in Sweden, there was an emphasis on research outcomes such as the adoption of ICT, efficient organisational processes and growth with a strong emphasis on green growth. The authors discuss a VINNOVA research impact study that highlighted the broader industrial, economic and social factors that affect research impact and are beyond the control of the researcher. These included the need for subsequent funding to effectively adopt innovative outcomes in practice, the impact of broader economic conditions and the long timescale required to capture some research impact effects.

Conclusions

This publication reveals that the aspirations presented by Barrett and others at the WBC13 in Brisbane are already embedded in the R&D activities and aspirations of those countries represented in this volume. Key challenges remain, however, in achieving these ambitions. These include building meaningful collaboration across the supply chain sustained by more durable R&D infrastructure, improving the diffusion of knowledge throughout the complex supply chain and better understanding and addressing the impact of research.

Developing metrics for understanding the benefit of R&D and its impact is a complex task. Country-based statistics regarding the performance of R&D in the construction industry often do not fully reveal its contribution, due to the expansive nature of the industry and the applied nature of much of the R&D activity. And while universities, funding agencies and individual organisations may have internal performance metrics, these do not typically permit benchmarking across the broader industry. Thus, carrying out cross-country comparisons is also problematic. Organisations such as the Global Reporting Initiative (GRI) and the Green Building Council are increasingly embracing this complexity and deliver global reporting, assessment and management systems that enable performance to be benchmarked, both locally and in the broader global context (GRI, 2013). The ability to benchmark the performance of construction industry similarly requires effective reporting mechanisms across the various value propositions discussed previously.

This publication thus brings a deeper understanding of the role of this industry to a broader audience. This is needed to better facilitate knowledge creation, uptake and dissemination across the industry supply chain, from micro businesses to large-scale trans-boundary organisations. This requires effective policy frameworks that promote stronger, more accessible and productive collaborations that maximise the opportunities presented by new technologies. All in the service of providing liveable infrastructure that is resilient to the significant economic, environmental and social changes evident across the globe.

Notes

1 CIB World Building Congress, Brisbane, 5–9 May 2013. More information Available at: http://worldbuildingcongress2013.com/.
2 Calculated by Roos (Roos, G. (2013)) 'The innovation ecosystem in Australia', CEDA, Australia (in press) from Table 2, page 30 in *Building the cluster commons. An evaluation of 12 cluster organisations in Sweden 2005–2012*; Sölvell, Ö. & Williams, M. (eds), Stockholm: Ivory Tower Publishers.
3 Additive manufacturing or 3D printing is an advanced manufacturing technology that provides a paradigm shift for manufacturing due to a number of advantages with this technology, including greater energy efficiency in part due to weight savings; savings in material costs; reduced cost associated with transportation, warehousing, and inventories; the ability to produce objects with three-dimensional characteristics not possible to produce using subtractive manufacturing techniques; and the ability to generate and produce in alloys not possible to produce using traditional techniques.
4 The term originates from Ashton, K. (2009) 'That "internet of things" thing, in the real world things matter more than ideas', *RFID Journal*. Available at: www.rfidjournal.com/articles/view?4986.
5 For more information, visit www.encord.org/.
6 For more information, visit cic.vtt.fi/vera/english.htm.
7 For more information, visit www.cbri.res.in/index.php?option=com_content&view=article&id=97&Itemid=99.

References

Adner, R. (2002) 'When are technologies disruptive? A demand-based view of the emergence of competition', *Strategic Management Journal*, 23: 667–688.

Afuah, A. & Tucci, C.L. (2003) 'A model of the internet as creative destroyer', *IEEE Transactions on Engineering Management*, 50: 395–402.

Ahlqvist, T., Valovirta, V. & Loikkanen, T. (2011) 'Innovation policy roadmapping as a systemic instrument for policy design', paper presented at the Fourth International Seville Conference on Future-Oriented Technology Analysis and Grand Societal Challenges – Shaping And Driving Structural And Systemic Transformations, Seville, 13–15 May.

Allen Consulting Group (2010) *Productivity in the buildings network – Assessing the impacts of building information models*, Canberra: Built Environment Industry Innovation Council.

Australian Research Council (2012) *ERA: National report*, Canberra: Commonwealth of Australia.

Barrett, P. (2013) 'Impact of research and development on construction', panel address at the CIB World Building Congress, Brisbane, 7 May.

Berman, S.J. & Hagan, J. (2006) 'How technology-driven business strategy can spur innovation and growth', *Strategy & Leadership*, 34: 28–34.

Bougrain F. & Haudeville B. (2002) 'Innovation, collaboration and SMEs internal research capacities', *Research Policy*, 31: 735–747.

Bower, J.L. & Christensen, C.M. (1995) 'Disruptive technologies: Catching the wave', *Harvard Business Review*, 73: 43–53.

Bruneel, J., D'Este, P. & Salter, A. (2010) 'Investigating the factors that diminish the barriers to university–industry collaboration', *Research Policy*, 39: 858–868.

Buildingvalue (2013) *Building and construction productivity partnership*. Available at: www.buildingvalue.co.nz/ (accessed 9 March 2013).

Built Environment Industry Innovation Council (BEIIC) (2012) *Final report to government*, Canberra: BEIIC.

Chubb, I. (2013) 'Impact of research and development on construction', panel address at the CIB World Building Congress, Brisbane, 7 May.

Chui, M., *et al.* (2012) *The social economy: Unlocking value and productivity through social technologies*, McKinsey Global Institute. Available at: www.mckinsey.com/insights/high_tech_telecoms_internet/the_social_economy (accessed 2 September 2013).

CIB (2013) *About CIB*. Available at: www.cibworld.nl/site/about_cib/index.html (accessed 30 September 2013).

Danneels, E. (2004) 'Disruptive technology reconsidered: A critique and research agenda', *Journal of Product Innovation Management*, 21: 246–258.

Dzisah, J. & Etzkowitz, H. (2011) 'The dynamics of universities, knowledge and society', in *The age of knowledge: The dynamics of universities, knowledge and society*, Dzisah, J. & Etzkowitz, H. (eds), Leiden: Brill.

Edquist, C. (1997) 'Systems of innovation approaches – Their emergence and characteristics', in *Systems of innovation: technologies, institutions and organisations*, Edquist, C. (ed.), London: Pinter.

Etzkowitz, H. & Leydesdorff, L. (2000) 'The dynamics of innovation: From national systems and "mode 2" to a triple helix of university–industry–government relations', *Research Policy*, 29: 109–123.

European Commission (2009) 'Preparing for our future: Developing a common strategy for key enabling technologies', in *The EU, communication from the Commission to the European Parliament*, The Council, Brussels: The European Economic and Social Committee and the Committee of the Regions.

European Commission (2013) *European innovation partnerships*. Available at: http://ec.europa.eu/research/innovation-union/index_en.cfm?pg=eip (accessed 4 September 2013).

European Network of Construction Companies for Research and Development (ENCORD) (2009) *Position & strategy paper*, ENCORD.

Foster, G. (1986) *Financial statement analysis*, Englewood Cliffs: Prentice-Hall.

Garnett, H.M., *et al.* (2008) *Outcomes of higher education: Quality relevance and impact: IMHE Programme on Institutional Management in Higher Education*, Paris: Organisation for Economic Cooperation and Development (OECD).

GRI (2013) *G4 sustainability reporting guideline: Reporting principals and standard disclosures*, Amsterdam: Global Reporting Initiative.

Hacklin, F. (2008) *Management of convergence in innovation – Strategies and capabilities for value creation beyond blurring industry boundaries*, Heidelberg: Physica-Verlag.

Hampson, K.D. & Kraatz, J.A. (2013) 'Modelling, collaboration and integration: A case study for the delivery of public buildings', paper presented at CIB World Building Congress, Brisbane, 5–9 May.

Keast, R. & Hampson, K.D. (2007) 'Building constructive innovation networks: Role of relationship management', *Journal of Construction Engineering and Management*, 133: 364–373.

Kraatz, J.A. & Hampson, K.D. (2013) 'Brokering innovation to better leverage R&D investment', *Building Research & Information*, 41: 187–197.

Kraatz, J.A. & Hampson, K.D. (2014) *Private sector R&D investment: A case study*, Perth: Sustainable Built Environment National Research Centre.

Kraatz, J.A., *et al.* (2012) *Leveraging R&D investment for the Australian built environment: Industry report*, Brisbane: Sustainable Built Environment National Research Centre.

Le Gall, C. (2013) 'Impact of research and development on construction', panel address at the CIB World Building Congress, Brisbane, 7 May.

Leitner, K.-H. & Guldenberg, S. (2010) 'Generic strategies and firm performance in SMEs: A longitudinal study of Austrian SMEs', *Small Business Economics*, 35: 169–189.

Miozzo, M. & Dewick, P. (2002) 'Building competitive advantage: Innovation and corporate governance in European construction', *Research Policy*, 31: 989–1008.

OECD (2002) *Frascati manual: Proposed standard practice for surveys on research and experimental development*, Paris: Organisation for Economic Cooperation and Development (OECD).

OECD (2011) *Global science forum: Effective modelling of urban systems to address the challenges of climate change and sustainability*, Paris: Organisation for Economic Cooperation and Development. Available at: www.oecd.org/sti/sci-tech/49352636.pdf (accessed 16 July 2013).

Paap, J. & Katz, R. (2004) 'Anticipating disruptive innovation', *Research-Technology Management*, 47: 13–22.

Parker, R. & Hine, D. (2013) 'The role of knowledge intermediaries in developing firm learning capabilities', *European Planning Studies*, 4: 1–23.

Peng, T.-J.A., Pike, S. & Roos, G. (2012) 'Is cooperation with competitors a good idea? An example in practice', *British Journal of Management*, 23: 532–560.

Research Councils UK (2012) *Excellence with impact*. Available at: www.rcuk.ac.uk/kei/impacts/Pages/meanbyimpact.aspx (accessed 30 September 2013).

Roos, G. (2013) 'The role of intellectual capital in business model innovation: An empirical study', in *Intellectual capital strategy management for knowledge-based organizations*, Ordoñez de Pablos, P., Tennyson, R.D. & Zhao, J. (eds), Hershey: Business Science Reference.

Roos, G. & Pike, S. (2011) 'The relationship between university research and firm innovation', in *Bridging the gap between academic accounting research and professional practice*, Evans, E., Burritt, R. & Guthrie, J. (eds), Adelaide: Institute of Chartered Accountants in Australia & University of South Australia.

Ruddock, L. & Ruddock, S. (2009) 'The scope of the construction sector – Determining its value', in *Economics for the modern built environment*, Ruddock, L. (ed.), Abingdon: Taylor & Francis.

Russom, P. (2011) *Big data analytics – TDWI best practices report*, Fourth Quarter, Renton, WA: TDWI.

Schiele, H. & Krummaker, S. (2011) 'Consortium benchmarking: Collaborative academic–practitioner case study research', *Journal of Business Research*, 64: 1137–1145.

Spithoven, A., Clarysse, B. & Knockaert, M. (2010) 'Building absorptive capacity to organise inbound open innovation in traditional industries', *Technovation*, 30: 130–141.

Tassey, G. (2010) 'Rationales and mechanisms for revitalizing US manufacturing R&D strategies', *Journal of Technology Transfer*, 35: 283–333.

Tassey, G. (2012) *Beyond the business cycle: The need for a technology-based growth strategy*, Gaithersburg, MD: National Institute of Standards and Technology, U.S. Department of Commerce.

van de Ven, A.H. (2007) *Engaged scholarship: A guide for organizational and social research*, Oxford and New York: Oxford University Press.

Vianova (2013) *BIM for infrastructure in Sweden*. Available at: www.vianovasystems.com/BIM/BIM-Today/BIM-for-Infrastructure-in-Sweden#.UcpfKPmLDgc (accessed 26 June 2013).

Winch, G. (2005) 'Managing complex connective processes: Innovation broking', in *Building tomorrow: Innovation in construction and engineering*, Manseau, A. & Shields, R. (eds), Aldershot: Ashgate Publishing Limited.

Wright, M., Clarysse, B., Lockett, A. & Knockaert, M. (2008) 'Mid-range universities' linkages with industry: Knowledge types and the role of intermediaries', *Research Policy*, 37: 1205–1223.

Index

Note: page numbers in **bold** are for figures, those in *italics* are for tables.